Logic Colloquium 2007

The Annual European Meeting of the Association for Symbolic Logic, also known as the Logic Colloquium, is among the most prestigious annual meetings in the field. The current volume, *Logic Colloquium 2007*, with contributions from plenary speakers and selected special session speakers, contains both expository and research papers by some of the best logicians in the world. This volume covers many areas of contemporary logic: model theory, proof theory, set theory, and computer science, as well as philosophical logic, including tutorials on cardinal arithmetic, on Pillay's conjecture, and on automatic structures.

This volume will be invaluable for experts as well as those interested in an overview of central contemporary themes in mathematical logic.

Françoise Delon was Directrice d'études at the Centre de Formation des PEGC of Reims and Humboldt Stipendiatin at Freiburg and is presently a Directrice de Recherche at Centre National de la Recherche Scientifique.

Ulrich Kohlenbach is a Professor of Mathematics at TU Darmstadt (Germany). He is the coordinating editor of *Annals of Pure and Applied Logic* and the president of the Deutsche Vereinigung für Mathematische Logik und für Grundlagen der Exakten Wissenschaften.

Penelope Maddy is a Distinguished Professor of Logic and Philosophy of Science at the University of California, Irvine. She is a Fellow of the American Academy of Arts and Sciences and is currently the president of the Association for Symbolic Logic.

Frank Stephan is an Associate Professor in the departments of mathematics and computer science at the National University of Singapore. He is the editor of the *Journal of Symbolic Logic*.

LECTURE NOTES IN LOGIC

A Publication of
The Association for Symbolic Logic

This series serves researchers, teachers, and students in the field of symbolic logic, broadly interpreted. The aim of the series is to bring publications to the logic community with the least possible delay and to provide rapid dissemination of the latest research. Scientific quality is the overriding criterion by which submissions are evaluated.

More information, including a list of the books in the series, can be found at http://www.aslonline.org/books-lnl.html.

Logic Colloquium 2007

Edited by

FRANÇOISE DELON
UFR de Mathématiques

ULRICH KOHLENBACH
Technische Universität Darmstadt

PENELOPE MADDY
University of California, Irvine

FRANK STEPHAN
National University of Singapore

ASSOCIATION FOR SYMBOLIC LOGIC

CAMBRIDGE
UNIVERSITY PRESS

CAMBRIDGE
UNIVERSITY PRESS

32 Avenue of the Americas, New York NY 10013-2473, USA

Cambridge University Press is part of the University of Cambridge.

It furthers the University's mission by disseminating knowledge in the pursuit of education, learning and research at the highest international levels of excellence.

www.cambridge.org
Information on this title: www.cambridge.org/9781107696778

First published 2010
Reprinted 2011
First paperback edition 2014

A catalogue record for this publication is available from the British Library

ISBN 978-0-521-76065-2 Hardback
ISBN 978-1-107-69677-8 Paperback

CONTENTS

INTRODUCTION

The Logic Colloquium 2007, the European Summer Meeting of the Association for Symbolic Logic, was held in Wrocław, Poland, from 14 to 19 July 2007. It was colocated with the following events: The thirty-fourth International Colloquium on Automata, Languages and Programming (ICALP), the twenty-second Annual IEEE Symposium on Logic in Computer Science (LICS) and the ninth ACM-SIGPLAN International Symposium on Principles and Practice of Declarative Programming (PPDP). There was an agreement with LICS on running joint sessions for one day.

More than 200 participants from all over the world took part in the Logic Colloquium. The programme consisted of 3 tutorials, 11 invited plenary talks, 6 joint talks with LICS (2 long, 4 short) and 21 talks in 5 special sessions on set theory, proof complexity and nonclassical logics, philosophical and applied logic at the JPL, logic and analysis and model theory. In addition to these invited talks, there were 63 contributed talks.

The programme committee consisted of Alessandro Andretta (Turin), Françoise Delon (Paris 7), Ulrich Kohlenbach (Darmstadt), Steffen Lempp (Madison, Chair), Penelope Maddy (UC Irvine), Jerzy Marcinkowski (Wrocław), Ludomir Newelski (Warsaw), Andrew Pitts (Cambridge), Pavel Pudlák (Prague), Sławomir Solecki (Urbana-Champaign), Frank Stephan (Singapore) and Göran Sundholm (Leiden). The local organizing committee consisted of Tobias Kaiser, Piotr Kowalski, Jan Kraszewski, Amador Martin-Pizarro, Serge Randriambololona and Roman Wencel.

The Logic Colloquium 2007 wants to acknowledge its sponsors for their generous support of the event: the Association for Symbolic Logic, the Polish Academy of Sciences and the University of Wrocław.

The next pages give an overview of the programme of the meeting; in addition to the talks listed, there were contributed talks from all fields of mathematical logic. Speakers invited to give a plenary or special session talk were also invited to contribute to this volume. The contributions handed in were refereed according to the high standards of logic journals. We would like to thank the authors for their excellent contributions as well as the referees for their diligent work to review and evaluate the submissions.

The Editors
Françoise Delon
Ulrich Kohlenbach
Penelope Maddy
Frank Stephan

SPEAKERS AND TITLES

Tutorials

Steve Jackson (North Texas): Cardinal arithmetic in $L(R)$.

Bakh Khoussainov (Auckland): Automatic structures.

Kobi Peterzil (Haifa): The infinitesimal subgroup of a definably compact group.

Invited Plenary Talks

Albert Atserias (Barcelona): Structured finite model theory.

Matthias Baaz (Vienna): Towards a proof theory of analogical reasoning.

Vasco Brattka (Cape Town): Computable analysis and effective descriptive set theory.

Zoé Chatzidakis (Paris 7): Model theory of difference fields and some applications.

Gabriel Debs (Paris 6): Coding compact spaces of Borel functions.

Fernando Ferreira (Lisbon): On a new functional interpretation.

Andrzej Grzegorczyk (Warsaw): Philosophical content of formal achievements.

Bjørn Kjos-Hanssen (Hawaii): Brownian motion and Kolmogorov complexity.

Piotr Kowalski (Wrocław): Definability in differential fields.

Paul Larson (Miami U): Large cardinals and forcing-absoluteness.

Tony Martin (UC Los Angeles): Sets and the concept of set.

LICS Joint Long Talks

Martin Hyland (Cambridge): Combinatorics and proofs.

Colin Stirling (Edinburgh): Higher-order matching, games and automata.

LICS Joint Short Talks

Cristiano Calcagno (Imperial College): Can logic tame systems programs?

Martín Escardó (Birmingham): Infinite sets that admit exhaustive search.

Rosalie Iemhoff (Utrecht): Skolemization in constructive theories.

Alex Simpson (Edinburgh): Non-well-founded proofs.

Special Session Set Theory

Organized by Ilijas Farah and Joel Hamkins.

Márton Elekes (Hungarian Academy of Sciences): Partitioning κ-fold covers into κ many subcovers.

Gunter Fuchs (Münster): Maximality principles for closed forcings.

Victoria Gitman (City University of New York): Scott's problem for proper Scott sets.

Lionel Nguyen Van The (Calgary): The Urysohn sphere is oscillation stable.

Asger Tornquist (Toronto): Classifying measure preserving actions up to conjugacy and orbit equivalence.

Matteo Viale (Vienna): The constructible universe for the anti-foundation axiom system ZFA.

Special Session Proof Complexity and Nonclassical Logics

Organized by Pavel Pudlák.

Emil Jerábek (Prague): Proof systems for modal logics.

George Metcalfe (Vanderbilt): Substructural fuzzy logics.

Alasdair Urquhart (Toronto): Complexity problems for substructural logics.

Special Session Philosophical and Applied Logic at the JPL

Organized by Penelope Maddy.

Aldo Antonelli (UC Davis): From philosophical logic to computer science.

Horacio Arlò Costa (Carnegie-Mellon): Philosophical logic meets formal epistemology.

Greg Restall (Melbourne): Proof theory and meaning: three case studies.

Albert Visser (Utrecht): Interpretations in philosophical logic.

Special Session Logic and Analysis

Organized by Itaï Ben Yaacov and Ulrich Kohlenbach.

Philipp Gerhardy (Pittsburgh/Oslo): Local stability of ergodic averages.

Peter Hertling (Univ. der Bundeswehr München): Computability and non-computability results for the topological entropy of shift spaces.

Julien Melleray (Urbana-Champaign): Geometry of the Urysohn space: A model-theoretic approach.

Andreas Weiermann (Ghent): Analytic combinatorics of the transfinite.

Special Session Model Theory

Organized by Françoise Delon and Ludomir Newelski.

Vera Djordjevic (Uppsala): Independence in structures and finite satisfiability.

Amador Martin-Pizarro (Lyon 1): Some thoughts on bad objects.

Ziv Shami (Tel Aviv): Countable imaginary simple unidimensional theories.

Marcus Tressl (Regensburg): Super real closed rings.

DECORATED LINEAR ORDER TYPES AND THE THEORY OF CONCATENATION

VEDRAN ČAČIĆ, PAVEL PUDLÁK, GREG RESTALL, ALASDAIR URQUHART, AND ALBERT VISSER

Abstract. We study the interpretation of Grzegorczyk's Theory of Concatenation TC in structures of decorated linear order types satisfying Grzegorczyk's axioms. We show that TC is incomplete for this interpretation. What is more, the first order theory validated by this interpretation interprets arithmetical truth. We also show that every extension of TC has a model that is not isomorphic to a structure of decorated order types. We provide a positive result, to wit a construction that builds structures of decorated order types from models of a suitable concatenation theory. This construction has the property that if there is a representation of a certain kind, then the construction provides a representation of that kind.

§1. Introduction. In his paper [2], Andrzej Grzegorczyk introduces a theory of concatenation TC. The theory has a binary function symbol $*$ for concatenation and two constants a and b. The theory is axiomatized as follows.

TC1. $\vdash (x * y) * z = x * (y * z)$

TC2. $\vdash x * y = u * v \to ((x = u \wedge y = v) \vee \exists w\, ((x*w = u \wedge y = w*v) \vee (x = u*w \wedge y*w = v)))$

TC3. $\vdash x * y \neq \mathsf{a}$

TC4. $\vdash x * y \neq \mathsf{b}$

TC5. $\vdash \mathsf{a} \neq \mathsf{b}$

Axioms TC1 and TC2 are due to Tarski [7]. Grzegorczyk calls axiom TC2 the *editor axiom*. We will consider two weaker theories. The theory TC_0 has the signature with just concatenation, and is axiomatized by TC1,2. The theory TC_1 is axiomatized by TC1,2,3. We will also use TC_2 for TC.

Some of the results of this note were obtained during the Excursion to mountain Ślęża of the inspiring Logic Colloquium 2007 in Wrocław and, in part, in the evening after the Excursion. We thank the organizers for providing this wonderful opportunity. We thank Dana Scott for his comments, insights and questions. We are grateful to Vincent van Oostrom for some perceptive remarks.

Pavel Pudlák was supported by grants A1019401 and 1M002162080.

Logic Colloquium '07
Edited by Françoise Delon, Ulrich Kohlenbach, Penelope Maddy, and Frank Stephan
Lecture Notes in Logic, 35

The theories we are considering have various interesting interpretations. First they are, of course, theories of strings with concatenation; in other words, theories of free semigroups. Secondly they are theories of wider classes of structures, to wit structures of *decorated linear order types*, which will be defined below[1].

The theories TC_i are theories for concatenation without the empty string, i.e., without the unit element. Adding a unit ε one obtains another class of theories $\mathsf{TC}_i^\varepsilon$, theories of free monoids, or theories of structures of decorated linear order types including the empty linear decorated order type. The basic list of axioms is as follows.

$\mathsf{TC}^\varepsilon 1.\ \vdash \varepsilon * x = x \wedge x * \varepsilon = x$

$\mathsf{TC}^\varepsilon 2.\ \vdash (x * y) * z = x * (y * z)$

$\mathsf{TC}^\varepsilon 3.\ \vdash x*y = u*v \rightarrow \exists w\ ((x*w = u \wedge y = w*v) \vee (x = u*w \wedge w*y = v))$

$\mathsf{TC}^\varepsilon 4.\ \vdash \mathsf{a} \neq \varepsilon$

$\mathsf{TC}^\varepsilon 5.\ \vdash x * y = \mathsf{a} \rightarrow (x = \varepsilon \vee y = \varepsilon)$

$\mathsf{TC}^\varepsilon 6.\ \vdash \mathsf{b} \neq \varepsilon$

$\mathsf{TC}^\varepsilon 7.\ \vdash x * y = \mathsf{b} \rightarrow (x = \varepsilon \vee y = \varepsilon)$

$\mathsf{TC}^\varepsilon 8.\ \vdash \mathsf{a} \neq \mathsf{b}.$

We take $\mathsf{TC}_0^\varepsilon$ to be the theory axiomatized by $\mathsf{TC}^\varepsilon 1, 2, 3$, $\mathsf{TC}_1^\varepsilon$ to be $\mathsf{TC}_0^\varepsilon +$ $\mathsf{TC}^\varepsilon 4, 5$ and $\mathsf{TC}^\varepsilon := \mathsf{TC}_2^\varepsilon$ to be $\mathsf{TC}_1^\varepsilon + \mathsf{TC}^\varepsilon 6, 7, 8$.

One can show that TC is *bi-interpretable* with TC^ε, in which a unit ε is added via one dimensional interpretations without parameters. The theory TC_1 is bi-interpretable with $\mathsf{TC}_1^\varepsilon$ via two-dimensional interpretations with parameters. The situation for TC_0 seems to be more subtle. See also [10]. In Section 6, we will study an extension of $\mathsf{TC}_0^\varepsilon$.

Andrzej Grzegorczyk and Konrad Zdanowski have shown that TC is essentially undecidable. This result can be strengthened by showing that Robinson's Arithmetic Q is mutually interpretable with TC. Note that TC_0 is undecidable —since it has an extension that parametrically interprets TC— but that TC_0 is not essentially undecidable: it is satisfied by a one-point model. Similarly TC_1 is undecidable, but it has as an extension the theory of finite strings of a's, which is a notational variant of Presburger Arithmetic and, hence, decidable.

We will call models of TC_0 *concatenation structures*, and we will call models of TC_i *concatenation i-structures*. The relation of isomorphism between concatenation structures will be denoted by $\cong$. We will be interested in concatenation structures, whose elements are decorated linear order types with the operation *concatenation of decorated order types*. Let a non-empty class A be given. An A-decorated linear ordering is a structure $\langle D, \leq, f\rangle$, where D is a non-empty domain, $\leq$ is a linear ordering on D, and f is a function from

[1]A special case of decorated linear order types is addition of sets as discovered by Tarski (see [8]). It is shown by Laurence Kirby in [3] that the structure of addition on sets is isomorphic to addition of well-founded order types with a proper class of decorating objects.

D to A. A mapping ϕ is an *isomorphism* between A-decorated linear order types $\langle D, \leq, f\rangle$ and $\langle D', \leq', f'\rangle$ iff it is a bijection between D and D' such that, for all d, e in D, $d \leq e \Leftrightarrow \phi d \leq' \phi e$, and $fd = f'\phi d$. Our notion of isomorphism gives us a notion of A-decorated linear order type. We have an obvious notion of concatenation between A-decorated linear orderings which induces a corresponding notion of concatenation for A-decorated linear order types. We use $\alpha, \beta, \ldots$ to range over such linear order types. Since, linear order types are classes we have to follow one of two strategies: either to employ Scott's trick to associate a set object to any decorated linear order type or to simply refrain from dividing out isomorphism but to think about decorated linear orderings modulo isomorphism. We will employ the second strategy.

We will call a concatenation structure whose domain consists of (representatives of) A-decorated order types, for some A, and whose concatenation *is* concatenation of decorated order types: a *concrete concatenation structure*. It seems entirely reasonable to stipulate that e.g. the interpretation of $\mathtt{a}$ in a concrete concatenation structure is a decorated linear order type of a one element order. However, for the sake of generality we will refrain from making this stipulation.

Grzegorczyk conjectured that every concatenation 2-structure is isomorphic to a concrete concatenation structure. We prove that this conjecture is false. (i) Every extension of TC_1 has a model that is not isomorphic to a concrete concatenation 1-structure and (ii) the set of principles valid in all concrete concatenation 2-structures interprets arithmetical truth.

The plan of the paper is as follows. We show, in Section 2, that we have, for all decorated order types α, β and γ, the following principle:

$$(\dagger) \qquad \beta * \alpha * \gamma = \alpha \Rightarrow \beta * \alpha = \alpha * \gamma = \alpha.$$

This fact was already known. It is due to Lindenbaum, credited to him in Sierpiński's book [6] on p. 248. It is also problem 6.13 of [4].

It is easy to see that every group is a concatenation structure and that (†) does not hold in the two element group. We show, in Section 5, that every concatenation structure can be extended to a concatenation structure with any number of atoms. It follows that there is a concatenation structure with at least two atoms in which (†) fails. Hence, TC is incomplete for concrete concatenation structures. In Section 3, we provide a counterargument of a different flavour. We provide a tally interpretation that defines the natural numbers (with concatenation in the role of addition) in every concrete concatenation 2-structure. It follows that every extension of TC_1 is satisfied by a concatenation 1-structure that is not isomorphic to any concrete concatenation 1-structure, to wit any model of that extension that contains a non-standard element. In Section 4, we strengthen the result of Section 3, by showing that in concrete concatenation 2-structures we can add multiplication to the natural numbers.

It follows that the set of arithmetically true sentences is interpretable in the concretely valid consequences of TC_2.

Finally, in Section 6 we prove a positive result. We provide a mapping from arbitrary models of a variant of an extension of TC_0 to structures of decorated order types. As we have shown such a construction cannot always provide a representation. We show that, for a restricted class of representations, we do have: if a model has a representation in the class, then the construction yields such a representation.

§2. A principle for decorated order types. In this section we prove a universal principle that holds in all concatenation structures, which is not provable in TC. There is an earlier proof of this principle [4, p. 187]. Our proof, however, is different from that of Komjáth and Totik.

THEOREM 1. *Let $\alpha_0, \alpha_1, \alpha_2$ be decorated order types. Suppose that $\alpha_1 = \alpha_0 * \alpha_1 * \alpha_2$. Then, $\alpha_1 = \alpha_0 * \alpha_1 = \alpha_1 * \alpha_2$.*

PROOF. Suppose $\alpha_1 = \alpha_0 * \alpha_1 * \alpha_2$. Consider a decorated linear ordering $\mathcal{A} := \langle A, \leq, f \rangle$ of type α_1, By our assumption, we may partition A into A_0, A_1, A_2, such that:

$$\langle A, \leq, f \rangle = \langle A_0, \leq\restriction A_0, f \restriction A_0 \rangle * \langle A_1, \leq\restriction A_1, f \restriction A_1 \rangle * \langle A_2, \leq\restriction A_2, f \restriction A_2 \rangle,$$

where $\mathcal{A}_i := \langle A_i, \leq\restriction A_i, f \restriction A_i \rangle$ is an instance of α_i, Let $\phi : \mathcal{A} \to \mathcal{A}_1$ be an isomorphism.

Let $\phi^n \mathcal{A}_{(i)} := \langle \phi^n[A_{(i)}], \leq\restriction \phi^n[A_{(i)}], f \restriction \phi^n[A_{(i)}] \rangle$. We have: $\phi^n \mathcal{A}_i$ is of order type α_i and $\phi^n \mathcal{A}$ is of order type α_1.

Clearly, $\phi\mathcal{A}_0$ is an initial substructure of $\phi\mathcal{A} = \mathcal{A}_1$. So, $\mathcal{A}_0$ and $\phi\mathcal{A}_0$ are disjoint and $\phi\mathcal{A}_0$ adjacent to the right of $\mathcal{A}_0$. Similarly, for $\phi^n \mathcal{A}_0$ and $\phi^{n+1}\mathcal{A}_0$. Take $A_0^\omega := \bigcup_{i \in \omega} \phi^i A_0$. We find that $\mathcal{A}_0^\omega := \langle A_0^\omega, \leq\restriction A_0^\omega, f \restriction A_0^\omega \rangle$ is initial in $\mathcal{A}$ and of decorated linear order type α_0^ω. So $\alpha_1 = \alpha_0^\omega * \rho$, for some ρ. It follows that $\alpha_0 * \alpha_1 = \alpha_0 * \alpha_0^\omega * \rho = \alpha_0^\omega * \rho = \alpha_1$. The other identity is similar. ⊣

So, every concrete concatenation structure validates that $\alpha_1 = \alpha_0 * \alpha_1 * \alpha_2$ implies $\alpha_1 = \alpha_0 * \alpha_1 = \alpha_1 * \alpha_2$. We postpone the proof that this principle is not provable in TC to Section 5.

§3. Definability of the natural numbers. In this section, we show that the natural numbers can be defined in every concrete concatenation 1-structure. We define:

- $x \subseteq y :\leftrightarrow x = y \vee \exists u\, (u * x = y) \vee \exists v\, (x * v = y) \vee \exists u, v\, (u * x * v = y)$.
- $x \subseteq_{\mathsf{ini}} y :\leftrightarrow x = y \vee \exists v\, (x * v = y)$.
- $x \subseteq_{\mathsf{end}} y :\leftrightarrow x = y \vee \exists u\, (u * x = y)$.
- $(n : \widetilde{\mathsf{N}}_{\mathsf{a}}) :\leftrightarrow \forall m \subseteq_{\mathsf{ini}} n\, (m = \mathsf{a} \vee \exists k\, (k \neq m \wedge m = k * \mathsf{a}))$.

The use of ':' in $n : \widetilde{\mathsf{N}}_{\mathsf{a}}$ is derived from the analogous use in type theory. We could read it as: n is of sort N_{a}. We write $m, n : \widetilde{\mathsf{N}}_{\mathsf{a}}$ for: $(m : \widetilde{\mathsf{N}}_{\mathsf{a}}) \wedge (n : \widetilde{\mathsf{N}}_{\mathsf{a}})$, etc. In the context of a structure we will confuse $\widetilde{\mathsf{N}}_{\mathsf{a}}$ with the extension of $\widetilde{\mathsf{N}}_{\mathsf{a}}$ in that structure.

We prove the main theorem of this section.

THEOREM 2. *In any concrete concatenation* 1*-structure, we have*:

$$\widetilde{\mathsf{N}}_{\mathsf{a}} = \{\mathsf{a}^{n+1} \mid n \in \omega\}.$$

In other words, $\widetilde{\mathsf{N}}_{\mathsf{a}}$ *is precisely the class of natural numbers in tally representation* (*starting with* 1). *Note that* $*$ *on this set is addition.*

PROOF. Consider any concrete concatenation 1-structure $\mathfrak{A}$. It is easy to see that every a^{n+1} is in $\widetilde{\mathsf{N}}_{\mathsf{a}}$. Clearly, every element x of $\widetilde{\mathsf{N}}_{\mathsf{a}}$ is either a or it has a predecessor, i.e., there is a y such that $x = y * \mathsf{a}$. The axioms of TC_1 guarantee that this predecessor is unique. This justifies the introduction of the partial predecessor function pd on $\widetilde{\mathsf{N}}_{\mathsf{a}}$. Let α be the order type corresponding to a. Let β_0 be any element of $\widetilde{\mathsf{N}}_{\mathsf{a}}$. If, for some n, $\mathsf{pd}^n\beta_0$ is undefined, then β_0 is clearly of the form α^{k+1}, for k in ω.

We show that the other possibility cannot obtain. Suppose $\beta_n := \mathsf{pd}^n\beta_0$ is always defined. Let $\mathcal{A}$ be a decorated linear ordering of type α and let $\mathcal{B}_i$ be a decorated linear ordering of type β_i. We assume that the domain A of $\mathcal{A}$ is disjoint from the domains B_i of the $\mathcal{B}_i$. Thus, we may implement $\mathcal{B}_{i+1} * \mathcal{A}$ just by taking the union of the domains.

Let ϕ_i be isomorphisms from $\mathcal{B}_{i+1} * \mathcal{A}$ to $\mathcal{B}_i$. Let $\mathcal{A}_i := (\phi_0 \circ \cdots \circ \phi_i)(\mathcal{A})$. Then, the $\mathcal{A}_i$ are all of type α and, for some $\mathcal{C}$, we have $\mathcal{B}_0 \cong \mathcal{C} * \cdots * \mathcal{A}_1 * \mathcal{A}_0$. Similarly $\mathcal{B}_1 \cong \mathcal{C} * \cdots * \mathcal{A}_2 * \mathcal{A}_1$. Let $\breve{\omega}$ be the opposite ordering of ω. It follows that $\beta_0 = \gamma * \alpha^{\breve{\omega}} = \beta_1 = \mathsf{pd}(\beta_0)$. Hence, β_0 is not in $\widetilde{\mathsf{N}}_{\mathsf{a}}$[2]. A contradiction. ⊣

We call a concatenation structure *standard* if $\widetilde{\mathsf{N}}_{\mathsf{a}}$ defines the tally natural numbers. Since, by the usual argument, any extension of TC_1 has a model with non-standard numbers, we have the following corollary.

COROLLARY 3. *Every extension of* TC_1 *has a model that is not isomorphic to a concrete concatenation* 1*-structure. In a different formulation*: *for every concatenation* 1*-structure there is an elementarily equivalent concatenation* 1*-structure that is not isomorphic to a concrete concatenation* 1*-structure.*

Note that the non-negative tally numbers with addition form a concrete concatenation 1-structure. Thus, the concretely valid consequences of $\mathsf{TC}_1 + \forall x\ (x : \widetilde{\mathsf{N}}_{\mathsf{a}})$, i.e., the principles valid in every concrete concatenation 1-structure satisfying $\forall x\ (x : \widetilde{\mathsf{N}}_{\mathsf{a}})$ are decidable.

[2]Note that we are not assuming that γ is in $\mathfrak{A}$.

§4. Definability of multiplication. If we have two atoms to work with, we can add multiplication to our tally numbers. This makes the set of concretely valid consequences of TC non-arithmetical. The main ingredient of the definition of multiplication is the theory of relations on tally numbers. In TC, we can develop such a theory. The development has some resemblance to the construction in the classic paper of Quine [5]. However, the ideas here are somewhat more intricate, since we are working in a more general context than that of [5]. We represent the relation $\{\langle x_0, y_0\rangle, \dots, \langle x_{n-1}, y_{n-1}\rangle\}$, by:

$$\mathsf{bb} * x_0 * \mathsf{b} * y_0 * \mathsf{bb} * x_1 * \dots \mathsf{bb} * x_{n-1} * \mathsf{b} * y_{n-1} * \mathsf{bb}.$$

We define:

- $r : \mathsf{REL} :\leftrightarrow \mathsf{bb} \subseteq_{\mathsf{end}} r$,
- $\emptyset := \mathsf{bb}$,
- $x[r]y :\leftrightarrow x, y : \widetilde{\mathsf{N}}_{\mathsf{a}} \wedge \mathsf{bb} * x * \mathsf{b} * y * \mathsf{bb} \subseteq r$.
- $\mathsf{adj}(r, x, y) := r * x * \mathsf{b} * y * \mathsf{bb}$.

Clearly, we have: $\mathsf{TC} \vdash \forall u, v \ \neg u[\emptyset]v$. To verify that this coding works we need the adjunction principle.

THEOREM 4. *We have*:

$$\mathsf{TC} \vdash (r : \mathsf{REL} \wedge x, y, u, v : \widetilde{\mathsf{N}}_{\mathsf{a}}) \to \\ \big(u[\mathsf{adj}(r, x, y)]v \leftrightarrow \big(u[r]v \vee (u = x \wedge v = y)\big)\big).$$

We can prove this result by laborious and unperspicuous case splitting. However, it is more elegant to do the job with the help of a lemma. Consider any model of TC_0. Fix an element w. We call a sequence $(w_0, \dots, w_k)$ a *partition* of w if we have that $w_0 * \dots * w_k = w$. The partitions of w form a category with the following morphisms. $f : (u_0, \dots, u_n) \to (w_0, \dots, w_k)$ iff f is a surjective and weakly monotonic function from $n + 1$ to $k + 1$, such that, for any $i \leq k$, $w_i = u_s * \dots * u_\ell$, where $f(j) = i$ iff $s \leq j \leq \ell$. We write $(u_0, \dots, u_n) \leq (w_0, \dots, w_k)$ for: $\exists f \ f : (u_0, \dots, u_n) \to (w_0, \dots, w_k)$. In this case we say that $(u_0, \dots, u_n)$ is *a refinement* of $(w_0, \dots, w_k)$.

LEMMA 5. *Consider any concatenation structure. Let w be an element of the structure. Then, any two partitions of w have a common refinement.*

PROOF. Fix any concatenation structure. We first prove that, for all w, all pairs of partitions $(u_0, \dots, u_n)$ and $(w_0, \dots, w_k)$ of w have a common refinement, by induction of $n + k$.

If either n or k is 0, this is trivial. Suppose that $(u_0, \dots, u_{n+1})$ and $(w_0, \dots, w_{k+1})$ are partitions of w. By the editor axiom, we have either (a) $u_0 * \dots * u_n = w_0 * \dots * w_k$ and $u_{n+1} = w_{k+1}$, or there is a v such that (b) $u_0 * \dots * u_n * v = w_0 * \dots * w_k$ and $u_{n+1} = v * w_{k+1}$, or (c) $u_0 * \dots * u_n = w_0 * \dots * w_k * v$ and $v * u_{n+1} = w_{k+1}$. We only treat case (b), the other cases being easier or similar. By the induction hypothesis, there is a common refinement $(x_0, \dots x_m)$ of $(u_0, \dots, u_n, v)$ and $(w_0, \dots, w_n)$. Let this be witnessed by f, resp. g. It is easily seen that $(x_0, \ldots x_m, w_{k+1})$ is the desired refinement with witnessing functions f' and g', where $f' := f[m+1 \mapsto n+1]$, $g' := g[m+1 \mapsto k+1]$. Here $f[m+1 \mapsto n+1]$ in the result of extending f to assign $n+1$ to $m+1$. ⊣

We turn to the proof of Theorem 4. The verification proceeds more or less as one would do it for finite strings.

Proof. Consider any concatenation 2-structure. Suppose $\mathsf{REL}(r)$. The right-to-left direction is easy, so we treat left-to-right. Suppose x, y, u and v are tally numbers. and $u[\mathsf{adj}(r, x, y)]v$. There are two possibilities. Either $r = \mathsf{bb}$ or $r = r_0 * \mathsf{bb}$. We will treat the second case. Let $s := \mathsf{adj}(r, x, y)$. One the following four partitions is a partition of s: (i) $(\mathsf{b}, \mathsf{b}, u, \mathsf{b}, v, \mathsf{b}, \mathsf{b})$, or (ii) $(w, \mathsf{b}, \mathsf{b}, u, \mathsf{b}, v, \mathsf{b}, \mathsf{b})$, or (iii) $(\mathsf{b}, \mathsf{b}, u, \mathsf{b}, v, \mathsf{b}, \mathsf{b}, z)$, or (iv) $(w, \mathsf{b}, \mathsf{b}, u, \mathsf{b}, v, \mathsf{b}, \mathsf{b}, z)$. We will treat cases (ii) and (iv).

Suppose $\sigma := (w, \mathsf{b}, \mathsf{b}, u, \mathsf{b}, v, \mathsf{b}, \mathsf{b})$ is a partition of s. We also have that $\tau := (r_0, \mathsf{b}, \mathsf{b}, x, \mathsf{b}, y, \mathsf{b}, \mathsf{b})$ is a partition of s. Let $(t_0, \ldots, t_k)$ be a common refinement of σ and τ, with witnessing functions f and g. The displayed b's in these partitions must have unique places among the t_i. We define m_σ to be the unique i such that $f(i) = m$, provided that $\sigma_m = \mathsf{b}$. Similarly, for m_τ. (To make this unambiguous, we assume that if $\sigma = \tau$, we take σ as the common refinement with f and g both the identity function.)

We evidently have $7_\sigma = 7_\tau = k$ and $6_\sigma = 6_\tau = k - 1$. Suppose $4_\sigma < 4_\tau$. It follows that $\mathsf{b} \subseteq v$. So, v would have an initial subsequence that ends in b, which is impossible. So, $4_\sigma \nless 4_\tau$. Similarly, $4_\tau \nless 4_\sigma$. So $4_\sigma = 4_\tau$. It follows that $v = y$. Reasoning as in the case of 4_σ and 4_τ, we can show that $2_\sigma = 2_\tau$ and, hence $u = x$.

Suppose $\rho := (w, \mathsf{b}, \mathsf{b}, u, \mathsf{b}, v, \mathsf{b}, \mathsf{b}, z)$ is a partition of s. We also have that $\tau := (r_0, \mathsf{b}, \mathsf{b}, x, \mathsf{b}, y, \mathsf{b}, \mathsf{b})$ is a partition of s. Let $(t_0, \ldots, t_k)$ be a common refinement of ρ and τ, with witnessing functions f and g. We consider all cases, where $1_\tau < 6_\rho$. Suppose $6_\rho = 1_\tau + 1 = 2_\tau$. Note that $7_\rho = 6_\rho + 1$, so we find: $\mathsf{b} \subseteq x$, quod non, since x is in $\widetilde{\mathsf{N}}_\mathsf{a}$. Suppose $2_\tau < 6_\rho < 4_\tau$. In this case we have a b as substring of x. Quod non. Suppose $6_\rho = 4_\tau$. Since $7_\rho = 6_\rho + 1$, we get a b in y. Quod non. Suppose $4_\tau < 6_\rho < 6_\tau$. In this case, we get a b in y. Quod impossible. Suppose $6_\rho \geq 6_\tau = k - 1$. In this place there is no place left for z among the t_i. So, in all cases, we obtain a contradiction. So the only possibility is $6_\rho \leq 1_\tau$. Thus, it follows that $u[r]v$. ⊣

We can now use our relations to define multiplication of tally numbers in the usual way. See e.g. Section 2.2 of [1]. In any concrete concatenation 2-structure, we can use induction to verify the defining properties of multiplication as defined. It follows that we can interpret all arithmetical truths in the set of concretely valid consequences of TC.

Corollary 6. *We can interpret true arithmetic in the set of all principles valid in concrete concatenation 2-structures.*

§5. The sum of concatenation structures. In this section we show that concatenation structures are closed under sums. This result will make it possible to verify the claim that the universal principle of Section 2 is not provable in TC. The result has some independent interest, since it provides a good closure property of concatenation structures.

Consider two concatenation structures $\mathfrak{A}_0$ and $\mathfrak{A}_1$. We write $\star$ for concatenation in the $\mathfrak{A}_i$. We may assume, without loss of generality, that the domains of $\mathfrak{A}_0$ and $\mathfrak{A}_1$ are disjoint. We define the sum $\mathfrak{B} := \mathfrak{A}_0 \oplus \mathfrak{A}_1$ as follows.

- The domain of $\mathfrak{B}$ consists of non-empty sequences $w_0 \cdots w_{n-1}$, where the w_j are alternating between elements of the domains of $\mathfrak{A}_0$ and $\mathfrak{A}_1$. In other words, if w_j is in the domain of $\mathfrak{A}_i$, then w_{j+1}, if it exists, is in the domain of $\mathfrak{A}_{1-i}$.
- The concatenation $\sigma * \tau$ of $\sigma := w_0 \cdots w_{n-1}$ and $\tau := v_0 \cdots v_{k-1}$ is $w_0 \cdots w_{n-1} v_0 \cdots w_{k-1}$, in case w_{n-1} and v_0 are in the domains of different structures $\mathfrak{A}_i$. The concatenation $\sigma * \tau$ is $w_0 \cdots (w_{n-1} \star v_0) \cdots w_{k-1}$, in case w_{n-1} and v_0 are in in the same domain.

In case $\sigma * \tau$ is obtained via the first case, we say that σ and τ are *glued together*. If the second case obtains, we say that σ and τ are *clicked together*.

THEOREM 7. *The structure* $\mathfrak{B} = \mathfrak{A}_0 \oplus \mathfrak{A}_1$ *is a concatenation structure.*

PROOF. Associativity is easy. We check the editor property TC2. Suppose $\sigma_0 * \sigma_1 = z_0 \cdots z_{m-1} = \tau_0 * \tau_1$. We distinguish a number of cases.

CASE 1. Suppose both of the pairs σ_0, σ_1 and τ_0, τ_1 are glued together. Then, for some $k, n > 0$, we have $\sigma_0 = z_0 \cdots z_{k-1}$, $\sigma_1 = z_k \cdots z_{m-1}$, $\tau_0 = z_0 \cdots z_{n-1}$, and $\tau_1 = z_n \cdots z_{m-1}$.

So, if $k = n$, we have $\sigma_0 = \tau_0$ and $\sigma_1 = \tau_1$.

If $k < n$, we have $\tau_0 = \sigma_0 * (z_k \cdots z_{n-1})$ and $\sigma_1 = (z_k \cdots z_{n-1}) * \tau_1$. The case that $n < k$ is similar.

CASE 2. Suppose σ_0, σ_1 is glued together and that τ_0, τ_1 is clicked together. So, there are $k, n > 0$, u_0, and u_1 such that $\sigma_0 = z_0 \cdots z_{k-2} u_0$, $\sigma_1 = u_1 z_k \cdots z_{m-1}$, $u_0 \star u_1 = z_{k-1}$, $\tau_0 = z_0 \cdots z_{n-1}$, and $\tau_1 = z_n \cdots z_{m-1}$.

Suppose $k \leq n$. Then, $\tau_0 = \sigma_0 * (u_1 z_k \cdots z_{n-1})$ and $\sigma_1 = (u_1 z_k \cdots z_{n-1}) * \tau_1$. Note that, in case $k = n$, the sequence $z_k \cdots z_{n-1}$ is empty. The case that $k \geq n$ is similar.

CASE 3. This case, where σ_0, σ_1 is clicked together and τ_0, τ_1 is glued together, is similar to Case 2.

CASE 4. Suppose that σ_0, σ_1 and τ_0, τ_1 are both clicked together. So, there are $k, n > 0$, u_0, u_1, v_0, v_1 such that $\sigma_0 = z_0 \cdots z_{k-2} u_0$, $\sigma_1 = u_1 z_k \cdots z_{m-1}$, $u_0 \star u_1 = z_{k-1}$, $\tau_0 = z_0 \cdots z_{n-2} v_0$, $\tau_1 = v_1 z_n \cdots z_{m-1}$ and $v_0 \star v_1 = z_{n-1}$.

Suppose $k = n$. We have $u_0 \star u_1 = z_{k-1} = v_0 \star v_1$. So, we have either (a) $u_0 = v_0$ and $u_1 = v_1$, or, for some w, either (b) $u_0 \star w = v_0$ and $u_1 = w \star v_1$, or (c) $u_0 = v_0 \star w$ and $w \star u_1 = v_1$. In case (b), we have: $\sigma_0 * w = \tau_0$ and $\sigma_1 = w * \tau_1$. We leave (a) and (c) to the reader.

Suppose $k < n$. We have:

$$\sigma_0 * (u_1 z_k \cdots z_{n-2} v_0) = \tau_0 \text{ and } \sigma_1 = (u_1 z_k \cdots z_{n-2} v_0) * \tau_1.$$

The case that $k > n$ is similar. ⊣

It is easy to see that $\oplus$ is a sum or coproduct in the sense of category theory. The following theorem is immediate.

THEOREM 8. *If a is an atom of $\mathfrak{A}_i$, then a is an atom of $\mathfrak{A}_0 \oplus \mathfrak{A}_1$.*

Finally, we have the following theorem.

THEOREM 9. *Let A be any set and let $\mathfrak{B} := \langle B, * \rangle$ be any concatenation structure. We assume that A and B are disjoint. Then, there is an extension of $\mathfrak{B}$ with at least A as atoms.*

PROOF. Let A^* be the free semi-group generated by A. We can take as the desired extension of $\mathfrak{B}$, the structure $A^* \oplus \mathfrak{B}$. ⊣

REMARK 5.1. The whole development extends with only minor adaptations, when we replace axiom TC2 by:

- $\vdash x * y = u * v \to ((x = u \wedge y = v) \dot\vee (\exists! w\, (x * w = u \wedge y = w * v) \vee \exists! w\, (x = u * w \wedge y * w = v))$

Here $\dot\vee$ is *exclusive or*.

§6. A canonical construction. Although we know that not every concatenation structure can be represented by decorated linear orderings, i.e., as a concrete concatenation structure, there may exist a canonical construction of a concrete concatenation structure which is a representation whenever there exists any concrete representation. In this section we shall propose such a construction, but we can only show that it is universal in a restricted subclass of all concrete representations.

It will be now more convenient to work with a theory for monoids, rather than for semigroups, as we did in the previous sections. We will work in the theory $\mathsf{TC}^\varepsilon_+$, which is $\mathsf{TC}^\varepsilon_0$ plus the following axiom.

$\mathsf{TC}^\varepsilon 9$. $\vdash x * y * z = y \to (x = \varepsilon \wedge z = \varepsilon)$.

We do not postulate the existence of irreducible elements, as they do not play any role in what follows, but they surely can be present. We shall call elements of a model $\mathcal{M}$ of $\mathsf{TC}^\varepsilon_+$: words. When possible, the concatenation symbol $*$ will be omitted.

When considering representations of structures with a unit element ε by decorated order types, one has to allow an empty decorated order structure.

Thus a representation of a model $\mathcal{M}$ of $\mathsf{TC}_+^\varepsilon$ is a mapping ρ that assigns a decorated order structure to every $w \in \mathcal{M}$ so that

1. $\rho(\varepsilon) = \emptyset$,
2. $\rho(uv) = \rho(u)\rho(v)$ and
3. $\rho(w) \cong \rho(v)$ implies $w = v$.

LEMMA 10. *In a model $\mathcal{M}$ of $\mathsf{TC}_+^\varepsilon$ the binary relation $\exists u(xu = y)$ defines an ordering on the elements of $\mathcal{M}$.*

DEFINITION 6.1. A *k-partition* of a word w is a k-tuple $(w_1, \dots, w_k)$ such that $w_1 \dots w_k = w$; we shall often abbreviate it by $w_1 \dots w_k$. An ordering relation is defined on 3-partitions of w by:

- $u_1u_2u_3 \leq v_1v_2v_3 \;:\Leftrightarrow\; \exists x_1, x_3\, (v_1x_1 = u_1 \wedge x_3v_3 = u_3)$.

The axioms ensure that for any two partitions there is a unique common refinement.

DEFINITION 6.2 (Word Ultrafilters). Let w be a word and S a set of 3-partitions of w. We shall call S a word ultrafilter on w if

1. $\varepsilon w \varepsilon \in S$
2. $x\varepsilon y \notin S$ for any x, y
3. if $U \in S$, V is a 3-partition of w and $U \leq V$, then $V \in S$
4. if $xyz \in S$ and $y = y_1y_2$, then exactly one of the following two cases holds: $(x, y_1, y_2z) \in S$ or $(xy_1, y_2, z) \in S$.

Let S be a word ultrafilter on w and $xyz \in S$. Then we define the natural restriction of S to y which is a word ultrafilter S_y on y defined by:

- $(r, s, t) \in S_y \;:\Leftrightarrow\; (xr, s, tz) \in S$.

We shall define an ordering on word ultrafilters on a fixed w and an equivalence on word ultrafilters on all words of M. Let S and T be word ultrafilters on w, then we define:

- $S < T \;:\Leftrightarrow\; \exists u, v\, ((\varepsilon, u, v) \in S \wedge (u, v, \varepsilon) \in T)$.

Let S and T be word ultrafilters on possibly different words, then we define:

- $S \sim T :\Leftrightarrow \exists x, x', y, z, z'\, (xyz \in S \wedge x'yz' \in T \wedge S_y = T_y)$.

Notice that $<$ is a strict ordering on word ultrafilters on w, but for $S < T$ it still may be $S \sim T$.

DEFINITION 6.3. Let w be a word. The *canonical decorated ordering* associated with w is the ordering of all word ultrafilters on w, where each word ultrafilter S is decorated by $[S]_\sim$, the equivalence class of $\sim$ containing S. This decorated ordering will be denoted by $C(w)$.

Here are some basic properties of $C(w)$.

- The topological space determined by the ordering is compact and totally disconnected. In particular, it has largest and smallest elements.

- For every proper prefix x of w, there is a uniquely determined pair of word ultrafilters which forms a gap (no word ultrafilter in between). Thus there is a natural embedding of the ordering of the prefixes into $C(w)$. More precisely, we have have two mappings ϕ_w^- and ϕ_w^+ such that for a proper prefix x the pair $\phi_w^-(x), \phi_w^+(x)$ is the gap corresponding to x. If $x = \varepsilon$ (or $x = w$), then only $\phi_w^+(x)$ (or $\phi_w^-(x)$) is defined and it is the least (largest) element of $C(w)$. Furthermore the images of these mappings are dense sets in $C(w)$.
- Vice versa, every gap in $C(w)$ corresponds to a prefix (or equivalently to a 2-partition).
- If a is an atom (irreducible element in $\mathcal{M}$), then it determines a principal word ultrafilter. For a given atom a all such principal word ultrafilters are equivalent.

C satisfies conditions 1. and 2. but, in general, condition 3. fails.

EXAMPLE 6.4. Let $\mathcal{M} = A^*$ be the monoid generated by A (the alphabet). Then, all word ultrafilters are principal and $C(w)$ is essentially the string w itself.

EXAMPLE 6.5. This is a 'pathological example'. Let $\mathcal{M}$ be the nonnegative real numbers with $+$. For a positive real r, the order type of $C(r)$ is the order type of: $\{(0,1)\} \cup (\mathcal{I} \times \{0,1\}) \cup \{(1,0)\}$, where $\mathcal{I}$ is the open unit interval, with the lexicographic order. The equivalence relation $\sim$ has two classes; elements of the form $(x, 0)$ are decorated by one type of word ultrafilters, elements $(x, 1)$ are decorated by the other type. Hence for every $r, s > 0$, $C(r) \cong C(s)$, thus C is not a representation.

DEFINITION 6.6. ρ is a *regular* representation of $\mathcal{M}$ by decorated orderings, if for every 2-partition $x_1x_2 = w$ of $w \in \mathcal{M}$, there exists a *unique* 2-partition $A_1A_2 = \rho(w)$ such that $A_1 \cong \rho(x_1)$ and $A_2 \cong \rho(x_2)$.

We do not know if every concatenation structure that has a concrete representation also has a concrete regular representation.

If ρ is regular, we have an analogous property for k-partitions for every k. For a k-partition $(x_1, \ldots, x_k)$ of w in $\mathcal{M}$, we shall write

$$\rho^k(x_1, \ldots, x_k) = (A_1, \ldots, A_k),$$

where $(A_1, \ldots, A_k)$ is the uniquely determined k-partition of $\rho(x_1 \ldots x_k)$ such that $A_i \cong \rho(x_i)$, for $i = 1, \ldots, k$.

THEOREM 11. 1. *If the canonical mapping C is a representation of $\mathcal{M}$, then it is a regular representation of $\mathcal{M}$.*

2. *If there exists a regular representation ρ of $\mathcal{M}$, then so is also C.*

PROOF. *Ad 1.* Let $uv = w$ be in $\mathcal{M}$, and suppose we have two different 2-partitions $AB = C(w)$, $A'B' = C(w)$, with $A = C(u) \cong A'$, $B = C(v) \cong B'$.

Suppose that A is a proper initial segment of A'. Since AB corresponds to a 2-partition uv, there is a gap between A and B. Since A is a proper initial segment of A', the gap is in A'. As every gap corresponds to a 2-partition of the preimage in C, there exists y and D, such that $A' = AD$ and $D \cong C(y)$. Hence $u = uy$, which is possible only if $y = \varepsilon$. But then D is empty, which is a contradiction.

Ad 2. Given a regular representation ρ we should show condition 3. for mapping C. Since ρ is a representation it suffices to show that $C(u) \cong C(v)$ implies $\rho(u) \cong \rho(v)$. Our strategy is

(i) to construct, for every $w \in M$, an order preserving mapping h with $h : \mathsf{Supp}(\rho(w)) \to \mathsf{Supp}(C(w))$, and then
(ii) to show that if $\imath : C(u) \to C(v)$ is an isomorphism, then for every $S \in \mathsf{Supp}(C(u))$ the fibers of S and $\imath(S)$, as decorated orderings, are isomorphic, i.e., $h^{-1}(S) \cong h^{-1}(\imath(S))$, or they are both empty.

If (i) and (ii) are true, then it is easy to construct an isomorphism $\rho(u) \cong \rho(v)$: it suffices to connect the isomorphisms of the fibers into one isomorphism.

Ad (i). Let $w \in M$, let $j \in \mathsf{Supp}(\rho(w))$. We define:

- $h(j) := \{(x, y, z) \mid \exists A, B, D\ (j \in B \text{ and } \rho^3(x, y, z) = (A, B, D))\}$.

One can readily verify that $h(j)$ is a word ultrafilter, and that h is order preserving.

Ad (ii). Let $S \in \mathsf{Supp}(C(u))$ and $T \in \mathsf{Supp}(C(v))$ such that $T = \imath(S)$. Then S and T have the same decoration, which means that $S \sim T$. By definition, there exist 3-partitions $(x, y, z) \in S$ and $(x', y, z') \in T$ such that $S_y = T_y$. Let $\rho^3(x, y, z) = (A, B, D)$ and $\rho^3(x', y, z') = (A', B', D')$. Then, $B \cong B'$, as ρ is a representation. Let us denote this isomorphism by κ.

Take an arbitrary 3-partition $y_1y_2y_3 = y$ and let

$$\rho^5(x, y_1, y_2, y_3, z) = (A, B_1, B_2, B_3, D),$$

and

$$\rho^5(x', y_1, y_2, y_3, z') = (A', B_1', B_2', B_3', D').$$

Then $B_i \cong B_i'$, for $i = 1, 2, 3$. By the regularity of ρ, the segments B_1, B_2, B_3 in B and the segments B_1', B_2', B_3' in B' are uniquely determined by their isomorphism types, whence:

$$\kappa(B_i) = B_i', \text{ for } i = 1, 2, 3. \tag{1}$$

The fiber $h^{-1}(S)$ is defined as the intersection of all segments B_2 that belong to 5-partitions (x, y_1, y_2, y_3, z) such that $(y_1, y_2, y_3) \in S_y = T_y$. Similarly, the fiber $h^{-1}(T)$ is defined as the intersection of all segments B_2' that belong to 5-partitions (x', y_1, y_2, y_3, z') such that $(y_1, y_2, y_3) \in S_y = T_y$. According to (1), for all such 3-partitions, $\kappa : (B_1, B_2, B_3) \cong (B_1', B_2', B_3')$. Hence κ is also an isomorphism of $h^{-1}(S)$ onto $h^{-1}(T)$, or both fibers are empty. ⊣

REFERENCES

[1] John Burgess, ***Fixing Frege***, Princeton Monographs in Philosophy, Princeton University Press, Princeton, NJ, 2005.

[2] Andrzej Grzegorczyk, *Undecidability without arithmetization*, ***Studia Logica. An International Journal for Symbolic Logic***, vol. 79 (2005), no. 2, pp. 163–230.

[3] Laurence Kirby, *Addition and multiplication of sets*, ***MLQ. Mathematical Logic Quarterly***, vol. 53 (2007), no. 1, pp. 52–65.

[4] P. Komjáth and V. Totik, ***Problems and Theorems in Classical Set Theory***, Problem Books in Mathematics, Springer, New York, 2006.

[5] W. V. Quine, *Concatenation as a basis for arithmetic*, ***The Journal of Symbolic Logic***, vol. 11 (1946), pp. 105–114.

[6] W. Sierpiński, ***Cardinal and Ordinal Numbers***, Polska Akademia Nauk, Monografie Matematyczne, vol. 34, Państwowe Wydawnictwo Naukowe, Warsaw, 1958.

[7] A. Tarski, *Der wahrheitsbegriff in den formalisierten Sprachen*, ***Studia Philosophica***, vol. 1 (1935), pp. 261–405, Reprinted as [9]. The paper is a translation of the Polish *Pojęcie prawdy w językach nauk dedukcyjnych*, Prace Towarzystwa Naukowego Warszawskiego, Wydział III matematyczno-fizycznych, no. 34, Warsaw 1933.

[8] ———, *The notion of rank in axiomatic set theory a some of its applications. abstract 628*, ***Bulletin of the American Mathematical Society***, vol. 61 (1955), p. 433.

[9] ———, *The concept of truth in formalised languages*, ***Logic, Semantics, Metamathematics*** (A. Tarski, editor), Oxford University Press, Oxford, 1956, (this paper is a translation of [7]), pp. 152–278.

[10] A. Visser, *Growing commas—a study of sequentiality and concatenation*, ***Notre Dame Journal of Formal Logic***, vol. 50 (2009), no. 1, pp. 61–85.

DEPARTMENT OF MATHEMATICS
UNIVERSITY OF ZAGREB, BIJENIČKA 30
10000 ZAGREB, CROATIA
E-mail: veky@math.hr

MATHEMATICAL INSTITUTE
ACADEMY OF SCIENCES OF THE CZECH REPUBLIC
ŽITNÁ 25, 115 67 PRAHA 1, CZECH REPUBLIC
E-mail: pudlak@math.cas.cz

SCHOOL OF PHILOSOPHY, ANTHROPOLOGY AND SOCIAL INQUIRY
THE UNIVERSITY OF MELBOURNE
PARKVILLE 3010, AUSTRALIA
E-mail: restall@unimelb.edu.au

DEPARTMENT OF COMPUTER SCIENCE
UNIVERSITY OF TORONTO, TORONTO
ONTARIO, CANADA M5S 1A4
E-mail: urquhart@cs.toronto.edu

DEPARTMENT OF PHILOSOPHY
UTRECHT UNIVERSITY, HEIDELBERGLAAN 8
3584 CS UTRECHT, THE NETHERLANDS
E-mail: albert.visser@phil.uu.nl

CARDINAL PRESERVING ELEMENTARY EMBEDDINGS

ANDRÉS EDUARDO CAICEDO

Abstract. Say that an elementary embedding $j : N \to M$ is *cardinal preserving* if $\mathrm{CAR}^M = \mathrm{CAR}^N = \mathrm{CAR}$. We show that if PFA holds then there are no cardinal preserving elementary embeddings $j : M \to V$. We also show that no ultrapower embedding $j : V \to M$ induced by a set extender is cardinal preserving, and present some results on the large cardinal strength of the assumption that there is a cardinal preserving $j : V \to M$.

§1. Introduction. This paper is the first of a series attempting to investigate the structure of (not necessarily fine structural) inner models of the set theoretic universe under assumptions of two kinds:

1. Forcing axioms, holding either in the universe V of all sets or in both V and the inner model under study, and
2. Agreement between (some of) the cardinals of V and the cardinals of the inner model.

I try to be as self-contained as is reasonably possible, given the technical nature of the problems under consideration. The notation is standard, as in Jech [8]. I assume familiarity with inner model theory; for fine structural background and notation, the reader is urged to consult Steel [19] and Mitchell [15].

In the remainder of this introduction, I include some general observations on large cardinal theory, forcing axioms, and fine structure, and state the main results of the paper.

Consider set theory with the axiom of choice as formalized by the Gödel-Bernays axioms GBC, so we can freely treat proper classes. An *inner model* (or simply, a *model*) is a transitive class model M of the Zermelo-Fraenkel ZFC axioms containing all the ordinals. If M is a model and φ is a statement, φ^M is the assertion that φ holds in M. If τ is a definable term, τ^M indicates the interpretation of τ inside M. Denote by ORD the class of ordinals and by CAR the class of cardinals. The cofinality of an ordinal α is denoted $\mathrm{cf}(\alpha)$. All our embeddings are elementary and non-trivial, and the classes involved are inner models; the critical point of such an embedding j is denoted $\mathrm{cp}(j)$.

Logic Colloquium '07
Edited by Françoise Delon, Ulrich Kohlenbach, Penelope Maddy, and Frank Stephan
Lecture Notes in Logic, 35

1.1. Large cardinal axioms. See Kanamori [10] as a general reference for large cardinals. It is well-known that the GBC axioms fall quite short of providing a complete picture of the universe of sets. Among the many independent extensions that have been studied, two sets of statements have been isolated as natural candidates to add to the basic axioms: large cardinal axioms and forcing axioms. By far, large cardinal axioms are better understood and more readily accepted. There is no formal definition of what a large cardinal is, but a few features can be distinguished. Typically, they are regular cardinals κ such that V_κ is itself a model of ZFC and, more importantly for present purposes, one can associate to κ a family of elementary embeddings $j : V \to M$ where M is a transitive class. The association is usually (as in the case of measurable, strong or supercompact cardinals) that κ is the *critical point* $\mathrm{cp}(j)$ of j, the first ordinal α such that $j(\alpha) > \alpha$, but it can take other shapes, as is the case with Woodin cardinals.

An important remark is that these notions can be stated in terms of the existence of certain *ultrafilters* or systems of ultrafilters called *extenders*, the connection being given by an analogue of the model theoretic *ultrapower* construction. An extender is essentially (a family of ultrafilters coding together) a fragment of an elementary embedding, and it is by now a standard device; good expositions and definitions can be found in Jech [8, Chapter 20], Kanamori [10, § 26], and Steel [19, § 2.1], among others. Briefly:

For a set X and a cardinal κ, let

$$[X]^\kappa = \{Y \subseteq X : |Y| = \kappa\}$$

and define $[X]^{<\kappa}$ and $[X]^{\leq\kappa}$ similarly.

Definition 1.1. Let κ be a cardinal and let $\lambda > \kappa$. A *non-trivial* (κ, λ)-*extender* E is a sequence $(E_a : a \in [\lambda]^{<\omega})$ such that there is some $\zeta \geq \kappa$ for which the following hold:

1. For each $a \in [\lambda]^{<\omega}$, E_a is a κ-complete ultrafilter over $[\zeta]^{|a|}$. We call $\kappa = \mathrm{cp}(E)$ the *critical point* of E and $\lambda = \mathrm{lh}(E)$ the *length* of E.
2. (Non-triviality) For at least one $a \in [\lambda]^{<\omega}$, E_a is not κ^+-complete.
3. For each $\xi \in \zeta$ there is $a \in [\lambda]^{<\omega}$ such that $\{s \in [\zeta]^{|a|} : \xi \in s\} \in E_a$.
4. (Coherence or Compatibility) Whenever $a \subseteq b \in [\lambda]^{<\omega}$ let $\pi_{ba} : [\zeta]^{|b|} \to [\zeta]^{|a|}$ be the *projection* map given by
$$\pi_{ba}(s) = \pi_s[a]$$
where $\pi_s : (b, <) \to (s, <)$ is the unique order isomorphism. Then, for $a \subseteq b \in [\lambda]^{<\omega}$ and $X \subseteq [\zeta]^{|a|}$,
$$X \in E_a \text{ iff } \{s : \pi_{ba}(s) \in X\} \in E_b.$$
5. (Normality) If $a \in [\lambda]^{<\omega}$, $i < |a|$, and $f : [\zeta]^{|a|} \to V$ is such that
$$\left\{s \in [\zeta]^{|a|} : f(s) < s_i\right\} \in E_a$$

(where s_i is the i-th element of s in increasing order) then there is $b \in [\lambda]^{<\omega}$ with $a \subseteq b$ and $k < |b|$ such that

$$\left\{s \in [\zeta]^{|b|} : f(\pi_{ba}(s)) = s_k\right\} \in E_b,$$

where π_{ba} is as above.

The relation between extenders and elementary embeddings is described in the following result:

LEMMA 1.2. 1. *Assume $j : V \to N$ is elementary. Let $\kappa = \mathrm{cp}(j)$ and $\lambda > \kappa$. Let ζ be (least) such that $j(\zeta) \geq \lambda$. For all $a \in [\lambda]^{<\omega}$, define*

$$E_a := \{X \subseteq [\zeta]^{|a|} : a \in j(X)\}.$$

Then $E = (E_a : a \in [\lambda]^{<\omega})$ is a (κ, λ)-extender.

2. *Conversely, given a (κ, λ)-extender E with each E_a over $[\zeta]^{|a|}$, there is an elementary embedding $j : V \to N$ such that*
   (a) *$\kappa = \mathrm{cp}(j)$, ζ is least such that $\lambda \leq j(\zeta)$, and for all $a \in [\lambda]^{<\omega}$, $E_a = \{X \subseteq [\zeta]^{|a|} : a \in j(X)\}$. Moreover,*
   (b) *$N = \{j(f)(s) : s \in [\lambda]^{<\omega}$ and $f : [\zeta]^{|s|} \to V\}$.* ⊣

For a proof see any of the references cited above; the model N and the embedding j in item 2 are obtained as the direct limit of the system that consists of the ultrapowers $\mathrm{ult}(V, E_a)$ and their associated embeddings; that this system is directed follows from the compatibility condition in Definition 1.1. It is customary to refer to this direct limit (or its transitive collapse) as $\mathrm{ult}(V, E)$. Our interest in Lemma 1.2 lies in the following corollary, an immediate consequence of Lemma 1.2 2(b):

COROLLARY 1.3. *Let E be a (κ, λ)-extender with each E_a over $[\zeta]^{|a|}$, and let $j_E : V \to N$ be the associated embedding. Then, for any ξ, $j_E(\xi) < (|\xi^\zeta| \cdot |\lambda|)^+$.* ⊣

We have mentioned examples of large cardinals. However, what is accepted as a large cardinal *axiom* is more general. For example, the existence of certain *mice* is a large cardinal axiom. They do not imply that large cardinals exist in V, but rather in certain inner models.

A compelling reason for accepting large cardinal axioms comes from the heuristic realization that natural statements that do not involve large cardinals in their formulation can be shown equiconsistent with certain large cardinal axioms. At the moment this is more a conjecture than a fact (hence the informal adjective "natural" in the description just given, intended to exclude, for example, pathological statements such as those generated by means of Gödel sentences), but it is a widely accepted state of affairs having been verified for a varied class of examples; to name just a few: reflection of stationary sets, determinacy hypotheses, or the failure of the singular cardinals hypothesis

SCH. Recall that SCH is the statement that for all cardinals κ,

$$\kappa^{\mathrm{cf}(\kappa)} = 2^{\mathrm{cf}(\kappa)} + \kappa^{+};$$

it follows from work of Woodin and Gitik (see Jech [8, Chapter 36] and references therein) that the failure of SCH is equiconsistent with the existence of a cardinal κ of Mitchell order $o(\kappa) = \kappa^{++}$. Baumgartner showed that the proper forcing axiom PFA (see subsection 1.2) is consistent relative to the existence of supercompact cardinals, see Shelah [18, Chapter VII]; it is widely expected that, once fine structural inner model theory has been developed enough, it will be shown that PFA is in fact equiconsistent with a strong large cardinal axiom. The best available result to date is from Jensen-Schimmerling-Schindler-Steel [9], where it is shown that PFA implies the existence of non-domestic mice (see subsection 1.3), see also Andretta-Neeman-Steel [1].

1.2. Forcing axioms. As said before, forcing axioms are not yet so widely understood. While some set theorists go as far as considering them natural statements, what seems to be the consensus is that they formalize a very desirable feature of the universe of sets. Namely, one would like the universe to be as "wide" or "saturated" as possible. The way in which forcing axioms formalize this desire is by stating that certain sets that can be added by certain forcing posets already exist. There are restrictions on the class of posets to be considered and on how generic these sets that would have been added can actually be. By loosening or increasing the restrictions, a variety of forcing axioms can be identified. The most widely used and best known is by far Martin's axiom MA, see Martin-Solovay [14], but this is not a strong enough statement to even decide the size of the continuum. Strong forcing axioms are much stronger than MA both in consequences and in consistency strength. Typical examples are the proper forcing axiom PFA introduced by Baumgartner and Shelah, see Baumgartner [2] and Shelah [18, Chapter VII], and Martin's maximum MM introduced in the groundbreaking paper Foreman-Magidor-Shelah [5].

For Γ a collection of forcing notions, let $\mathsf{FA}(\Gamma)$ be the statement that for any $\mathbb{P} \in \Gamma$ and any ω_1-many given dense subsets of $\mathbb{P}$ there is (in V) a filter $G \subseteq \mathbb{P}$ sufficiently generic in the sense that G meets each of these dense filters.

Assume that X is an uncountable set. A set $S \subseteq [X]^{\omega}$ is said to be *stationary* if for any function $f : [X]^{<\omega} \to X$ there is a set $y \in S$ closed under f. A forcing $\mathbb{P}$ is *proper* iff for any stationary set S, it is still the case in $V^{\mathbb{P}}$ that S is stationary. In particular, uncountable sets in V are still uncountable in $V^{\mathbb{P}}$ and thus $\omega_1^V = \omega_1^{V^{\mathbb{P}}}$. The proper forcing axiom PFA is the statement $\mathsf{FA}(\mathsf{Proper})$, where Proper is the class of proper posets.

Given a poset $\mathbb{P}$, it can be shown, see Foreman-Magidor-Shelah [5], that $\mathsf{FA}(\{\mathbb{P}\})$ fails if there is a stationary subset S of ω_1 that is no longer stationary in $V^{\mathbb{P}}$ (one says that the stationarity of S was not preserved). Martin's maximum MM is the statement that $\mathsf{FA}(\Gamma)$ holds, where Γ is the class of posets that preserve stationary subsets of ω_1. Obviously, MM is the strongest forcing axiom possible, and in particular it implies PFA. As with PFA, it is consistent relative to the existence of a supercompact cardinal.

Many consequences of PFA can be considered natural features of the universe of sets (for example, the failure of the continuum hypothesis, the failure of square principles, the singular cardinal hypothesis, determinacy in $L(\mathbb{R})$, and generic absoluteness of $L(\mathbb{R})$; see for example Bekkali [3], Todorčević [20], Viale [22], and Jensen-Schimmerling-Schindler-Steel [9]) thus providing evidence for its acceptance as a natural extension of ZFC. However, even if one does not consider PFA to be "natural", it seems reasonable that some common features of forcing axioms and other similar *strong reflection principles* will eventually be considered as natural as large cardinal axioms.

Recall:

DEFINITION 1.4. Let κ be an uncountable cardinal.

1. κ is *strongly compact* iff for any set S, any κ-complete filter over S can be extended to a κ-complete ultrafilter over S.
2. Given a set X, a filter F over $[X]^{<\kappa}$ is *fine* iff it is κ-complete and, for all $x \in X$,
$$\{\sigma \in [X]^{<\kappa} : x \in \sigma\} \in F.$$
3. Given a cardinal $\lambda \geq \kappa$, say that κ is *λ-strongly compact* or, simply, *λ-compact* iff there is a fine ultrafilter over $[\lambda]^{<\kappa}$.

The following characterization of strong compactness in terms of elementary embeddings will be useful in Section 3, see Kanamori [10, Theorem 22.17]:

THEOREM 1.5. *Let $\omega < \kappa \leq \lambda$. Then the following are equivalent*:

1. *κ is λ-compact.*
2. *There is an elementary $j : V \to N$ with $\mathrm{cp}(j) = \kappa$ and such that for any $X \subseteq N$ with $|X| \leq \lambda$, there is $Y \in N$ with $X \subseteq Y$ and $N \models |Y| < j(\kappa)$.*
3. *For any set S, any κ-complete filter over S generated by at most λ sets can be extended to a κ-complete ultrafilter over S.* ⊣

We have occasion to use some covering properties introduced in the highly recommended Viale [23]. We proceed to recall the relevant notions and results:

DEFINITION 1.6 (Viale). Let $\lambda < \kappa$ be regular cardinals. A collection of sets $\mathcal{D} = (K(\alpha, \beta) : \alpha < \lambda, \beta < \kappa)$ is a *λ-covering matrix for κ* iff the following conditions are met:

1. $\beta \subseteq \bigcup_{\alpha<\lambda} K(\alpha, \beta)$ for all $\beta < \kappa$.
2. $K(\alpha, \beta) \subsetneq K(\eta, \beta)$ for all $\beta < \kappa$ and all $\alpha < \eta < \lambda$.
3. For all $\alpha < \lambda$ and all $\gamma < \beta < \kappa$ there is $\eta < \lambda$ such that $K(\alpha, \gamma) \subseteq K(\eta, \beta)$.
4. For all $X \in [\kappa]^{\leq\lambda}$ there is $\gamma_X < \kappa$ such that for all $\beta < \kappa$ and $\eta < \lambda$ there is $\alpha < \lambda$ such that $K(\eta, \beta) \cap X \subseteq K(\alpha, \gamma_X)$.

DEFINITION 1.7 (Viale). If $\kappa > \lambda$ are regular cardinals, the *covering property* $\mathsf{CP}(\kappa, \lambda)$ holds iff for every λ-covering matrix $\mathcal{D}$ for κ there is an $A \in [\kappa]^\kappa$ such that

$$[A]^\lambda \subseteq \bigcup \left\{ [K(\alpha, \beta)]^\lambda : \alpha < \lambda, \beta < \kappa \right\}.$$

The definition of the combinatorial principle $\square(\kappa)$ mentioned below can be found, for example, in Moore [16].

THEOREM 1.8 (Viale). 1. $\mathsf{CP}(\kappa, \omega)$ *implies that* $\square(\kappa)$ *fails.*

2. *If* κ *is singular,* $\mathrm{cf}(\kappa) = \lambda$, $2^\lambda < \kappa$, *and* $\mathsf{CP}(\kappa^+, \lambda)$ *holds, then* $\kappa^\lambda = \kappa^+$.
3. $\mathsf{CP}(\kappa^+, \omega)$ *for all singular* κ *of cofinality* ω *implies the singular cardinal hypothesis* SCH.
4. PFA *implies* $\mathsf{CP}(\kappa^+, \omega)$ *for all singular* κ *of cofinality* ω.
5. *If* λ *is strongly compact, then* $\mathsf{CP}(\kappa, \theta)$ *holds for all regular* $\theta < \lambda$ *and all regular* $\kappa \geq \lambda$.
6. *If* $\mathsf{CP}(\kappa^+, \theta)$ *holds and* M *is an inner model such that* κ *is regular in* M *and* $(\kappa^+)^M = \kappa^+$, *then* $\mathrm{cf}(\kappa) \neq \theta$.
7. *If* PFA *holds and* M *is an inner model with the same cardinals, then* M *computes correctly all ordinals of cofinality* ω. $\dashv$

1.3. Fine structure. See Steel [19], Mitchell [15], and references within for historical references and details on the notions and results mentioned here. What follows owes much to expositions by Ketchersid. Inner model theory is a rather technical area in set theory, and by necessity the presentation I make of it here is somewhat of a caricature. Gödel defined the constructible universe L; it is not really a good model of set theory as currently understood since it does not contain any significant large cardinals. On the other hand, its theory admits a very detailed analysis (known as fine structure theory and introduced by Jensen) by studying how new sets are added in the inductive construction of L in terms of the complexity of their definitions. One can generalize L to models of the form $L[\mathcal{E}]$, the constructible universe built from an additional predicate $\mathcal{E}$, and by allowing $\mathcal{E}$ to code (one or several) elementary embeddings, $L[\mathcal{E}]$ can be made to model a substantial fragment of the large cardinal hierarchy. However, to replicate the fine structural analysis of L in this larger generality requires substantially new ideas, since the predicate $\mathcal{E}$ should be chosen very carefully in order to obtain some kind of canonicity.

The (partial) solution devised by Mitchell and Steel, following a long line of research including work of Jensen, Solovay, Dodd and Mitchell, among others, is to let $\mathcal{E} = \langle E_\alpha : \alpha < \beta \rangle$ be a *coherent* sequence of extenders. By reference to just a extender E, i.e., without need of knowing the domain of the embedding being coded, one can recover the critical point of this embedding, which justifies talking of $\mathrm{cp}(E)$. Also, as long as some agreement conditions are satisfied, an *ultrapower* $\mathrm{ult}(M, E)$ of a model M by an extender E can be formed, with critical point $\mathrm{cp}(E)$, even if M was not the domain of the original embedding coded by E or if $E \notin M$. Coherency is a technical requirement; each E_α is either $\emptyset$ or else it is an extender over $L_\alpha[\mathcal{E} \restriction \alpha]$; moreover, if $j : L_\alpha[\mathcal{E} \restriction \alpha] \to \mathrm{ult}(L_\alpha[\mathcal{E} \restriction \alpha], E_\alpha)$ is the ultrapower embedding by E_α, then $j(\mathcal{E})\restriction\alpha = \mathcal{E}\restriction\alpha$ and $j(\mathcal{E})_\alpha = \emptyset$. (In this description I have ignored a few details, in particular, some technical remarks related to the well-foundedness of the target model.)

A *potential premouse* (ppm) is a structure $\mathcal{M} = \langle L_\alpha[\mathcal{E}\restriction\alpha], E_\alpha\rangle$ where $\mathcal{E}$ is a coherent sequence of extenders. These are structures that "look like" initial segments of one of the canonical models one wants to build. Part of what the canonicity of the construction requires is an analogue of some of the nice *condensation* properties of L. This is called the *initial segment condition*, and a ppm satisfying this is called a premouse.

For $\mathcal{M}$ as above and $\beta \leq \alpha$, let $\mathcal{M}|\beta = L_\beta[\mathcal{E}\restriction\beta]$ and $\mathcal{M}\|\beta = \langle L_\beta[\mathcal{E}\restriction\beta], E_\beta\rangle$. A key issue when studying premice is the question of *iterability*. Two premice $\mathcal{M}$ and $\mathcal{N}$ are *lined up* if one is an initial segment of the other, i.e., if for some β, $\mathcal{N} = \mathcal{M}\|\beta$ or $\mathcal{M} = \mathcal{N}\|\beta$. To *compare* two premice $\mathcal{M}$ and $\mathcal{N}$ means to produce from them two (other) lined up premice $\mathcal{M}^*$ and $\mathcal{N}^*$. To do this, the notion of *iteration tree*, due to Martin and Steel, is required. In short, nice inner models for small large cardinals can be compared by iterating ultrapowers, see Jech [8, Chapter 19]: One looks at the first ordinal β where $\mathcal{M}\|\beta \neq \mathcal{N}\|\beta$. This disagreement must come from their *top* measures being different, $E^{\mathcal{M}}_\beta \neq E^{\mathcal{N}}_\beta$. Forming $\mathcal{M}' = \mathrm{ult}(\mathcal{M}|\beta, E^{\mathcal{M}}_\beta)$ and $\mathcal{N}' = \mathrm{ult}(\mathcal{N}|\beta, E^{\mathcal{N}}_\beta)$, the coherency property of the extender sequences implies that $\mathcal{M}'\|\beta = \mathcal{N}'\|\beta$, i.e., we have effectively removed a disagreement. The process is continued (taking direct limits at limit stages) until (if) lined up models are produced. It is a remarkable result of Kunen that this is indeed the case for models for one measurable cardinal, and this result can be generalized. The iterations so obtained are essentially *linear* iterations, and this linearity seriously bounds both the complexity of the reals that can belong to such inner models and their large cardinals. Martin and Steel found a non-linear method of iterating ultrapowers of inner models in the region of Woodin cardinals. These models give then rise to iteration trees, trees of structures with embeddings between the models appearing along their branches. If two models are compared this way, at limit stages of the comparison process, different possibilities on how to continue the trees may arise, and the existence of these choices increases the complexity of their comparison process and explains why these models allow more complicated reals and large cardinals than those appearing in linearly iterable models. From the existence of enough large cardinals one can deduce the existence of nice models M, i.e., *iterable* (in an appropriate sense that allows us to carry out the comparisons mentioned above) models of enough set theory with roughly the same large cardinals. Exactly how far in the large cardinal hierarchy this process can go is open, but models with many Woodin cardinals can be obtained this way.

A premouse $\mathcal{M}$ is *collapsing* if every element of $\mathcal{M}$ is Σ_1-definable over $\mathcal{M}$ from no parameters—in particular, $\mathcal{M}$ is countable. Collapsing (sufficiently iterable) premice are already lined up. The most famous example of a collapsing mouse (and the smallest one) is *zero sharp*, $0^\sharp$, this is an iterable model of the form $0^\sharp = M_0^\sharp = \langle L_\alpha, E_\alpha \rangle$ where $0^\sharp \models$"E_α is a measure on $\text{cp}(E_\alpha)$", with α chosen as least as possible. Hence, $0^\sharp$ resembles an inner model with a measurable cardinal, except that by considering the next level of the constructible hierarchy, the universe is collapsed to ω. Other examples include $0^\dagger$, which is a model with a genuine measure and a top extender, and $M_l^\sharp$, $l \leq \omega$, models with l Woodin cardinals and a top extender. Since these models are lined up, one can talk of a mouse being "below a Woodin cardinal", for example. Work of Steel, Martin, Woodin and Neeman has shown that $\utilde{\Sigma}^1_{n+1}$-determinacy is equivalent to the existence and iterability of $M_n^\sharp(x)$ for all reals x, where these models are defined as $M_n^\sharp$, but the real x is added as an element at the bottom stage (for $n = 0$ this is a result of Martin and Harrington). Similarly, $\mathsf{AD}^{L(\mathbb{R})}$ in all set generic extensions is equivalent to the existence and iterability of $M_\omega^\sharp$ and this is in turn equivalent to the generic absoluteness of $L(\mathbb{R})$. Below, we mention non-domestic premice; these are premice $\mathcal{M}$ such that there is an ordinal $\nu \leq \text{ORD}^{\mathcal{M}}$ such that in $\mathcal{M}\|\nu$, $\kappa = \text{cp}(E_\nu)$ is a limit of Woodin cardinals and of cardinals strong up to κ, see Andretta-Neeman-Steel [1].

The model K, the *core model*, is an iterable model built from a coherent sequence of extenders as above. It is intended to faithfully represent the large cardinal structure of the universe. Its existence is in general an open problem (and requires some additional technical assumptions to ensure its iterability), but it has been successfully identified as long as this large cardinal structure is not too complicated. For example, it is just L if $0^\sharp$ does not exist. Much like L, K is quite canonical; for example, it is definable, and its definition is invariant under forcing. It satisfies the *weak covering property*, namely that $(\lambda^+)^K = \lambda^+$ for all V-singular cardinals λ; when $K = L$ (which is to say, when $0^\sharp$ does not exist) this is a consequence of Jensen's celebrated *covering lemma*. An important technical feature of K is its *rigidity*. This means that there are no elementary embeddings $j : K \to K$. In fact, a standard technique to show that an assumption φ implies the existence of mice capturing certain large cardinals, consists on attempting to build K under the assumption that φ holds and the mice in question do not exist, and then proceeding to show that K is non-rigid.

1.4. Results. We study embeddings $j : V \to M$ or $j : M \to V$ where the model M computes cardinals correctly, i.e., $\text{CAR}^M = \text{CAR}$. We call these embeddings *cardinal preserving*. The expectation is that they do not exist, and the results of this paper can be seen as a step towards confirming this expectation. In Section 2 we show:

THEOREM 1.9. *If* PFA *holds, then there are no cardinal preserving embeddings* $j : M \to V$.

Theorem 1.9 follows Viale's Theorem 1.8 via a structural restriction we identify, see Theorem 2.5.

In Section 3 we study cardinal preserving embeddings $j : V \to M$, and prove:

THEOREM 1.10. *If E is a set extender and $j : V \to M$ is the corresponding ultrapower embedding, then j is not cardinal preserving.*

Theorem 1.10 also makes use of (the proof of) Viale's Theorem 1.8. Next, we show that cardinal preserving embeddings $j : V \to M$ have significant large cardinal strength, both *locally* (e.g., $\text{cp}(j)$ is $\text{cp}(j)^+$-compact and much more) and *globally* (e.g., V is closed under sharps, and significantly more).

All the lower bounds in consistency strength obtained in this paper, Theorems 2.11, 3.2 and 3.3, except for the combinatorial Theorem 3.7 (joint with H. Woodin), follow from fine structural considerations that trace back to violations of the appropriate covering lemma, see Mitchell [15]; Steel [19] and Mitchell [15] provide all unexplained notation and background results used in these proofs.

§2. Embeddings into V.

QUESTION 2.1. Assume $j : M \to N$ is an elementary embedding. Can we have $\text{CAR}^M = \text{CAR}^N = \text{CAR}$?

We suspect the answer is no. In this paper we take the first step towards this question by considering the case where either M or N is V, and analyzing the structure of such *cardinal preserving elementary embeddings*.

Notice that such embeddings seem rather difficult to attain by standard means. For example:

THEOREM 2.2 (Hamkins [6]). *No embedding $j : V \to V[G]$ can be produced by set forcing.* ⊣

Embeddings into V *can* be produced by class forcing, see below, or by means of indiscernibles, see Vickers-Welch [24].

Also, if E is a set extender and $j : V \to M$ is the corresponding ultrapower embedding, then j is not cardinal preserving. We present here a quick argument under the additional assumption of the singular cardinal hypothesis SCH: Let E be a (κ, β)-extender with each E_a over $[\zeta]^{|a|}$, let $j : V \to M$ be the corresponding embedding, and let λ be a sufficiently large strong limit cardinal of cofinality κ. Then $\lambda < \lambda^+ = \lambda^{|\xi|}$ by SCH, and $\lambda < j(\lambda)$, since $\mathrm{cf}(\lambda) = \kappa$. But

$$\lambda^{|\xi|} \leq j(\lambda^{|\xi|}) < (\lambda^{|\xi|}|\beta|)^+,$$

by Corollary 1.3. So, if j is cardinal preserving, then $j(\lambda^{|\xi|}) = \lambda^{|\xi|}$, and $\lambda < j(\lambda) < j(\lambda^{|\xi|}) = \lambda^+$, contradiction. We present a proof without the assumption of SCH in Theorem 3.6.

Initial impulse for this line of research came from the following conjecture:

Conjecture 2.3. Assume PFA. Let M be an inner model that computes cardinals correctly and contains all the reals. Then $\mathrm{ORD}^\omega \subset M$.

This is not the place to explain why we expect Conjecture 2.3 to be the case; see Viale [22, 23] and Caicedo-Veličović [4]. As a quick motivation, recall that Moore has shown that MRP, a consequence of PFA (see Moore [16]), fails after adding a Prikry sequence, and Viale [21] has generalized this result.

We can help clarify the role of the assumption of agreement of cardinals:

1. Suppose for example that M has enough large cardinals. Then there are forcing extensions V of M where PFA holds, but V and M have little else in common so, clearly, additional agreement between V and M is required.
2. Or suppose that κ is measurable. Let M be the result of iterating ω-many times a given measure on κ and let $j : V \to M$ be the corresponding embedding. Then $\mathrm{cf}(j(\kappa)) = \omega$, $2^\kappa < j(\kappa) < (2^\kappa)^+$ and, in M, $j(\kappa)$ is inaccessible (by elementarity of j). It follows that M is not closed under ω-sequences and $\mathbb{R} \subset M$. Of course, we can also assume in this situation that PFA holds in V.
3. Also, if M has a proper class of completely Jónsson cardinals, $\mathbb{P}_\infty$ is the class stationary tower (see Larson [13]), G is $\mathbb{P}_\infty$-generic over M, and $V = M[G]$, then there is (definably in the structure (V, M, G)) an elementary
$$j : M \to V,$$
and we can arrange that $\mathrm{cp}(j)$ is arbitrarily high and $\mathrm{cf}^V(\mathrm{cp}(j)) = \omega$. However, in the situation of this example, cardinals are always collapsed: The critical point of such a j is in M an uncountable regular cardinal λ. Let $a = S^\lambda_\omega$, the set of ordinals below λ of cofinality ω. By elementary properties of $\mathbb{P}_\infty$, $\mathrm{cp}(j) = \lambda$ and $\mathrm{cf}(\lambda) = \omega$ iff $a \in G$. For example, if $a \in G$ then $j[\bigcup a] \in j(a)$, so $j[\lambda]$ will be an ordinal of cofinality ω below $j(\lambda)$. Of course, this means that $j[\lambda]$ must equal λ, and λ is therefore the critical point of j. If λ is a successor cardinal in M, then it is collapsed in V but if it is inaccessible in M, then it is preserved even though its cofinality changes. Decoding definitions, λ^+ is preserved in $M[G]$ iff the intersection of a with the set of subsets of λ^+ of order type λ^+ is stationary, and this intersection is in G. In particular, λ^+ is a Jónsson

cardinal in M, see Kanamori [10, § 8]. On the other hand, no successor of a regular cardinal is Jónsson, so λ^+ has to be collapsed.

The following result indicates that under PFA the existence of a cardinal preserving $j : M \to V$ would contradict Conjecture 2.3; this motivated Theorem 1.9 stating that no such embeddings can occur under PFA. Theorem 1.9 is proven in the following subsection; we show that if $j : M \to V$ is cardinal preserving, then M computes incorrectly many cofinalities (which is significantly stronger than $\mathrm{ORD}^\omega \not\subseteq M$).

THEOREM 2.4 (Foreman, see Vickers-Welch [24]). *If $j : M \to V$ is elementary, then $\mathrm{ORD}^\omega \not\subseteq M$.* ⊣

Thus, one expects that Question 2.1 has a negative answer under PFA (or even provably in GBC).

2.1. Discontinuities. In the case of cardinal preserving embeddings $j : M \to V$ a rather bizarre picture is known.

THEOREM 2.5. *Suppose that $j : M \to V$ is cardinal preserving. Let $\kappa = \mathrm{cp}(j)$. Then for all $\lambda > \kappa$, $j(\lambda) > \lambda$. In particular, if $j[\lambda] \subseteq \lambda$, then $\mathrm{cf}^M(\lambda) \geq \kappa$.*

PROOF. We proceed by contradiction, assuming $j : M \to V$ is a counterexample. Let $\kappa = \mathrm{cp}(j)$. We want to show that if $\lambda > \kappa$ then $j(\lambda) > \lambda$. Let λ be the first counterexample. Then λ is a cardinal: If $M \models \tau = |\lambda|$, then $j(\tau) = |j(\lambda)| = |\lambda|$, since $j(\lambda) = \lambda$. Since $j(\tau)$ is a cardinal and in M there is a bijection between λ and τ, $j(\tau) = |\lambda| = \tau$.

The argument now splits into cases, depending on whether λ is singular or not.

CASE 1. λ is singular.

The following notions and result are essential to pcf theory, see Shelah [17, Chapter II].

DEFINITION 2.6. 1. Let I be an ideal on a set X and $f, g : X \to \mathrm{ORD}$. Then $f <_I g$ iff

$$\{x \in X : g(x) \leq f(x)\} \in I.$$

2. Given a cardinal τ, let $\mathcal{J}_\tau^{bd}$ be the ideal of bounded subsets of τ.

DEFINITION 2.7. Let μ be singular. A *scale for μ is a tuple $(\vec{\mu}, \vec{f})$ such that*

1. $\vec{\mu} = (\mu_i : i < \mathrm{cf}(\mu))$ *is an increasing sequence of regular cardinals cofinal in μ.*
2. $\vec{f} = (f_\alpha : \alpha < \mu^+)$ *is a sequence of functions such that*
   (a) $f_\alpha \in \prod_{i<\mathrm{cf}(\mu)} \mu_i$ *for all $\alpha < \mu^+$,*
   (b) *If $\beta < \gamma < \mu^+$, then $f_\beta <_{\mathcal{J}^{bd}_{\mathrm{cf}(\mu)}} f_\gamma$, and*
   (c) *If $f \in \prod_{i<\mathrm{cf}(\mu)} \mu_i$, then there is $\alpha < \mu^+$ such that $f <_{\mathcal{J}^{bd}_{\mathrm{cf}(\mu)}} f_\alpha$.*

THEOREM 2.8 (Shelah [17, Chapter II]). *Let λ be singular. Then there is a scale for λ.* ⊣

Let $(\vec{\lambda}, \vec{f})$ be a scale for λ in M. We may assume that $\lambda_0 > \kappa$, so $j(\lambda_i) > \lambda_i$ for all $i < \mathrm{cf}^M(\lambda)$.

Let $\delta = \mathrm{cf}^M(\lambda)$. Then $j(\delta) = \mathrm{cf}(j(\lambda)) = \mathrm{cf}(\lambda) \leq \mathrm{cf}^M(\lambda) = \delta$, so $\delta = j(\delta)$ and therefore, since λ is the least fixed point of j above κ, $\delta < \kappa$ and $j(\vec{\lambda}) = j[\vec{\lambda}]$.

Now we use an argument of Zapletal [25]. We have that $(j(\vec{\lambda}), j(\vec{f}))$ is a scale for λ in V. Notice that $\lambda^+ = j(\lambda^+)$, so $j[\vec{f}] = (j(f_\alpha) : \alpha < \lambda^+)$ is a scale for λ as well.

Now consider $g \in \prod_{i<\delta} j(\lambda_i)$ given by

$$g(i) = \sup(j[\lambda] \cap j(\lambda_i)) = \sup(j[\lambda_i]).$$

Notice that $g(i) < j(\lambda_i)$ since $j(\lambda_i)$ is regular in V, so g is well defined. Notice also that g dominates each $j(f_\alpha)$, since

$$j(f_\alpha)(i) = j(f_\alpha)(j(i)) = j(f_\alpha(i)) < g(i)$$

for all i, so $j[\vec{f}]$ is not a scale. Contradiction.

CASE 2. λ is regular.

Then $\lambda^{+\lambda}$ is a fixed point of j, singular, and limit of regular cardinals τ such that $j(\tau) > \tau$. We can then easily modify the argument of Case 1 so it applies to $\lambda^{+\lambda}$.

We have shown that for all $\lambda > \kappa$, $j(\lambda) > \lambda$. Now we prove the last assertion of the theorem. If $j[\lambda] \subseteq \lambda$ and $\mathrm{cf}^M(\lambda) < \kappa$, let $A \subseteq \lambda$ be in M cofinal and of order type $\mathrm{cf}^M(\lambda)$. Then $j(A) = j[A] \subseteq \lambda$ and $j(\lambda) = \lambda$, a contradiction. This completes the proof. ⊣

It follows from this result that the critical point of a cardinal preserving embedding into V is Π_1-indescribable in a very strong sense. For example:

COROLLARY 2.9. *If there is a cardinal preserving $j : M \to V$ then there is a proper class of weakly inaccessible cardinals.*

PROOF. Let $\kappa = \mathrm{cp}(j)$. Any weakly inaccessible cardinal λ in V is also weakly inaccessible in M and therefore $j(\lambda)$ is (another) weakly inaccessible cardinal. κ is weakly inaccessible in M, so there are (in V, thus in M) weakly inaccessible cardinals above κ. If there are only set many of them, their supremum would be a fixed point of j. ⊣

From Theorem 2.5 and Viale's Theorem 1.8, we immediately obtain:

COROLLARY 2.10. *If $j : M \to V$ is cardinal preserving,* PFA *fails and any strongly compact cardinal is larger than* $\mathrm{cp}(j)$.

PROOF. That PFA fails follows from Theorem 1.8.7 (itself a consequence of items 4 and 6 of Theorem 1.8), since we can find (arbitrarily large) cardinals

μ closed under j and of V-cofinality ω, but any such μ has M-cofinality at least $\mathrm{cp}(j)$, by Theorem 2.5; for example,

$$\mu = j^{\omega}(\mathrm{cp}(j)) := \sup\{\,\mathrm{cp}(j), j(\mathrm{cp}(j)), j(j(\mathrm{cp}(j))), \dots\}.$$

If $\lambda < \mathrm{cp}(j)$ is strongly compact, then $\mathsf{CP}(\tau, \omega)$ holds for all regular cardinals $\tau \geq \lambda$, by Theorem 1.8.5. In particular, no M-regular cardinal above (or equal to) $\mathrm{cp}(j)$ can have V-cofinality ω, by Theorem 1.8.6. But then we reach a contradiction exactly as above. ⊣

It also follows in the same way that if $j : M \to V$ is cardinal preserving and τ is strongly compact, then for any $\lambda \geq \tau$, $\mathrm{cf}^M(\lambda) \geq \tau$ iff $\mathrm{cf}(\lambda) \geq \tau$.

2.2. Consistency strength. Let $j : M \to V$ be cardinal preserving. Since such an M computes incorrectly many cofinalities, covering fails very badly for M, and it should be no surprise that such an embedding would require considerable consistency strength. For example:

Theorem 2.11. *Assume that there is a cardinal preserving embedding $j : M \to V$. Then there are inner models with strong cardinals.*

Proof. Assume otherwise. Then K exists, is rigid, and satisfies the weak covering property. In particular, K computes cofinally many successor cardinals correctly.

Universality of a class $W = L[\mathcal{E}]$ is a technical assumption, see Steel [19] for a definition, but below strong cardinals, it follows by the covering lemma, see Mitchell [15], that if $(\mu^+)^W = \mu^+$ for cofinally many cardinals μ, then W is universal. By Mitchell [15], any universal proper class mouse W is an iterate of K, so in particular there is an elementary $\pi : K \to W$.

Since we are below strong cardinals, K^M is iterable in V (since, below Woodin cardinals, iterability can be expressed as a Π^1_2 condition) and it follows that it is universal by the weak covering property (in M) and the fact that $\mathrm{CAR}^M = \mathrm{CAR}$. But then there is an embedding $\pi : K \to K^M$, so $j \circ \pi : K \to K$ is nontrivial. Contradiction. ⊣

§3. Embeddings of V.

Now we turn our attention to the case of embeddings $j : V \to M$.

3.1. Nice cardinals. We start by showing that cardinal preserving embeddings must have significant large cardinal strength.

Definition 3.1. Given a cardinal preserving $j : V \to M$, we say that a regular cardinal μ is *j-nice* (or *nice*, if j is clear from context) iff

$$\delta_\mu := \sup j[\mu] < j(\mu).$$

Throughout this section (even if not explicitly stated), $j : V \to M$ is cardinal preserving and $\kappa = \mathrm{cp}(j)$. If μ is j-nice, then we can define a uniform

κ-complete ultrafilter $\mathcal{U}_\mu$ on μ by setting

$$\mathcal{U}_\mu := \{X \subseteq \mu : \delta_\mu \in j(X)\}.$$

The point is that there are many j-nice cardinals μ. For example, μ is j-nice if $\mu = \rho^+$ and $\rho < j(\rho)$, because $\sup j[\rho^+]$ has cofinality ρ^+ while $j(\rho^+) = j(\rho)^+$ is regular and larger than ρ^+ since $j(\rho) > \rho$.

THEOREM 3.2. *If there is a cardinal preserving $j : V \to M$, then V is closed under sharps.*

PROOF. Let A be a set, which we may assume transitive. If $A^\sharp$ does not exist, by the covering lemma, see Mitchell [15], whenever λ is a singular strong limit cardinal larger than $\text{rk}(A)$, then $\lambda^+ = (\lambda^+)^{L(A)}$. Fix such λ of cofinality κ. Then λ^+ is nice. As in the Vopěnka-Hrbáček argument from strongly compact cardinals, see Kanamori [10, Theorem 5.9], we can then consider the two embeddings $j_{\lambda^+} : V \to M_{\lambda^+} \cong \text{ult}(V, \mathcal{U}_{\lambda^+})$ and $k : V \to N_{\lambda^+} \cong \text{ult}^-(V, \mathcal{U}_{\lambda^+})$, where ult^- is formed by only considering those functions $f : \lambda^+ \to V$ such that $|\text{ran}(f)| < \lambda^+$. We then also have an embedding $i : N_{\lambda^+} \to M_{\lambda^+}$ given by $i([f]^-) = [f]$, where $[f]^-$ is the collapse of the equivalence class of f in the ultrapower $\text{ult}^-(V, \mathcal{U}_{\lambda^+})$ and $[f]$ is the collapse of the corresponding class in $\text{ult}(V, \mathcal{U}_{\lambda^+})$. We then have (see Kanamori [10, Theorem 5.9]):

- $j_{\lambda^+} = i \circ k$.
- $\text{cp}(i) \geq k(\lambda^+)$ and $j_{\lambda^+}(A) = k(A)$.
- $k(\lambda^+) < j_{\lambda^+}(\lambda^+)$.

But then it follows that $j_{\lambda^+}(\lambda) = k(\lambda)$ and

$$\begin{aligned} j_{\lambda^+}(\lambda^+) &= j_{\lambda^+}((\lambda^+)^{L(A)}) = (j_{\lambda^+}(\lambda)^+)^{L(j_{\lambda^+}(A))} \\ &= (k(\lambda)^+)^{L(k(A))} = k((\lambda^+)^{L(A)}) \\ &= k(\lambda^+), \end{aligned}$$

contradiction. ⊣

Schindler pointed out that essentially the same argument gives more information:

THEOREM 3.3. *If there is a cardinal preserving $j : V \to M$, then for all n and all X, $M_n^\sharp(X)$ exists.* ⊣

We omit the (technical) proof of Theorem 3.3, but include a (very brief) sketch for the experts: The result follows by induction on n. Assuming, for example, that $M_n^\sharp(X)$ exists for all X but $M_{n+1}^\sharp$ does not, if one lets E be a long extender coding a sufficiently large fragment of j, then one can build K inside $M_n^\sharp(E)$. In $M_n^\sharp(E)$ there is then an elementary embedding (coming from E, as above) sending K to a universal weasel such that the map is discontinuous at some successor cardinal where covering holds (just as before). This gives a contradiction, exactly as above.

The cardinal $\kappa = \mathrm{cp}(j)$ possesses very strong large cardinal properties just shy of strong compactness. For example, cofinally often, μ is nice and therefore the ultrafilter $\mathcal{U}_\mu$ is defined; in particular, this holds for all regular cardinals $\mu \geq \kappa$ below λ_j, where

$$\lambda_j = \min\left\{\text{first weakly inaccessible above } \kappa, \text{ first fixed point of } j \text{ above } \kappa\right\},$$

since all these cardinals μ are of the form ρ^+ for some ρ such that $\rho < j(\rho)$. Ketonen [12] has shown that an uncountable regular cardinal ν is τ-compact for regular $\tau \geq \nu$ iff all regular μ with $\nu \leq \mu \leq \tau$ carry a uniform ν-complete ultrafilter. Thus:

COROLLARY 3.4. *If there is a cardinal preserving $j : V \to M$ with $\kappa = \mathrm{cp}(j)$, and λ_j is defined as above, then κ is $< \lambda_j$-strongly compact.* $\dashv$

In particular, κ is κ^+-strongly compact, so κ is a measurable cardinal such that $\square_\kappa$ fails, see Kanamori-Magidor [11]. It follows from Andretta-Neeman-Steel [1] that there is a non-domestic premouse and in particular there are inner models of $\mathsf{ZF} + \mathsf{AD}_{\mathbb{R}}$ containing all the reals.

The following is immediate from the proof of Viale's Theorem 1.8, see Viale [23]:

THEOREM 3.5. *Given a cardinal preserving $j : V \to M$, if λ is j-nice, then the covering property $\mathsf{CP}(\lambda, \theta)$ holds for all regular $\theta < \kappa = \mathrm{cp}(j)$ and therefore*

1. *$\square(\lambda)$ fails.*
2. *If $\lambda = \rho^+$ and $\mathrm{cf}(\rho) = \theta < \kappa$, then $\lambda^\theta = \lambda$.* $\dashv$

In particular, if $\lambda = \rho^+$ is nice, then either ρ is singular in M or else $\mathrm{cf}(\rho) \geq \kappa$.

THEOREM 3.6. *If $j : V \to M$ is cardinal preserving, then j is not the ultrapower embedding by a set extender.*

PROOF. We proceed as before: Let E be a (κ, β)-extender with each E_a over $[\zeta]^{|a|}$, and let $j : V \to M$ be the corresponding embedding. Towards a contradiction, suppose that j is cardinal preserving. Let $\lambda > \zeta, \beta$ be a singular strong limit cardinal of cofinality κ so, in particular, $\lambda > 2^\kappa$. Then $\lambda < j(\lambda)$ and since $j(\lambda)$ is a singular cardinal of M-cofinality $j(\kappa)$, $j(\lambda) \geq \lambda^{+\kappa}$. Since $2^\lambda = \lambda^\kappa = j(\lambda^\kappa) > j(\lambda)$, λ violates the singular cardinal hypothesis SCH; here, $\lambda^\kappa = j(\lambda^\kappa)$ follows as before: $j(\lambda^\kappa) < (\lambda^{\kappa|\zeta|}|\beta|)^+$, by Corollary 1.3. It follows that $\lambda^{+\kappa}$ also violates SCH, since $\lambda^{+\kappa} < (\lambda^{+\kappa})^\kappa = \lambda^\kappa$ and, since λ^κ is fixed by j, then

$$\lambda^{+\kappa+1} < j(\lambda)^{+j(\kappa)} = j(\lambda^{+\kappa}) < j(\lambda^\kappa) = \lambda^\kappa.$$

By Silver's theorem or the Galvin-Hajnal results, see for example Holz-Steffens-Weitz [7, Corollary 2.3.4], $\{\mu < \lambda^{+\kappa} : \mu^\omega > \mu^+\}$ contains an ω-club. For μ singular of cofinality ω, recall that $\tau = \mu^+$ is nice if $j(\mu) > \mu$, and that therefore $\mathsf{CP}(\tau, \omega)$ holds, so $\tau^\omega = \tau$, or $\mu^\omega = \mu^+$. Hence, $j(\mu) = \mu$

for an ω-club of cardinals below $\lambda^{+\kappa}$. However, all cardinals in the interval $[\lambda, \lambda^{+\kappa})$ are moved by j. Contradiction. ⊣

3.2. Cofinality preserving embeddings. We close the paper by showing how consistency strength can be extracted in a cleaner way from an embedding $j : V \to M$ if we impose the formally stronger requirement that M computes cofinalities correctly.

THEOREM 3.7 (Caicedo, Woodin). *Assume* $j : V \to M$ *is such that if* $\kappa = \mathrm{cp}(j)$, *then* $\mathrm{cf}(\lambda) = \mathrm{cf}^M(\lambda)$ *for any* $\lambda \leq \sup j[j(\kappa)]$. *Then*

1. $j(\kappa)$ *is strongly inaccessible and*
2. $V_{j(\kappa)} \models \kappa$ *is strongly compact.*

PROOF. The assumption guarantees that λ is nice whenever $\kappa \leq \lambda \leq j(\kappa)$ and λ is regular, and therefore κ is $\leq j(\kappa)$-strongly compact. We start by showing that j witnesses the $< j(\kappa)$-strong compactness of κ in the sense that for any $\lambda < j(\kappa)$ there is in M a set Y such that $j[\lambda] \subset Y$ and $|Y| < j(\kappa)$. For this, fix in V a sequence

$$C = (C_\alpha : \alpha < \lambda)$$

such that each C_α is a club subset of α of order type $\mathrm{cf}(\alpha)$. Then $j(C)_{\delta_\lambda}$ is club in δ_λ and has order type λ. Let $D = \{\beta < \lambda : j(\beta) \in j(C)_{\delta_\lambda}\}$. Then D is $< \kappa$-club in λ and $j[D] \subseteq j(C)_{\delta_\lambda}$.

Now use a bijection $\pi : D \to \lambda$ to lift the covering of $j[D]$ to a covering of $j[\lambda]$: Notice that $j(\pi)$ is a bijection between $j(D)$ and $j(\lambda)$, and that $j(\pi)[j[D]] = j[\lambda]$.

The same argument shows that $j[j(\kappa)]$ is covered by a set in M of size $j(\kappa)$.

We use this covering property of j to establish the strong inaccessibility of $j(\kappa)$. For suppose that there is a regular γ, $\kappa \leq \gamma < j(\kappa)$, such that $2^\gamma \geq j(\kappa)$. Fix $X \subseteq \mathcal{P}(\gamma)$ of size $j(\kappa)$. Then there is $Y \in M$ of size $j(\kappa)$ covering $j[X]$. Clearly $a \cap j[\gamma] \neq b \cap j[\gamma]$ whenever $a \neq b \in j[X]$. Let $S \in M$ cover $j[\gamma]$ and have size $< j(\kappa)$. It follows that $Y \cap \mathcal{P}(S)$ has size $j(\kappa)$ in M and therefore $j(\kappa)$ is not strongly inaccessible in M. Contradiction.

Finally, since any ultrafilter on $\mathcal{P}_\kappa(\lambda)$, $\lambda < j(\kappa)$, lives in $V_{j(\kappa)}$, clearly $V_{j(\kappa)} \models \mathsf{ZFC} + \kappa$ is strongly compact. ⊣

Notice that the assumptions of Theorem 3.7 are not (expected to be) vacuous, since for example a 2-huge cardinal induces an embedding as required in Theorem 3.7. What is interesting is that we have recovered a very strong hypothesis from an assumption that rather than directly imposing closure on the target model only requires of it some degree of "correctness." Recall that κ is 2-huge iff there is an embedding $j : V \to M$ with $\mathrm{cp}(j) = \kappa$ and such that $M^{j(j(\kappa))} \subseteq M$. In fact, the weaker assumption of a 2-superstrong cardinal suffices. Recall that κ is n-superstrong iff there is an embedding $j : V \to M$ with $\mathrm{cp}(j) = \kappa$ and $V_{j^n(\kappa)} \subseteq M$, where the superscript indicates iteration. These are in any case significant assumptions; while superstrong cardinals are "weak" in the sense that any supercompact cardinal is limit of superstrong cardinals, 2-superstrength is (consistency-wise) above supercompactness.

QUESTION 3.8. Is it consistent to have an embedding $j : V \to M$ such that the first $\lambda^{+\kappa+1}$ cardinals of M and of V coincide? Here, $\kappa = \mathrm{cp}(j)$ and λ is the first fixed point of j above κ.

Acknowledgments. Some results in this paper strengthen results first announced in the invited talk *Cardinal preserving elementary embeddings* given in Oaxaca, México, on August 2006 during the XIII SLALM; I want to thank the organizing committee of the SLALM for allowing me the opportunity to speak, and the NSF for support in attending the meeting through grant DMS-0605727. The new results were to form part of the talk I was to present during the Logic Colloquium 2007, but health issues prevented me from attending; I want to thank the ASL and the organizing committee of the Colloquium for the invitation and the opportunity to include this paper in its proceedings. I also want to thank Grigor Sargsyan for conversations regarding Section 3, and the referees for their useful suggestions.

REFERENCES

[1] A. ANDRETTA, I. NEEMAN, and J. STEEL, *The domestic levels of K^c are iterable*, ***Israel Journal of Mathematics***, vol. 125 (2001), pp. 157–201.

[2] J. BAUMGARTNER, *Applications of the proper forcing axiom*, ***Handbook of Set-Theoretic Topology*** (K. Kunen and J. Vaughan, editors), North-Holland, Amsterdam, 1984, pp. 913–959.

[3] M. BEKKALI, ***Topics in Set Theory***, Lecture Notes in Mathematics, vol. 1476, Springer-Verlag, Berlin, 1991, Lebesgue measurability, large cardinals, forcing axioms, rho-functions, Notes on lectures by Stevo Todorčević.

[4] A. CAICEDO and B. VELIČKOVIĆ, ***Properness and Reflection Principles in Set Theory***, in preparation.

[5] M. FOREMAN, M. MAGIDOR, and S. SHELAH, *Martin's maximum, saturated ideals, and nonregular ultrafilters. I*, ***Annals of Mathematics. Second Series***, vol. 127 (1988), no. 1, pp. 1–47.

[6] J. HAMKINS, ***Forcing and Large Cardinals***, in preparation.

[7] M. HOLZ, K. STEFFENS, and E. WEITZ, ***Introduction to Cardinal Arithmetic***, Birkhäuser Advanced Texts: Basler Lehrbücher. [Birkhäuser Advanced Texts: Basel Textbooks], Birkhäuser Verlag, Basel, 1999.

[8] T. JECH, ***Set Theory***, Springer Monographs in Mathematics, Springer-Verlag, Berlin, 2003, The third millennium edition, revised and expanded.

[9] R. JENSEN, E. SCHIMMERLING, R. SCHINDLER, and J. STEEL, *Stacking mice*, ***The Journal of Symbolic Logic***, vol. 74 (2009), no. 1, pp. 315–335.

[10] A. KANAMORI, ***The Higher Infinite: Large Cardinals in Set Theory from Their Beginnings***, second ed., Springer Monographs in Mathematics, Springer-Verlag, Berlin, 2003.

[11] A. KANAMORI and M. MAGIDOR, *The evolution of large cardinal axioms in set theory*, ***Higher Set Theory (Proc. Conf., Math. Forschungsinst., Oberwolfach, 1977)***, Lecture Notes in Mathematics, vol. 669, Springer, Berlin, 1978, pp. 99–275.

[12] J. KETONEN, *Strong compactness and other cardinal sins*, ***Annals of Pure and Applied Logic***, vol. 5 (1972/73), pp. 47–76.

[13] P. LARSON, ***The Stationary Tower: Notes on a Course by W. Hugh Woodin***, University Lecture Series, vol. 32, American Mathematical Society, Providence, RI, 2004.

[14] D. A. MARTIN and R. M. SOLOVAY, *Internal Cohen extensions*, ***Annals of Pure and Applied Logic***, vol. 2 (1970), no. 2, pp. 143–178.

[15] W. MITCHELL, *The Covering Lemma*, to appear in ***Handbook of Set Theory***, Foreman and Kanamori (Editors).

[16] J. MOORE, *Set mapping reflection*, ***Journal of Mathematical Logic***, vol. 5 (2005), no. 1, pp. 87–97.

[17] S. SHELAH, ***Cardinal Arithmetic***, Oxford Logic Guides, vol. 29, The Clarendon Press Oxford University Press, New York, 1994.

[18] ———, ***Proper and Improper Forcing***, second ed., Perspectives in Mathematical Logic, Springer-Verlag, Berlin, 1998.

[19] J. STEEL, *An Outline of Inner Model Theory*, to appear in ***Handbook of Set Theory***, Foreman and Kanamori (Editors).

[20] S. TODORČEVIĆ, *A note on the proper forcing axiom*, ***Axiomatic Set Theory (Boulder, Colorado, 1983)*** (J. Baumgartner and D. Martin, editors), Contemporary Mathematics, vol. 31, American Mathematical Society, Providence, RI, 1984, pp. 209–218.

[21] M. VIALE, ***Applications of the proper forcing axiom to cardinal arithmetic***, Ph.D. Dissertation, Université Paris 7, Denis Diderot, Paris, 2006.

[22] ———, *The proper forcing axiom and the singular cardinal hypothesis*, ***The Journal of Symbolic Logic***, vol. 71 (2006), no. 2, pp. 473–479.

[23] ———, *A family of covering properties*, ***Mathematical Research Letters***, vol. 15 (2008), no. 2, pp. 221–238.

[24] J. VICKERS and P. D. WELCH, *On elementary embeddings from an inner model to the universe*, ***The Journal of Symbolic Logic***, vol. 66 (2001), no. 3, pp. 1090–1116.

[25] J. ZAPLETAL, *A new proof of Kunen's inconsistency*, ***Proceedings of the American Mathematical Society***, vol. 124 (1996), no. 7, pp. 2203–2204.

CALIFORNIA INSTITUTE OF TECHNOLOGY
DEPARTMENT OF MATHEMATICS
MAIL CODE 253-37
PASADENA, CA 91125, USA

Current address: Department of Mathematics, Boise State University, 1910 University Drive, Boise, ID 83725, USA

E-mail: caicedo@math.boisestate.edu

URL: `http://math.boisestate.edu/~caicedo/`

PROOF INTERPRETATIONS AND MAJORIZABILITY

FERNANDO FERREIRA

Abstract. In the last fifteen years, the traditional proof interpretations of modified realizability and functional (*dialectica*) interpretation in finite-type arithmetic have been adapted by taking into account majorizability considerations. One of such adaptations, the monotone functional interpretation of Ulrich Kohlenbach, has been at the center of a vigorous program in applied proof theory dubbed *proof mining*. We discuss some of the traditional and majorizability interpretations, including the recent bounded interpretations, and focus on the main *theoretical* techniques behind proof mining.

§1. **Introduction.** Functional interpretations were introduced half a century ago by Kurt Gödel in [17]. Gödel's interpretation uses functionals of finite type and is an *exact* interpretation. It is exact in the sense that it provides precise witnesses of existential statements. Another example of an exact (functional) interpretation is Georg Kreisel's modified realizability [40, 41]. In the last fifteen years or so, there has been an interest in interpretations which are not exact, but only demand *bounds* for existential witnesses. These interpretations, when dealing with bounds for functionals of every type, are based on the majorizability notions of William Howard [20] and Marc Bezem [4]. This is the case with Ulrich Kohlenbach's monotone modified realizability [29] and monotone functional interpretation [27], the bounded modified realizability [11] of Fernando Ferreira and Ana Nunes, or the bounded functional

---

This work is based on four lectures on proof interpretations given in *Days in Logic'06* in Coimbra (Portugal), January 2006. Meanwhile, these lectures were published under the title *Proof Interpretations* in "Days in Logic'06 (Two Tutorials)," eds. R. Kahle & I. Oitavem, Textos de Matemática (Série B), Departamento de Matemática, Universidade de Coimbra, vol. 36, pp. 1-42 (2006). The lectures have few proofs, except for the ones that are somewhat new, not well-known or not easily available (other proofs are given suitable bibliographic pointers). The present article is a revision of the published tutorial, with some changes and corrections, omitting some issues, and inserting new material presented in our address to the *Logic Colloquium'07*. An introduction was also specially written for the present publication. The Coimbra lectures are rather hard to find (and came out with some annoying misprints). We hope that with the present publication our lectures become more widely available. We also want to thank the two anonymous referees for their suggestions of improvement. This work was partially supported by Fundação para a Ciência e Tecnologia (Financiamento Base 2008 - ISFL/1/209).

Logic Colloquium '07
Edited by Françoise Delon, Ulrich Kohlenbach, Penelope Maddy, and Frank Stephan
Lecture Notes in Logic, 35

interpretation [11] of Ferreira and Paulo Oliva. There are, to be sure, other interpretations which also incorporate majorizability notions to a certain extent: for instance, the seminal Diller-Nahm interpretation [6], Wolfgang Burr's interpretation of $\mathsf{KP}\omega$ [5] or the very recent interpretation of Jeremy Avigad and Henry Towsner [3] which is able to provide a proof theoretic analysis of ID_1. The reader should consult [2], [54] and [43] for more information on proof interpretations. Specifically for Gödel's interpretation, I also suggest the articles which appeared in a recent issue of *dialectica* [48] commemorating the 50th anniversary of Gödel's paper.

In these lectures, we give an overview of the various functional interpretations based on the notion of majorizability of Howard/Bezem. The starting points are the exact interpretations of Gödel and Kreisel. We have tried to organize the results around certain main theorems: soundness, extraction, conservation and characterization theorems. We hope that this organization makes it easier to appreciate the differences between the various interpretations, their advantages and limitations, and also the techniques involved in proving the theorems (even though, as observed in the introductory footnote, the lectures have few proofs). We also pay special attention to certain issues. For instance, we discuss the advantages of the monotone functional interpretation vis-à-vis Gödel's interpretation (Section 4.2), and discuss the role of extensionality in relation to *intensional* majorizability and the uniform boundedness principles (Section 5.6). The lectures finish with an extraction result for the fully extensional classical theory $\mathsf{E\text{-}PA}_0^\omega + \mathsf{AC}_{\mathrm{qf}}^{1,1}$ (the result entails that this theory has the provably total functions of Peano Arithmetic) via a *false* theory by applying the techniques developed in Part 5. It is not known whether there is a direct route. In order not to distract the reader, we give almost no references during the exposition of the material. Each of the four lectures closes with a section on suggested readings and historical notes where we try to give the proper references.

The monotone functional interpretation and associated theorems have been used to guide the extraction of effective data from given ordinary proofs in mathematics. In more picturesque words, it has been used to guide the "proof mining" of ordinary mathematical arguments. This paper does not attempt to even give a brief description of the applied work apart from saying that proof mining techniques (functional style) have been applied to approximation theory, fixed points of non-expansive mappings and, more recently, to ergodic theory. There are very good places to get into these applications and the recent book of Kohlenbach [35] is a very good place to start. Nevertheless, we would like to convey here a sense of the excitement of possible applications. Our example is taken from the research blog of Terence Tao. In a page entitled "Soft analysis, hard analysis, and the finite convergence principle" (see [51]), Tao speaks of an informal distinction between "soft" and "hard" principles in mathematics. Typically, "soft" principles are abstract, infinitary and lack computational content. They are extremely useful in modern mathematics and, given the usual training of a mathematician, they are simple to state and easy to remember and apply. An example of a "soft" principle is (in Tao's terminology) the *infinite convergence theorem*: every bounded monotone sequence of real numbers is convergent. On the other hand, "hard" principles are concrete, finitary and with computational content. They seem to be more difficult to grasp than the corresponding "soft" principles. In his blog, Tao sets himself the task of finding the "hard" counterpart of the infinite convergence theorem. Actually, Tao considers the following modification of the infinite convergence principle: every bounded nondecreasing sequence of reals is a Cauchy sequence. (The reader of like mind knows that this modified statement is weaker than the original one in terms of set existence: see, for instance, Section 13.3 of [35].) Tao proceeds in an informal, tentative manner, and is eventually satisfied with a principle dubbed the *finite convergence principle*. Afterwards — in an impressive application — Tao formulates a variant of the infinite convergence theorem for Hilbert spaces and shows that Szemerédi's regularity lemma, a major combinatorial tool in graph theory, follows from the corresponding finite convergence principle for Hilbert spaces.

As pointed by Towsner in a comment of the blog, the finite convergence principle can be seen as an application of Gödel's *dialectica* interpretation. In a detailed treatment, Kohlenbach shows in his recent book [35] that the finite convergence principle is the no-counterexample interpretation (cf. Section 4.5) of the modified infinite convergence principle reinforced with a uniformity observation (which is given an explicit quantitative meaning in Kohlenbach's analysis). Note that, for the case at hand, the no-counterexample interpretation coincides with the *dialectica* interpretation (after a double-negation translation). In the lines below, we show that the finite convergence principle is essentially the bounded functional interpretation (for the classical case) of the modified infinite convergence principle:[1,2] in contrast with the *dialectica* interpretation, the bounded functional interpretation takes care of the uniformities automatically. Let us see why this is so. We state the modified infinite convergence theorem in a normalized form as follows:

$$\forall k \in \mathbb{N} \forall x \in [0,1]^{\mathbb{N}} \exists N \in \mathbb{N} \forall n \in \mathbb{N} \, |x_{N+n} - x_N| \leq \frac{1}{2^k},$$

[1] A similar case can also be made for the monotone functional interpretation.

[2] In [8], we describe a direct bounded functional interpretation for the classical case. In the present lectures, the (intuitionistic) bounded functional interpretation is applied after a double-negation translation. The proviso "essentially" is made because, for the bounded functional interpretation to work as desired, the convergence principles would have to be formulated with *intensional* majorizability signs (these reformulations are nevertheless equivalent to the original statements by extensionality reasons, as discussed in the sequel and in Section 5.6). These are technical matters that we sidestep in this introduction.

under the assumption that $x_i \leq x_{i+1}$, for all $i \in \mathbb{N}$. Alternatively, one can drop the assumption and work instead with the sequence $\tilde{x}_n = \max_{i \leq n} x_i$. This is what we will do. Rather than go through the moves of applying the syntactic transformation of the bounded functional interpretation, we will put the above statement in the form $\forall\exists$ (with an appropriate matrix) using the so-called *characteristic principles* associated with the bounded functional interpretation (for the classical case). These are the principles $\mathsf{bBAC}^{\omega}_{\trianglelefteq}$ and $\mathsf{MAJ}^{\omega}_{\trianglelefteq}$ of Theorem 5.12. The first principle is a version of choice, a consequence of which is:

$$\text{(I)}: \quad \forall i \in \mathbb{N} \exists j \in \mathbb{N}\, A_\exists(i, j) \rightarrow \tilde{\exists} f \in \mathbb{N}^{\mathbb{N}} \forall i \in \mathbb{N} \exists j \leq f(i)\, A_\exists(i, j),$$

where $A_\exists$ is an existential formula, and the tilde on the quantifier "$\tilde{\exists} f$" means that f is nondecreasing. Another consequence of the first principle is the following "collection" principle (in our application, it reduces to a "compactness" property):

$$\text{(II)}: \quad \forall w \in [0,1]^{\mathbb{N}} \exists i \in \mathbb{N} A_\exists(w, i) \rightarrow \exists l \in \mathbb{N} \forall w \in [0,1]^{\mathbb{N}} \exists i \leq l A_\exists(w, i),$$

where $A_\exists$ is an existential formula. The quantification "$\forall w \in [0,1]^{\mathbb{N}}$" is a *bounded* quantification under a suitable representation of the compact separable metric space $[0,1]^{\mathbb{N}}$ (its topology is the product topology of $\mathbb{N}$-copies of the closed unit interval, and a natural metric is forthcoming). One should consult [35] for an exposition concerning representations of Polish spaces. According to the characteristic principles, the bounded quantification mentioned above should have been *intensional*, but in our application we can work with the regular bounded quantification because we apply it to a matrix $A_\exists(w, i)$ which is *extensional* in w (these issues are discussed in Section 5.6).[3]

Assume the modified infinite convergence principle. Consider the negation of the formula $\exists N \in \mathbb{N} \forall n \in \mathbb{N}\, |\tilde{x}_{N+n} - \tilde{x}_N| \leq \frac{1}{2^k}$. Using (I), this negation is equivalent to $\tilde{\exists} F \in \mathbb{N}^{\mathbb{N}} \forall N \in \mathbb{N} \exists n \leq F(N)\, |\tilde{x}_{N+n} - \tilde{x}_N| > \frac{1}{2^k}$ (note that the relation $>$ between real numbers is existential). We get,

$$\forall k \in \mathbb{N} \tilde{\forall} F \in \mathbb{N}^{\mathbb{N}} \forall x \in [0,1]^{\mathbb{N}} \exists N \in \mathbb{N} \forall n \leq F(N)\, |\tilde{x}_{N+n} - \tilde{x}_N| \leq \frac{1}{2^k}.$$

It is clear that the matrix "$\forall n \leq F(N)\, |\tilde{x}_{N+n} - \tilde{x}_N| \leq \frac{1}{2^k}$" can be replaced by "$\forall n \leq F(N)\, |\tilde{x}_{N+n} - \tilde{x}_N| < \frac{1}{2^k}$." The latter matrix can be considered existential. Hence, we can apply (II) and conclude that

$$\forall k \in \mathbb{N} \tilde{\forall} F \in \mathbb{N}^{\mathbb{N}} \exists M \in \mathbb{N} \forall x \in [0,1]^{\mathbb{N}} \exists N \leq M \forall n \leq F(N)\, |\tilde{x}_{N+n} - \tilde{x}_N| < \frac{1}{2^k}.$$

[3] In more familiar mathematical terms, our application of (II) boils down to the fact that the compact space $[0,1]^{\mathbb{N}}$ is covered by the family of open sets $U_i = \{w \in [0,1]^{\mathbb{N}} : A_\exists(w, i)\}$, with $i \in \mathbb{N}$. Hence, it has a finite sub-cover.

Observe that only the values of x_i, for $i < M + F(M) + 1$, do matter. Hence the above can the restated as,

$$\begin{cases} \forall k \in \mathbb{N} \tilde{\forall} F \in \mathbb{N}^{\mathbb{N}} \exists M \in \mathbb{N} \forall x \in [0,1]^{M+F(M)+1} \\ \qquad \exists N \leq M \forall n \leq F(N)\, |\tilde{x}_{N+n} - \tilde{x}_N| < \dfrac{1}{2^k}. \end{cases}$$

This is a straightforward reformulation of Tao's finite convergence principle. As usual, this "hard" principle is more difficult to read than the "soft" monotone convergence principle. We quote a description given by Tao of this principle: the finite convergence principle "asserts that any sufficiently long (but finite) bounded monotone sequence will experience arbitrarily high-quality amounts of *metastability* with a specified error tolerance $\frac{1}{2^k}$, in which the duration $F(N)$ of the metastability exceeds the time N of onset of the metastability by an arbitrary function F which is specified in advance." As Tao says, this is significantly more verbose than the "soft" formulation.

In a more recent blog [50] entitled "The correspondence principle and finitary ergodic theory," Tao speaks of a *correspondence principle* between qualitative (or "soft") results in infinite dynamical systems and quantitative (or "hard") results in finite dynamical systems. He illustrates such correspondence principle with eight examples, all of them mediated by compactness properties. It would be interesting to see how Tao's examples are relate to the majorizability interpretations: are they essentially instances of the bounded functional interpretation? Moreover, the majorizability interpretations are based on majorizability properties, not on compactness properties (consult Sections 17.7 and 17.8 of [35] to appreciate the difference, as well as the groundwork with new base types in [33] and [15]). Are there examples of "correspondences" waiting to be discovered based on the majorizability interpretations?

I want to thank my students Patrícia Engrácia, Gilda Ferreira and Jaime Gaspar for discussions and observations concerning the majorizability interpretations. I would also like to thank the two anonymous referees for their helpful criticism and suggestions of improvement.

§2. Basic theory and some models.

2.1. The language of finite-type arithmetic. The finite types $\mathcal{T}$ are syntactic expressions defined inductively: 0 (the base type) is a finite type; if τ and σ are finite types then $\tau \to \sigma$ is a finite type. It is useful to have the following interpretation in mind: the base type 0 is the type constituted by the natural numbers, whereas $\tau \to \sigma$ is the type of (total) functions of objects of type τ to objects of type σ.

To make the reading easier, we often omit brackets and associate the arrows to the right. E.g., $0 \to 0 \to 0$ means $0 \to (0 \to 0)$. The *pure types* are defined inductively for each natural number: the pure type corresponding to the natural number 0 is the base type 0; the pure type $n+1$ is $n \to 0$.

The language of Heyting arithmetic in all finite types, denoted by $\mathcal{L}_0^\omega$, is a sorted language with a sort for each finite type. There is a denumerable set of variables x^σ, y^σ, z^σ, etc. for each type σ. When convenient, we omit the type superscript. There are two kinds of constants:

(a) *Logical constants* or *combinators*. For each pair of types ρ, τ there is a combinator of type $\rho \to \tau \to \rho$ denoted by $\Pi_{\rho,\tau}$; for each triple of types δ, ρ, τ there is a combinator of type

$$(\delta \to \rho \to \tau) \to (\delta \to \rho) \to (\delta \to \tau)$$

denoted by $\Sigma_{\delta,\rho,\tau}$.

(b) *Arithmetical constants*. The constant 0 of type 0; the *successor* constant S of type 1; for each type ρ, a *recursor* constant of type

$$0 \to \rho \to (\rho \to 0 \to \rho) \to \rho$$

denoted by R_ρ.

Constants and variables of type ρ are terms of type ρ. If t is a term of type $\rho \to \tau$ and q is a term of type ρ then $App(t, q)$ is a term of type τ. These are all the terms there are. A term with no variables is a *closed term*. We usually write tq or $t(q)$ for $App(t, q)$. When writing tqr without brackets we associate to the *left* (note the difference with the previous convention): $(t(q))(r)$. We also write $t(q, r)$ instead of $(t(q))(r)$. In general, $t(q, r, \ldots, s)$ stands for $(\ldots((t(q))(r))\ldots)(s)$.

Atomic formulas are formulas of the form $t = q$ where t and q are terms of type 0. Note that, in the present setting, there is only one primitive equality symbol (infixing between terms of type 0). *Formulas* are obtained from atomic formulas by means of the usual propositional connectives $\wedge$, $\vee$, $\to$, $\perp$ (*falsum*) and universal and existential quantifiers $\forall x^\sigma$ and $\exists x^\sigma$ for each type σ. As usual, $\neg A$ abbreviates $A \to \perp$ and $A \leftrightarrow B$ abbreviates $(A \to B) \wedge (B \to A)$.

2.2. Heyting arithmetic in all finite types. The theory HA_0^ω is based on *intuitionistic logic*. It also has the following axioms for combinators and equality:

(a) *Axioms for combinators*. $A[\Pi(x, y)/w] \leftrightarrow A[x/w]$ and $A[\Sigma(x, y, z)/w] \leftrightarrow A[x(z, yz)/w]$, where A is an atomic formula with a distinguished variable w and $A[t/w]$ is obtained from A by replacing the occurences of w by t.

(b) *Equality axioms*. $x = x$ (reflexivity); $x = y \wedge A[x/w] \to A[y/w]$, where A is an atomic formula with a distinguished (type zero) variable w.

and the arithmetical axioms:

(c) *Successor axioms*. $Sx \neq 0$ and $Sx = Sy \to x = y$.

(d) *Axioms for recursors.* $A[R(0, y, z)/w] \leftrightarrow A[y/w]$ and $A[R(Sx, y, z)/w] \leftrightarrow A[z(R(x, y, z), x)/w]$, where A is an atomic formula with distinguished variable w.
(e) *Induction scheme.* $A(0) \wedge \forall x^0(A(x) \rightarrow A(Sx)) \rightarrow \forall x A(x)$, for each formula A of the language.

It can be shown that equality is symmetric and transitive and, moreover, the conditional $x = y \wedge A[x/w] \rightarrow A[y/w]$ also holds for every formula of the language (provided that there is no clash of variables). Similarly, the axioms for combinators and recursors extend to every formula A.

Equality is decidable in the following sense: $\mathsf{HA}_0^\omega \vdash \forall x^0(x = 0 \vee x \neq 0)$. This is easily proven by induction. Equality for higher types is defined inductively: $s =_{\rho\rightarrow\tau} t$ is $\forall x^\rho(sx =_\tau tx)$. Equality in higher types is not decidable anylonger. *Full extensionality* is the scheme of axioms $\forall z \forall x, y(x = y \rightarrow zx = zy)$. It should be noted that we are *not* assuming full extensionality in the theory HA_0^ω. The theory obtained by its inclusion is denoted by $\mathsf{E}\text{-}\mathsf{HA}_0^\omega$. In the sequel, we prefix a theory by E when adding full extensionality to it.

2.3. Combinatorial completeness. The combinators Π and Σ are instrumental in proving the following important property:

THEOREM 2.1 (Combinatorial completeness). *For each term $t[x^\rho]$ of type τ with a distinguished variable x of type ρ, we can construct a term q of type $\rho \rightarrow \tau$ whose variables are those of t except for x such that, for every term s of type ρ and atomic formula A with a distinguished variable w^τ,*

$$\mathsf{HA}_0^\omega \vdash A[t[s/x]/w] \leftrightarrow A[qs/w].$$

For instance, if t is of type 0, then $qs = t[s/x]$. The term q is usually denoted by $\lambda x.t[x]$ and the above equation can now be written $(\lambda x.t[x])s = t[s/x]$ (β-reduction). Of course, the proposition extends to all formulas A (provided that there is no clash of variables), i.e., we may substitute the term $t[s/x]$ by $(\lambda x.t[x])s$ in any formula.

Using the recursors, it is possible to associate to each description of a primitive recursive function a closed term that (with the proper understanding) satisfies the defining conditions of the description. Therefore, the theory HA_0^ω contains, in a natural sense, *primitive recursive arithmetic*. Actually, only the recursor R_0 is needed to define the primitive recursive functions. The presence of recursors of higher types has the effect of making possible the definition of functions beyond the primitive recursive functions (e.g., the Ackermann function).

We reserve the subscript "qf" for quantifier-free formulas:

PROPOSITION 2.2. *For each quantifier-free formula $A_{\mathrm{qf}}(\underline{x})$ there is a closed term t of appropriate type such that* $\mathsf{HA}_0^\omega \vdash A_{\mathrm{qf}}(\underline{x}) \leftrightarrow t\underline{x} = 0$.

As a consequence, $\mathsf{HA}_0^\omega \vdash A_{qf} \vee \neg A_{qf}$. The theory obtained from HA_0^ω by adjoining the unrestricted law of excluded middle $A \vee \neg A$ is the classical theory PA_0^ω. If full extensionality is present, we have $\mathsf{E\text{-}PA}_0^\omega$.

2.4. Main models. 1. *The full set-theoretical model* S^ω. Let $\mathrm{S}_0 = \mathbb{N}$ and $\mathrm{S}_{\rho\to\tau} = (\mathrm{S}_\tau)^{\mathrm{S}_\rho}$, where $(\mathrm{S}_\tau)^{\mathrm{S}_\rho}$ is the set of *all* functions from S_ρ to S_τ. Let S^ω be $\langle \mathrm{S}_\sigma \rangle_{\sigma \in \mathcal{T}}$. With the proper understanding, it is clear that S^ω is a model of HA_0^ω. It is actually a model of $\mathsf{E\text{-}PA}_0^\omega$. The model S^ω is called the *standard* structure of finite-type arithmetic. When we call a sentence of $\mathcal{L}_0^\omega$ true or false, we always mean true or false with respect to the *standard* model.

2. *The hereditarily recursive operations* HRO^ω. For each type σ we define a subset HRO_σ of the natural numbers in the following way: $\mathrm{HRO}_0 = \mathbb{N}$ and

$$\mathrm{HRO}_{\rho\to\tau} = \{n \in \mathbb{N} : \forall k \in \mathrm{HRO}_\rho\, \exists m \in \mathrm{HRO}_\tau\, (\{n\}(k) \simeq m)\},$$

where $\{\cdot\}$ denotes the Kleene bracket of recursion theory. Let HRO^ω be $\langle \mathrm{HRO}_\sigma \rangle_{\sigma \in \mathcal{T}}$. If $n \in \mathrm{HRO}_{\rho\to\tau}$ and $k \in \mathrm{HRO}_\rho$ then $App^{\mathrm{HRO}^\omega}(n, k)$ is defined as $\{n\}(k)$. With a proper interpretation of the constants, HRO^ω is a model of PA_0^ω. By internalizing the above definitions within first-order Heyting arithmetic HA, it is possible to prove the following conservation result:

PROPOSITION 2.3. *The theory* HA_0^ω *is conservative over* HA.

3. *The intensional continuous functionals* ICF^ω. In order to motivate some definitions below, let us briefly discuss continuous functionals of type 2. The set S_1 $(=\mathbb{N}^\mathbb{N})$ endowed with the product topology of the discrete space $\mathbb{N}$ with itself $\mathbb{N}$-times is known as the *Baire space*. It is easy to see that a functional $\Phi : \mathrm{S}_1 \mapsto \mathbb{N}$ is continuous at a point $\beta \in \mathrm{S}_1$ if, and only if,

$$\exists n \in \mathbb{N} \forall \gamma \in \mathrm{S}_1 (\overline{\gamma}(n) = \overline{\beta}(n) \to \Phi(\gamma) = \Phi(\beta)).$$

where $\overline{\gamma}(n)$ stands for the finite sequence $\langle \gamma(0), \dots, \gamma(n-1) \rangle$. In other words, the value of Φ at a point β only depends on a finite initial segment of β. Therefore, Φ is determined by what happens in a countable set of finite sequences. This can be made explicit in the following way. An *associate* of a continuous functional Φ is an element $\alpha \in \mathrm{S}_1$ with the following two properties:

i. If $s = \langle s_0, \dots, s_{n-1} \rangle$ is a finite sequence[4] of natural numbers, $\alpha(s) \neq 0$ and s is an initial segment of β, then $\Phi(\beta) = \alpha(s) - 1$.
ii. For all $\beta \in \mathrm{S}_1$, there is n such that $\alpha(\overline{\beta}(n)) \neq 0$.

Of course, such associates exist. It is clear that for every continuous functional Φ^2 with associate α, the following holds:

$$\Phi(\beta) = \alpha(\overline{\beta}(\mu k(\alpha(\overline{\beta}(k)) \neq 0))) - 1,$$

where μ is the minimization operator of recursion theory. Therefore, each continuous functional of type 2 is determined by a function of type 1. By

[4] We are identifying the finite sequence s with its numerical coding, as it is usually done in recursion theory.

means of this *type lowering* procedure, we obtain a structure for $\mathcal{L}_0^\omega$ in the following manner (for simplicity, we restrict ourselves to pure types). First, for $\alpha, \beta \in \mathrm{S}_1$ we define

$$\alpha(\beta) \simeq \alpha(\overline{\beta}(\mu k(\alpha(\overline{\beta}(k)) \neq 0))) - 1.$$

Second, we let $\mathrm{ICF}_0 = \mathbb{N}$, $\mathrm{ICF}_1 = \mathrm{S}_1$ and, for the *remaining* pure cases:

$$\mathrm{ICF}_{\rho \to 0} = \{\alpha \in \mathrm{S}_1 : \forall \beta \in \mathrm{ICF}_\rho \exists k \in \mathbb{N}\,(\alpha(\beta) = k)\}.$$

The above definition can be extended to all finite types. Let $\mathrm{ICF}^\omega = \langle \mathrm{ICF}_\sigma \rangle_{\sigma \in \mathcal{T}}$. Application between functionals is defined in the natural way: for $\alpha \in \mathrm{ICF}_{\rho \to 0}$ and $\beta \in \mathrm{ICF}_\rho$, with $\rho \neq 0$, then $App^{\mathrm{ICF}^\omega}(\alpha, \beta)$ is $\alpha(\beta)$. With a proper interpretation of the constants, ICF^ω is a model of PA_0^ω.

This construction can be generalized in the following way. Take $U \subseteq \mathbb{N}^\mathbb{N}$ closed under Turing reducibility. Define $\mathrm{ICF}_0(U) = \mathbb{N}$, $\mathrm{ICF}_1(U) = U$ and, for $\sigma \neq 0, 1$, put $\mathrm{ICF}_\sigma(U)$ following the blueprint above (*mutatis mutandis*). Then $\mathrm{ICF}^\omega(U) := \langle \mathrm{ICF}_\sigma(U) \rangle_{\sigma \in \mathcal{T}}$ is a model of PA_0^ω. A particularly nice source of (counter-)examples is $\mathrm{ICF}^\omega(Rec)$, where Rec is the set of (total) recursive functions from $\mathbb{N}$ to $\mathbb{N}$. This is due to the fact that recursive sets do not necessarily separate disjoint r.e. sets (a result of Kleene).

4. *The extensional counterparts* HEO^ω *and* ECF^ω. The models HRO^ω and ICF^ω are called *intensional* because the scheme of extensionality fails to hold for them. For instance, take $e_1, e_2 \in \mathbb{N}$ two different indices for the constant zero function. Clearly, $e_1, e_2 \in \mathrm{HRO}_1$. Let $e \in \mathbb{N}$ be an index for the identity function. Of course, $e \in \mathrm{HRO}_2$. Even though $\{e\}(e_1) \neq \{e\}(e_2)$, one has $\forall n \in \mathrm{HRO}_0\,(\{e_1\}(n) = \{e_2\}(n))$. Therefore, the form of extensionality

$$\forall \Phi^2 \forall \alpha^1, \beta^1 (\forall k^0 (\alpha k = \beta k) \to \Phi\alpha = \Phi\beta),$$

already fails in HRO^ω. The above form of extensionality holds in ICF^ω, but it is easy to find a counter-example a type above, i.e. to:

$$\forall \mathfrak{G}^3 \forall \Phi^2, \Psi^2 (\forall \alpha^1 (\Phi\alpha = \Psi\alpha) \to \mathfrak{G}\Phi = \mathfrak{G}\Phi).$$

We now define the extensional counterparts of HRO^ω and ICF^ω. These are called, respectively, the *hereditarily effective operations* HEO^ω and the *(extensional) countinuous functionals* ECF^ω. The elements of non-zero type of HEO^ω and ECF^ω are functions (functionals). For HEO^ω we define by induction on the type σ both the functionals in HEO_σ and the *indices* of these functionals (these indices are natural numbers). We let HEO_0 be the set of natural numbers and declare that each natural number is an index of itself. We say that a natural number e is an index for a function (functional) $F : \mathrm{HEO}_\sigma \mapsto \mathrm{HEO}_\tau$ if for each index x of a function h of HEO_σ, $\{e\}(x)$ is defined and is an index for the function $F(h)$. We take $\mathrm{HEO}_{\rho \to \tau}$ as the set of functions $F : \mathrm{HEO}_\rho \mapsto \mathrm{HEO}_\tau$ which have an index. The structure HEO^ω is defined as $\langle \mathrm{HEO}_\sigma \rangle_{\sigma \in \mathcal{T}}$.

Regarding ECF^ω, we let ECF_0 be the set of natural numbers and, for each non-zero type σ, we define by induction both the functionals in ECF_σ and the *associates* of these functionals (these are elements of S_1). We put ECF_1 as S_1 and declare each $\alpha \in S_1$ an associate of itself. We restrict to pure types in order to simplify. We say that $\alpha \in S_1$ is an associate for a function $F : ECF_\sigma \mapsto \mathbb{N}$, with σ a non-zero pure type, if for each associate β of a function h of ECF_σ, $\alpha(\beta)$ is defined and is the natural number $F(h)$. We take $ECF_{\rho\to 0}$ as the set of functions $F : ECF_\rho \mapsto \mathbb{N}$ which have an associate. This definition can be extended to all finite types. The structure ECF^ω is defined as $\langle ECF_\sigma \rangle_{\sigma\in\mathcal{T}}$.

The structure ECF^ω is also called the structure of the *countable* functionals (Kleene's terminology). In a way similar to the intensional case, we can also start with any subset U of S_1 closed under Turing reducibility, and get the struture $ECF^\omega(U)$.

5. *The strongly majorizable functionals* M^ω. Let $M_0 = \mathbb{N}$ and let $\leq_0^*$ be the usual "less than or equal" relation between natural numbers. Given types ρ and τ, and given x and y elements of $M_\tau^{M_\rho}$, we say that $x \leq_{\rho\to\tau}^* y$ (and read "y strongly majorizes x") if

$$\forall u, v \in M_\rho \, (u \leq_\rho^* v \to x(u) \leq_\tau^* y(v) \wedge y(u) \leq_\tau^* y(v)).$$

Define $M_{\rho\to\tau}$ as the set $\{x \in M_\tau^{M_\rho} : \exists y \in M_\tau^{M_\rho}(x \leq_{\rho\to\tau}^* y)\}$. Let M^ω be $\langle M_\sigma \rangle_{\sigma\in\mathcal{T}}$. Bezem showed that M^ω forms a model of E-PA_0^ω under function application and the natural interpretation of the constants. It is also important to observe that if $x, y \in M_\tau^{M_\rho}$ and $x \leq_{\rho\to\tau}^* y$, then $y \leq_{\rho\to\tau}^* y$. Therefore, not only is it the case that x is in $M_{\rho\to\tau}$ but y is also there.

For $x, y \in S_1$, $x \leq_1^* y$ iff $x \leq_1 y$ and y is non-decreasing. Note that $\leq_1^*$ is not reflexive. Given $\alpha : \mathbb{N} \mapsto \mathbb{N}$, let α^M be defined by $\alpha^M(n) = \max_{k\leq n} \alpha(k)$. It is clear that $\alpha \leq_1^* \alpha^M$. Therefore, $M_1 = S_1$. However, M_2 is *properly* contained in S_2. E.g., consider the functional $\Sigma \in S_2$ defined as follows:

$$\Sigma(\alpha) = \begin{cases} n & \text{if } n \text{ is the least value such that } \alpha(n) \neq 0, \\ 0 & \text{if } \forall k \, (\alpha(k) = 0). \end{cases}$$

Suppose, in order to get a contradiction, that there is Ψ with $\Sigma \leq_2^* \Psi$. In particular, $\forall \alpha \in S_1(\alpha \leq_1^* 1^1 \to \Sigma(\alpha) \leq \Psi(1^1))$. This is a contradiction: just consider the function α which takes the value 0 for numbers $n \leq \Psi(1^1)$ and is 1 afterwards. As a consequence, the following form of choice fails in M^ω: $\forall \alpha^1 \exists n^0 A(\alpha, n) \to \exists \Phi^2 \forall \alpha^1 A(\alpha, \Phi(\alpha))$. To see this, just take for A the formula:

$$(\alpha(n) \neq 0 \wedge \forall k < n(\alpha(k) = 0)) \vee (n = 0 \wedge \forall k(\alpha(k) = 0)).$$

The failure of choice in M^ω can be further improved by noticing that the *discontinuous* functional E defined thus:

$$E(\alpha) = \begin{cases} 1 & \text{if } \forall k\,(\alpha(k) = 0), \\ 0 & \text{otherwise} \end{cases}$$

is majorizable. Hence, we can replace the formula $\forall k(\alpha(k) = 0)$ by the quantifier-free formula $E(\alpha) = 1$ and, as a consequence, get the failure of the above form of choice with a quantifier-free matrix (although with the parameter E).

Continuous functionals have been playing an important role in discussions concerning finite-type functionals. Since the notion of majorizability (as opposed to continuity) plays the crucial role in some of the functional interpretations discussed in this paper, we believe that it is illuminating to make two observations regarding the relationship between continuity and majorizability. First, the continuous functionals of type 2 are in M_2. The proof of this fact uses a compactness argument. Given Φ a continuous functional of type 2, it makes sense to define the functional $\Phi^{M} \in S_2$ according to the equation $\Phi^{M}(\alpha) = \max_{\beta \leq_1 \alpha^{M}} \Phi(\beta)$ because, for a given $\alpha \in S_1$, the set $\{\beta \in S_1 : \beta \leq_1 \alpha^{M}\}$ is a *compact* subspace of the Baire space and, hence, $\Phi[\{\beta \in S_1 : \beta \leq_1 \alpha^{M}\}]$ is compact in the discrete topology of $\mathbb{N}$. In other words, it is finite and we can take its maximum. By construction, $\Phi \leq_2^* \Phi^{M}$. Additionally, it is not difficult to show that Φ^{M} is continuous. Clearly, by the continuity of Φ one has

$$S_1 = \cup_{s \in D}\{\beta \in S_1 : \beta \text{ extends } s\},$$

where D is constituted by the finite sequences of natural numbers that determine values for Φ (i.e., such that elements of S_1 that extend an element of D have the same values according to Φ). Given $\alpha \in S_1$, by compactness there is a finite $F_\alpha \subseteq D$ such that

$$\{\beta \in S_1 : \beta \leq_1 \alpha^{M}\} \subseteq \cup_{s \in F_\alpha}\{\beta \in S_1 : \beta \text{ extends } s\}.$$

Let $n \in \mathbb{N}$ be the maximal length of the sequences in F_α. It can now be easily argued that if $\overline{\alpha}(n)$ is an initial segment of γ then $\Phi^{M}(\gamma) = \Phi^{M}(\alpha)$.

The second observation is that, nevertheless, there is a type 3 functional in ECF^{ω} which is not majorizable. Kreisel showed that there is a FAN functional $\mathfrak{F}^3$ in ECF_3 such that

$$\forall \Phi \in ECF_2 \forall \alpha \leq_1 1 \forall \beta \leq_1 1\ (\alpha(\overline{\mathfrak{F}(\Phi)}) = \beta(\overline{\mathfrak{F}(\Phi)}) \rightarrow \Phi\alpha = \Phi\beta).$$

In other words, $\mathfrak{F}(\Phi)$ is a length witnessing the uniform continuity of Φ restricted to the Cantor space (i.e., to the compact subset of the Baire space constituted by the elements α such that $\alpha \leq_1 1^1$). However, the set $\{\mathfrak{F}(\Phi) : \Phi \in ECF_2 \text{ and } \Phi \leq_2^* 1^2\}$ is clearly unbounded. Therefore, $\mathfrak{F}$ is not majorizable. (I thank Dag Normann for this observation.)

It will be important in the sequel to formalize the majorizability relation. We finish our discussion of the strongly majorizable functionals with this issue. Fix a suitable formula "$x \leq y$" of $\mathcal{L}_0^\omega$ saying that the natural number x is less than or equal to the natural number y. The (strong) majorizability formulas "$x \leq_\sigma^* y$" are defined inductively on the types according to the following clauses:

(a) $x \leq_0^* y := x \leq y$
(b) $x \leq_{\rho\to\sigma}^* y := \forall u^\rho, v^\rho\,(u \leq_\rho^* v \to xu \leq_\sigma^* yv \wedge yu \leq_\sigma^* yv)$

The following easy, but important, absoluteness property holds: when interpreted in M^ω the formulas "$x \leq_\sigma^* y$" coincide with the strong majorizability relations defined in the beginning of this example. It follows that M^ω is a model of the *majorizability axioms* MAJ^ω: $\forall x \exists y (x \leq^* y)$. By previous discussions, MAJ^ω already fails in the full set-theoretic model S^ω at type 2, and fails in the structure of (extensional) continuous functionals ECF^ω at type 3.

LEMMA 2.4. *For each finite type σ, the theory HA_0^ω proves:*

(i) $x \leq_\sigma^* y \to y \leq_\sigma^* y$;
(ii) $x \leq_\sigma^* y \wedge y \leq_\sigma^* z \to x \leq_\sigma^* z$;
(iii) $x \leq_\sigma y \wedge y \leq_\sigma^* z \to x \leq_\sigma^* z$;

where the relation $\leq_\sigma$ is the pointwise "less than or equal to" relation: it is $\leq$ for type 0, and $x \leq_{\rho\to\tau} y$ is defined recursively by $\forall u^\rho\,(xu \leq_\tau yu)$.

The following theorem of Howard is a basic ingredient of the soundness proof of many interpretations discussed in this paper:

THEOREM 2.5. *For each* closed *term t there is a closed term q such that* $\mathsf{HA}_0^\omega \vdash t \leq^* q$.

6. *The term model.* We say that a term t *contracts* to a term q if one of the following clauses is satisfied:

(i) t is Πrs and q is r.
(ii) t is Σrst and q is $r(t, st)$.
(iii) t is $R0st$ and q is s.
(iv) t is $R(Sr)st$ and q is $t(Rrst, r)$.

A term t reduces to another term q in *one step* if q is obtained from t by contracting a single sub-term of t. A term is said to be in *normal form* if it does not admit reductions in one step. A *reduction sequence* is a sequence of terms $t_0, \dots, t_n$ such that each term reduces in one step to the next. In this case, we say that t_0 *reduces* to t_n. If t_n cannot be reduced further, then t_n is in normal form and the sequence is called *terminating*.

THEOREM 2.6 (Confluence and strong normalization). *Every term reduces to a unique term in normal form. Moreover, every reduction sequence eventually terminates.*

The uniqueness result (confluence) is the Church-Rosser theorem for this reduction calculus. Normalization *strictu sensu* is the fact that every term has a terminating reduction sequence (*strong* normalization is the fact that *every* reduction sequence eventually terminates). Proofs of normalization are bound to use strong forms of induction (viz. induction on non-arithmetical predicates), because it is known that it (elementarily) implies the consistency of first-order Peano arithmetic. This fact can be proved with the aid of Gödel's *dialectica* interpretation (see Section 4.6).

A corollary of the normalization theorem is that every closed term r of type 0 reduces to a numeral $\overline{n}$ and, clearly, $\mathsf{HA}_0^\omega \vdash r = \overline{n}$. Therefore, we can assign to each closed term t of type 1 a number theoretical function that maps each $k \in \mathbb{N}$ to the unique natural number n such that $t\overline{k}$ reduces to $\overline{n}$. The normalization process ensures that this function is recursive. As we will comment in Part 3, closed terms of type 1 give exactly the *provably total* Σ_1^0-functions of PA_0^ω (and, actually, of PA).

Let CT_σ be the set of *closed terms* of type σ, and let CT^ω be $\langle \mathrm{CT}_\sigma \rangle_{\sigma \in \mathcal{T}}$. Given closed terms t, q of type zero, we say that $t =^{\mathrm{CT}^\omega} q$ if the normal forms of t and q are the same term. If $t \in \mathrm{CT}_{\rho \to \tau}$ and $q \in \mathrm{CT}_\rho$ then $App^{\mathrm{CT}^\omega}(t, q)$ is defined as the *term* $App(t, q)$. With these specifications and the interpretation of constants by themselves, CT^ω is a model of HA_0^ω.

A modification of the reduction calculus above has an interesting application. Let T be a closed term of type 2. The interpretation of T in the full set-theoretical model S^ω, denoted by T^{S^ω}, is a function from S_1 to $\mathbb{N}$. We claim that this function is continuous. In fact, we claim more: T^{S^ω} is a computable function (such functions are, of necessity, continuous since each computation only depends on a finite initial segment of the input, which is considered an oracle for effecting the computation). Consider a distinguished variable $\check{x}$ of type 1. Fix $\alpha \in \mathrm{S}_1$. We complement the previous normalization calculus with a new contraction rule:

$(\mathrm{v})_\alpha$ t is $\check{x}(\overline{n})$ and q is $\overline{\alpha(n)}$.

This α-normalization calculus enjoys the property of confluence and strong normalization. Therefore, the type 0 term $T\check{x}$ has a (unique) normal form which (it may easily be argued) must be a numeral $\overline{n}$. Due to the semantical soundness of the contraction procedure, it is clear that $T^{\mathrm{S}^\omega}(\alpha) = n$. It is also clear that the process of normalization yields an oracle computation (the oracle is only invoked to effect contractions of the form $(\mathrm{v})_\alpha$).

2.5. Suggested reading and historical notes. The structure of the hereditarily continuous functionals was independently discovered by Stephen Kleene in [22] and Kreisel in [40]. The notion of majorizability was introduced by Howard in [20]. The structure of the *strongly* majorizable functionals was defined by Bezem in [4]. The failure of choice in M^ω is due to Kohlenbach in [25]. The existence of the continuous FAN functional mentioned at the end of the discussion of the structure M^ω appeared in [41], and the proof that this functional is not majorizable is discussed in [23]. The strong normalization theorem for finite-type functionals (as well as the oracle modification) is due to William Tait in [49].

Anne Troelstra's book [52] is still very much rewarding for studying the topics of this section. Volume II of [55] and the recent [35] are alternatives. More specifically: HRO^ω and ICF^ω and their extensional counterparts are covered in [52] ([55] has less material). Our treatment of the extensional structures is slightly unusual since we work directly with set theoretic functionals instead of working with their indexes or associates endowed with a suitable notion of equality. Kohlenbach also studies the extensional counterparts in [35], as well as the structure of the majorizable functionals in detail. Both [52] and [55] study the term model and the normalization theorems.

The treatment of equality in finite types in this paper comes from a suggestion in [18], elaborated by Troelstra in [53] (the subscript '0' in HA_0^ω comes from Troelstra given notation).

§3. Modified and bounded realizability.

3.1. Modified realizability. The method of realizability is reminiscent of the BHK (Brouwer-Heyting-Kolmogorov) interpretation of the intuitionistic connectives. We introduce a version of realizability, called *modified realizability*, due to Kreisel in the setting of finite-type arithmetic.

We present Kreisel's modified realizability in a slightly unfamilar way. Instead of saying what realizing tuples of functionals are, we associate to each formula of the language an existential formula. We need a preliminary definition:

DEFINITION 3.1. A formula of $\mathcal{L}_0^\omega$ is called $\exists$-free if it is built from atomic formulas by means of conjuntions, implications and universal quantifications.

Do notice that $\exists$-free also means free of disjunctions (compare with some definitions in the sequel). We are ready to define the assignment of modified realizability:

DEFINITION 3.2. To each formula A of the language $\mathcal{L}_0^\omega$ we assign formulas $(A)^{\mathrm{mr}}$ and A_{mr} so that $(A)^{\mathrm{mr}}$ is of the form $\exists \underline{x} A_{\mathrm{mr}}(\underline{x})$ with $A_{\mathrm{mr}}(\underline{x})$ a $\exists$-free formula, according to the following clauses:

1. $(A)^{\mathrm{mr}}$ and A_{mr} are simply A, for atomic formulas A.

If we have already interpretations for A and B given by $\exists \underline{x} A_{\mathrm{mr}}(\underline{x})$ and $\exists \underline{y} B_{\mathrm{mr}}(\underline{y})$ (respectively), then we define:

2. $(A \wedge B)^{\mathrm{mr}}$ is $\exists \underline{x}, \underline{y}(A_{\mathrm{mr}}(\underline{x}) \wedge B_{\mathrm{mr}}(\underline{y}))$,
3. $(A \vee B)^{\mathrm{mr}}$ is $\exists n^0 \exists \underline{x}, \underline{y}((n = 0 \rightarrow A_{\mathrm{mr}}(\underline{x})) \wedge (n \neq 0 \rightarrow B_{\mathrm{mr}}(\underline{y})))$,
4. $(A \rightarrow B)^{\mathrm{mr}}$ is $\exists \underline{f} \forall \underline{x}(A_{\mathrm{mr}}(\underline{x}) \rightarrow B_{\mathrm{mr}}(\underline{f}(\underline{x})))$,

5. $(\forall z A(z))^{\mathrm{mr}}$ is $\exists \underline{f} \forall z A_{\mathrm{mr}}(\underline{f}(z), z)$,
6. $(\exists z A(z))^{\mathrm{mr}}$ is $\exists z, \underline{x} A_{\mathrm{mr}}(\underline{x}, z)$.

In the established literature, we say that $\underline{x}$ mr-*realizes* A instead of $A_{\mathrm{mr}}(\underline{x})$. Notice that the tuple $\underline{x}$ may be empty. The realizers of a disjunction include a flag n of type 0 that decides which way to fork. Similarly, the realizers of an existential quantifier include an existential witness. It is easy to check that the interpretation of negation $(\neg A)^{\mathrm{mr}}$ is $\forall \underline{x} \neg A_{\mathrm{mr}}(\underline{x})$. Notice that $(\neg A)^{\mathrm{mr}}$ is always an $\exists$-free formula and, therefore, demands an empty realizer. Realizability is unsuitable for extracting constructive information from negated formulas.

There are two important principles in connection with modified realizability:

I. *Axiom of Choice* AC^{ω}: $\forall \underline{x} \exists \underline{y} A(\underline{x}, \underline{y}) \to \exists \underline{f} \forall \underline{x} A(\underline{x}, \underline{f}\underline{x})$, where A is any formula.

II. *Independence of Premises* $\mathsf{IP}^{\omega}_{\exists\mathrm{free}}$: $(A \to \exists \underline{z} B(\underline{z})) \to \exists \underline{z}(A \to B(\underline{z}))$, where A is $\exists$-free and B is an arbitrary formula.

The last principle is a classical, but not intuitionistic, law.

THEOREM 3.3 (Soundness of modified realizability). *Suppose that*

$$\mathsf{HA}^{\omega}_0 + \mathsf{AC}^{\omega} + \mathsf{IP}^{\omega}_{\exists\mathrm{free}} + \Delta \vdash A(\underline{z}),$$

where Δ is a set of $\exists$-free sentences and $A(\underline{z})$ is an arbitrary formula (with the free variables as shown). Then there are closed terms $\underline{t}$ of appropriate types such that

$$\mathsf{HA}^{\omega}_0 + \Delta \vdash \forall \underline{z} A_{\mathrm{mr}}(\underline{t}(\underline{z}), \underline{z}).$$

The proof is by induction on the length of formal derivations, and the terms are effectively constructed from the formal derivations. The principles AC^{ω} and $\mathsf{IP}^{\omega}_{\exists\mathrm{free}}$ disappear because they are trivially realizable. Full extensionality is not automatically interpretable but the next best thing happens: It is constituted by $\exists$-free sentences and, therefore, it is self-interpretable. Therefore, in the above theorem, we may replace in *both* places the theory HA^{ω}_0 by $\mathsf{E}\text{-}\mathsf{HA}^{\omega}_0$.

3.2. Extraction and all that (I). The following is an important consequence of the soundness theorem:

PROPOSITION 3.4 (Extraction and conservation, modified realizability). *Suppose that*

$$\mathsf{HA}^{\omega}_0 + \mathsf{AC}^{\omega} + \mathsf{IP}^{\omega}_{\exists\mathrm{free}} + \Delta \vdash \forall x \exists y A(x, y),$$

where Δ is a set of $\exists$-free sentences and A is a $\exists$-free formula with free variables among x and y. Then there is a closed term t of appropriate type such that

$$\mathsf{HA}^{\omega}_0 + \Delta \vdash \forall x A(x, tx).$$

Letting A be the sentence $0 = 1$ and Δ be empty, we conclude that the theory $\mathsf{HA}_0^\omega + \mathsf{AC}^\omega + \mathsf{IP}_{\exists\text{free}}^\omega$ is consistent *relative* to HA_0^ω. This is not, however, terribly interesting.

THEOREM 3.5 (Characterization). *For any formula A,*

$$\mathsf{HA}_0^\omega + \mathsf{AC}^\omega + \mathsf{IP}_{\exists\text{free}}^\omega \vdash A \leftrightarrow (A)^{\text{mr}}.$$

The two principles AC^ω and $\mathsf{IP}_{\exists\text{free}}^\omega$ are called the *characteristic principles* of modified realizability. The characterization theorem also ensures that we are not missing any principles besides AC^ω and $\mathsf{IP}_{\exists\text{free}}^\omega$ in the statement of the soundness theorem. To see this, suppose that we could state the soundness theorem with a further principle (sentence) P. Since P is a consequence of itself, from soundness it would follow that there are closed terms $\underline{t}$ such that $\mathsf{HA}^\omega \vdash \mathsf{P}_{\text{mr}}(\underline{t})$. *A fortiori*, $\mathsf{HA}^\omega \vdash \exists \underline{x}\mathsf{P}_{\text{mr}}(\underline{x})$, i.e., $\mathsf{HA}^\omega \vdash (\mathsf{P})^{\text{mr}}$. By the characterization theorem, we get $\mathsf{HA}^\omega + \mathsf{AC}^\omega + \mathsf{IP}_{\exists\text{free}}^\omega \vdash \mathsf{P}$. In conclusion, P is superfluous.

COROLLARY 3.6. *The following two properties hold*:

(a) (*Disjunction Property*) *Suppose that* $\mathsf{HA}_0^\omega + \mathsf{AC}^\omega + \mathsf{IP}_{\exists\text{free}}^\omega \vdash A \vee B$, *for* sentences *$A$ and B. Then, either* $\mathsf{HA}_0^\omega + \mathsf{AC}^\omega + \mathsf{IP}_{\exists\text{free}}^\omega \vdash A$ *or* $\mathsf{HA}_0^\omega + \mathsf{AC}^\omega + \mathsf{IP}_{\exists\text{free}}^\omega \vdash B$.

(b) (*Existence Property*) *Suppose that* $\mathsf{HA}_0^\omega + \mathsf{AC}^\omega + \mathsf{IP}_{\exists\text{free}}^\omega \vdash \exists x A(x)$, *for the* sentence $\exists x A(x)$. *Then there is a closed term t such that* $\mathsf{HA}_0^\omega + \mathsf{AC}^\omega + \mathsf{IP}_{\exists\text{free}}^\omega \vdash A(t)$.

The above corollary is consequence of the soundness and characterization theorems. For instance, suppose that $A \vee B$ is provable in $\mathsf{HA}_0^\omega + \mathsf{AC}^\omega + \mathsf{IP}_{\exists\text{free}}^\omega$. By the soundness theorem, there are closed terms s^0 and $\underline{t}, \underline{q}$ such that,

$$\mathsf{HA}_0^\omega \vdash (s = 0 \rightarrow A_{\text{mr}}(\underline{t})) \wedge (s \neq 0 \rightarrow B_{\text{mr}}(\underline{q})).$$

Since s is a closed term, there is a numeral $\overline{n}$ such that $\mathsf{HA}_0^\omega \vdash s = \overline{n}$. Suppose $n = 0$ (the other case is similar). Then $\mathsf{HA}_0^\omega \vdash A_{\text{mr}}(\underline{t})$ and, therefore, $\mathsf{HA}_0^\omega \vdash (A)^{\text{mr}}$. By the characterization theorem, we get the desired conclusion.

It is important to notice that the above argument works because, in the presence of AC^ω and $\mathsf{IP}_{\exists\text{free}}^\omega$, it is possible to *come back* from the formula $(A)^{\text{mr}}$ to the original formula A. The existence property has the following natural generalization:

PROPOSITION 3.7. *Let A be an* arbitrary *formula with free variables among x and y. Suppose that*

$$\mathsf{HA}_0^\omega + \mathsf{AC}^\omega + \mathsf{IP}_{\exists\text{free}}^\omega \vdash \forall x \exists y A(x, y).$$

Then there is a closed term t of appropriate type such that

$$\mathsf{HA}_0^\omega + \mathsf{AC}^\omega + \mathsf{IP}_{\exists\text{free}}^\omega \vdash \forall x A(x, tx).$$

This is not a conservation result anymore. However, it is still a *sound* extraction result since the conclusion $\forall x A(x, tx)$ is *true*.

THEOREM 3.8 (FAN rule). *Let A be an arbitrary formula containing only the variables x^1 and n^0. Then the following rule holds: If*

$$\mathsf{HA}_0^\omega + \mathsf{AC}^\omega + \mathsf{IP}_{\exists\text{free}}^\omega \vdash \forall x \leq_1 1 \exists n^0\, A(x,n)$$

then there is a natural number m such that

$$\mathsf{HA}_0^\omega + \mathsf{AC}^\omega + \mathsf{IP}_{\exists\text{free}}^\omega \vdash \forall x \leq_1 1 \exists n \leq \overline{m}\, A(x,n).$$

The proof of this result uses Howard's majorizability relation. Assume that $\forall x \leq_1 1 \exists n^0\, A(x,n)$ is provable in $\mathsf{HA}_0^\omega + \mathsf{AC}^\omega + \mathsf{IP}_{\exists\text{free}}^\omega$. We get,

$$\mathsf{HA}_0^\omega + \mathsf{AC}^\omega + \mathsf{IP}_{\exists\text{free}}^\omega \vdash \forall x (x \leq_1 1 \to \exists n \exists \underline{z} A_{\text{mr}}(\underline{z}, x, n)).$$

By $\mathsf{IP}_{\exists\text{free}}^\omega$, $\forall x \exists n \exists \underline{z} (x \leq_1 1 \to A_{\text{mr}}(\underline{z}, x, n))$ is provable. As a consequence of the soundness theorem, there is a closed term t of type 2 such that

$$\mathsf{HA}_0^\omega \vdash \forall x (x \leq_1 1 \to \exists \underline{z} A_{\text{mr}}(\underline{z}, x, tx)).$$

By Howard's majorizability result, take a closed term q such that $t \leq_2^* q$. It is clear that

$$\mathsf{HA}_0^\omega + \mathsf{AC}^\omega + \mathsf{IP}_{\exists\text{free}}^\omega \vdash \forall x \leq_1 1 \exists n \leq q(1^1)\, A(x,n).$$

3.3. Digression on intuitionistic extraction. Modified realizability does not quite achieve the disjunction and existence property for HA_0^ω. For instance, if $\mathsf{HA}_0^\omega \vdash A \vee B$ then we can only guarantee that either A or B is provable in the *stronger* theory $\mathsf{HA}_0^\omega + \mathsf{AC}^\omega + \mathsf{IP}_{\exists\text{free}}^\omega$. One would want instead that A or B were already provable in the original HA_0^ω. This is in fact the case, and the proof uses a variant of modified realizability: *modified realizability with truth*. The clauses of this variant of realizability are the same of modified realizability except that the conditional has an extra clause:

$$(A \to B)^{\text{mrt}} \textit{ is } \exists \underline{f} \forall \underline{x} (A_{\text{mrt}}(\underline{x}) \to B_{\text{mrt}}(\underline{f}(\underline{x}))) \wedge (A \to B).$$

THEOREM 3.9 (Soundness of modified realizability with truth). *Suppose that*

$$\mathsf{HA}_0^\omega \vdash A(\underline{z}),$$

where $A(\underline{z})$ is an arbitrary formula (with the free variables as shown). Then there are closed terms $\underline{t}$ of appropriate types such that

$$\mathsf{HA}_0^\omega \vdash \forall \underline{z} A_{\text{mrt}}(\underline{t}(\underline{z}), \underline{z}).$$

For the purpose at hand, we crucially have:

LEMMA 3.10. *For every formula A, $\mathsf{HA}_0^\omega \vdash (A)^{\text{mrt}} \to A$.*

Now it is clear how to prove the disjunction and existence properties for plain HA_0^ω.

3.4. Bounded modified realizability. The new bounded interpretations rely heavily on the Howard-Bezem majorizability notions. In view of this fact, it is convenient to work with an extension of the language $\mathcal{L}_0^\omega$ (with exactly the same terms). Firstly, we extend the language $\mathcal{L}_\leq^\omega$ with a *primitive* binary relation symbol $\leq$ that infixes between terms of type 0. There are now new atomic formulas, and the syntactic notions extend in the natural way. Actually, the language that we get is an extension *by definitions* of $\mathcal{L}_0^\omega$ because $x \leq y$ may be defined by a natural quantifier-free formula. In this setting, we use the primitive binary symbol to define the majorizability formulas. Secondly, it is convenient to introduce the primitive syntactical device of *bounded quantifications*, i.e., quantifications of the form $\forall x \leq^* t$ and $\exists x \leq^* t$, for terms t not containing the variable x. *Bounded formulas* are formulas in which every quantifier is bounded.

The theory $\mathsf{HA}_\leq^\omega$ is HA_0^ω together with the (universal) defining axioms of $\leq$ and the following schemes:

$$\mathsf{B}_\forall : \forall x \leq^* tA(x) \leftrightarrow \forall x(x \leq^* t \to A(x)),$$
$$\mathsf{B}_\exists : \exists x \leq^* tA(x) \leftrightarrow \exists x(x \leq^* t \wedge A(x)).$$

It is clear that $\mathsf{HA}_\leq^\omega$ is a conservative extension of HA_0^ω. We say that a functional f is monotone if $f \leq^* f$. In the sequel, we often quantify over monotone functionals. We abbreviate the quantifications $\forall x(x \leq^* x \to A(x))$ and $\exists x(x \leq^* x \wedge A(x))$ by $\tilde{\forall} x A(x)$ and $\tilde{\exists} x A(x)$, respectively.

Definition 3.11. A formula of $\mathcal{L}_\leq^\omega$ is called $\tilde{\exists}$*-free* if it is built from atomic formulas by means of conjunctions, disjunctions, implications, bounded quantifications and monotone universal quantifications, i.e., quantifications of the form $\tilde{\forall} a$.

This notion resembles the notion of $\exists$-free formula, but notice that *disjunctions* are allowed.

Definition 3.12. To each formula A of the language $\mathcal{L}_\leq^\omega$ we assign formulas $(A)^{\mathrm{br}}$ and A_{br} so that $(A)^{\mathrm{br}}$ is of the form $\tilde{\exists}\underline{b} A_{\mathrm{br}}(\underline{b})$, with $A_{\mathrm{br}}(\underline{b})$ a $\tilde{\exists}$-free formula, according to the following clauses:

1. $(A)^{\mathrm{br}}$ and $(A)_{\mathrm{br}}$ are simply A, for atomic formulas A.

If we have already interpretations for A and B given by $\tilde{\exists}\underline{b} A_{\mathrm{br}}(\underline{b})$ and $\tilde{\exists}\underline{d} B_{\mathrm{br}}(\underline{d})$ (respectively) then, we define

2. $(A \wedge B)^{\mathrm{br}}$ is $\tilde{\exists}\underline{b}, \underline{d}(A_{\mathrm{br}}(\underline{b}) \wedge B_{\mathrm{br}}(\underline{d}))$,
3. $(A \vee B)^{\mathrm{br}}$ is $\tilde{\exists}\underline{b}, \underline{d}(A_{\mathrm{br}}(\underline{b}) \vee B_{\mathrm{br}}(\underline{d}))$,
4. $(A \to B)^{\mathrm{br}}$ is $\tilde{\exists}\underline{f}\tilde{\forall}\underline{b}(A_{\mathrm{br}}(\underline{b}) \to B_{\mathrm{br}}(\underline{f}(\underline{b})))$.

For bounded quantifiers we have:

5. $(\forall x \leq^* t\, A(x))^{\mathrm{br}}$ is $\tilde{\exists}\underline{b}\forall x \leq^* t\, A_{\mathrm{br}}(\underline{b}, x)$,
6. $(\exists x \leq^* t\, A(x))^{\mathrm{br}}$ is $\tilde{\exists}\underline{b}\exists x \leq^* t\, A_{\mathrm{br}}(\underline{b}, x)$.

And for unbounded quantifiers we define

7. $(\forall x A(x))^{\text{br}}$ is $\tilde{\exists} \underline{f} \tilde{\forall} a \forall x \leq^* a\, A_{\text{br}}(\underline{f}(a), x)$.
8. $(\exists x A(x))^{\text{br}}$ is $\tilde{\exists} a, \underline{b} \exists x \leq^* a\, A_{\text{br}}(\underline{b}, x)$.

Notice that the realizers of a disjunction do not include a flag deciding which way to fork, and that only a *bound* for the existential witness is included in the realizers of an existential statement. As usual, negation is a particular case of the implication: $(\neg A)^{\text{br}}$ is $\tilde{\forall} \underline{b} \neg A_{\text{br}}(\underline{b})$.

Three principles are important in connection with the above assignment:

I. *Bounded Choice* bAC^{ω}:

$$\forall \underline{x} \exists \underline{y} A(\underline{x}, \underline{y}) \to \tilde{\exists} \underline{f} \tilde{\forall} \underline{b} \forall \underline{x} \leq^* \underline{b} \exists \underline{y} \leq^* \underline{f}\, \underline{b}\, A(\underline{x}, \underline{y}),$$

where A is an arbitrary formula.

II. *Bounded Independence of Premises* $\mathsf{bIP}^{\omega}_{\tilde{\exists}\text{free}}$:

$$(A \to \exists \underline{y} B(\underline{y})) \to \tilde{\exists} \underline{b} (A \to \exists \underline{y} \leq^* \underline{b}\, B(\underline{y})),$$

where A is a $\tilde{\exists}$-free formula and B is an arbitrary formula.

III. *Majorizability Axioms* MAJ^{ω}:

$$\forall x \exists y (x \leq^* y).$$

It must be remarked that each principle above is *false* in the full set theoretical model S^{ω}. The majorizability axioms MAJ^{ω} (as well as $\mathsf{bIP}^{\omega}_{\tilde{\exists}\text{free}}$) are, of course, true in the structure M^{ω} of the majorizable functionals. However, as we saw in the first part, the bounded choice principle already fails in M^{ω} for x of type 1 and y of type 0.

PROPOSITION 3.13. *The theory* $\mathsf{HA}^{\omega}_{\leq} + \mathsf{bAC}^{\omega} + \mathsf{bIP}^{\omega}_{\tilde{\exists}\text{free}}$ *proves the* Collection Principle bC^{ω}:

$$\forall \underline{z} \leq^* \underline{c}\, \exists \underline{y} A(\underline{y}, \underline{z}) \to \tilde{\exists} \underline{b} \forall \underline{z} \leq^* \underline{c} \exists \underline{y} \leq^* \underline{b} A(\underline{y}, \underline{z}),$$

where A is an arbitrary formula and $\underline{c}$ is a tuple of monotone functionals.

PROOF. Suppose that $\forall \underline{z} \leq^* \underline{c}\, \exists \underline{y} A(\underline{y}, \underline{z})$. By $\mathsf{bIP}^{\omega}_{\tilde{\exists}\text{free}}$, we get

$$\forall \underline{z} \tilde{\exists} \underline{b} (\underline{z} \leq^* \underline{c} \to \exists \underline{y} \leq^* \underline{b} A(\underline{y}, \underline{z})).$$

By bAC^{ω}, we may conclude that $\tilde{\exists} \underline{f} \forall \underline{z} \leq^* \underline{c} \tilde{\exists} \underline{b} \leq^* \underline{f}\, \underline{c} \exists \underline{y} \leq^* \underline{b}\, A(\underline{x}, \underline{y})$. The desired conclusion follows from the transitivity of the majorizability relation. ⊣

The case where the types of z and y are 0 extends the familiar *collection principle* of arithmetic. In the context of intuitionistic analysis, Brouwer's FAN theorem is the case where z is of type 1 and y is of type 0. The formulation for the Cantor space is:

$$\forall z \leq_1 1 \exists n^0 A(n, z) \to \exists m^0 \forall z \leq_1 1 \exists n \leq m A(n, z),$$

for arbitrary formulas A. The above formulation of the FAN theorem must be distinguished from the following, which also appears in the literature:

$$\forall z \leq_1 1 \exists n^0 A(n, z) \rightarrow \exists k^0 \forall z \leq_1 1 \exists n^0 \forall w \leq_1 1(\overline{w}(k) = \overline{z}(k) \rightarrow A(n, w)),$$

for arbitrary formulas A. The latter formulation explicitly includes a *continuity principle*, whereas the former does not. It is important to keep in mind that bounded interpretations concern majorizability notions, not continuity notions.

THEOREM 3.14 (Soundness of the bounded modified realizability). *Suppose that*

$$\mathsf{HA}^\omega_\leq + \mathsf{bAC}^\omega + \mathsf{bIP}^\omega_{\tilde{\exists}\text{free}} + \mathsf{MAJ}^\omega + \Delta \vdash A(\underline{z}),$$

where Δ is a set of $\tilde{\exists}$-free sentences and $A(\underline{z})$ is an arbitrary formula (with the free variables as shown). Then there are closed monotone terms $\underline{t}$ of appropriate types such that

$$\mathsf{HA}^\omega_\leq + \Delta \vdash \tilde{\forall}\underline{a} \forall \underline{z} \leq^* \underline{a}\, A_{\mathrm{br}}(\underline{t}(\underline{a}), \underline{z}).$$

Above, and in similar situations in the sequel, by a closed *monotone* term t we mean a closed term such that the monotonicity condition ($t \leq^* t$, in this case) is provable in the base theory ($\mathsf{HA}^\omega_\leq$, in this case).

3.5. Extraction and all that (II). The following is a consequence of the soundness theorem:

PROPOSITION 3.15. (Extraction and conservation, bounded modified realizability) *Suppose that*

$$\mathsf{HA}^\omega_\leq + \mathsf{bAC}^\omega + \mathsf{bIP}^\omega_{\tilde{\exists}\text{free}} + \mathsf{MAJ}^\omega + \Delta \vdash \forall x \exists y A(x, y),$$

where Δ is a set of $\tilde{\exists}$-free sentences and A is a $\tilde{\exists}$-free formula with free variables among x and y. Then there is a closed monotone term t of appropriate type such that

$$\mathsf{HA}^\omega_\leq + \Delta \vdash \tilde{\forall} a \forall x \leq^* a \exists y \leq^* ta\, A(x, y).$$

Strictly speaking, we do not have a conservation result. However, if the type of x is 0 or 1 then we do have such a result:

COROLLARY 3.16. *Suppose that*

$$\mathsf{HA}^\omega_\leq + \mathsf{bAC}^\omega + \mathsf{bIP}^\omega_{\tilde{\exists}\text{free}} + \mathsf{MAJ}^\omega + \Delta \vdash \forall x^{0/1} \exists y A(x, y),$$

where Δ is a set of $\tilde{\exists}$-free sentences, A a $\tilde{\exists}$-free formula with free variables among x and y. Then there is a closed monotone term t of appropriate type such that

$$\mathsf{HA}^\omega_\leq + \Delta \vdash \forall x \exists y \leq^* tx\, A(x, y).$$

In particular, this corollary yields a relative consistency result of the theory $\mathsf{HA}^{\omega}_{\leq} + \mathsf{bAC}^{\omega} + \mathsf{bIP}^{\omega}_{\tilde{\exists}\text{free}} + \mathsf{MAJ}^{\omega}$ over $\mathsf{HA}^{\omega}_{\leq}$. This is interesting in the present setting because the former theory is *classically inconsistent*: It refutes the classically true Markov's principle. To see this, suppose that

$$\forall x^1(\neg\neg\exists n^0(xn = 0) \to \exists n^0(xn = 0)).$$

By intuitionistic logic and $\mathsf{bIP}^{\omega}_{\tilde{\exists}\text{free}}$, $\forall x^1 \exists n^0(\neg\forall k^0(xk \neq 0) \to \exists i \leq n(xi = 0))$. Now, by the collection principle bC^{ω}, there is a natural number m^0 such that $\forall x \leq_1 1(\neg\forall k^0(xk \neq 0) \to \exists n \leq m\,(xn = 0))$. This is a contradiction (just consider the number-theoretic primitive recursive function that takes the value 1 for values less than $m + 1$ and is 0 afterwards).

The reader should take notice that the world of $\mathsf{HA}^{\omega}_{\leq} + \mathsf{bAC}^{\omega} + \mathsf{bIP}^{\omega}_{\tilde{\exists}\text{free}} + \mathsf{MAJ}^{\omega}$ is a world with some principles related to Brouwerian intuitionism and, in the terminology of intuitionism, has *strong counterexamples* to classical logic.

We can also prove a characterization theorem for bounded modified realizability:

THEOREM 3.17 (Characterization). *For any formula A,*

$$\mathsf{HA}^{\omega}_{\leq} + \mathsf{bAC}^{\omega} + \mathsf{bIP}^{\omega}_{\tilde{\exists}\text{free}} + \mathsf{MAJ}^{\omega} \vdash A \leftrightarrow (A)^{\mathrm{br}}.$$

Of course, bAC^{ω}, $\mathsf{bIP}^{\omega}_{\tilde{\exists}\text{free}}$ and MAJ^{ω} are called the *characteristic* principles of the bounded modified realizability. As noticed, in contrast with the traditional case, these charateristic principles are not true. If we prove $\forall x^1 \exists y A(x, y)$ in the extended theory $\mathsf{HA}^{\omega}_{\leq} + \mathsf{bAC}^{\omega} + \mathsf{bIP}^{\omega}_{\tilde{\exists}\text{free}} + \mathsf{MAJ}^{\omega}$ for an *arbitrary* A then, in complete analogy with the traditional case, we can find a closed monotone term t such that $\forall x^1 \exists y \leq^* tx\, A(x, y)$ is provable in $\mathsf{HA}^{\omega}_{\leq} + \mathsf{bAC}^{\omega} + \mathsf{bIP}^{\omega}_{\tilde{\exists}\text{free}} + \mathsf{MAJ}^{\omega}$. However, the extended theory is not a *true* theory and, therefore, we may not conclude that the statement $\forall x^1 \exists y \leq^* tx\, A(x, y)$ is true. In other words, the extraction of the term t is *not sound* when the matrix A is arbitrary. However, as we saw above, the extraction is indeed sound if $A(x, y)$ is an $\tilde{\exists}$-free formula (in particular, if $A(x, y)$ is quantifier-free).

3.6. Benign principles. In the context of bounded modified realizability, if we are interested in extracting sound bounding information from proofs, we must restrict ourselves to conclusions of the form $\forall\exists$ whose matrix is $\tilde{\exists}$-free. If such statements are proved in the theory $\mathsf{HA}^{\omega}_{\leq} + \mathsf{bAC}^{\omega} + \mathsf{bIP}^{\omega}_{\tilde{\exists}\text{free}} + \mathsf{MAJ}^{\omega}$, then it is possible to extract truthful bounding information. Of course, we may use in our proofs principles that *follow* from the above theory. This is the case, e.g., with Kohlenbach's *uniform boundedness principles* UB_{ρ}, a combination of the FAN theorem (extended to higher types) with choice:

$$\forall k^0 \forall x \leq yk\, \exists z^0 A(x, y, k, z) \to \exists \chi^1 \forall k^0 \forall x \leq yk\, \exists z \leq \chi k\, A(x, y, k, z),$$

for arbitrary A and y of type $0 \to \rho$.

There are, however, principles that do not follow from the extended theory but whose use as premises does yield sound bounding information (since that information is checked in a *true* theory). We call these principles *benign.* Note that benign principles may be *false* (because they could follow from suitable true principles with the aid of the false characteristic principles). We present a list of eight benign principles:

1. The axioms of extensionality $\forall z^{\sigma\to\delta}\forall x^{\sigma}, y^{\sigma}(x =_{\sigma} y \to zx =_{\delta} zy)$, for $\sigma = 0, 1$ or 2 and δ arbitrary, are benign.

2. The classical, but not intuitionistic, truth

$$\forall x\forall y(A(x) \vee B(y)) \to \forall x A(x) \vee \forall y B(y),$$

where A and B are $\tilde{\exists}$-free formulas, is benign. When the types of x and y are 0 and A and B are quantifier-free formulas, we have the *lesser limited principle of omniscience* LLPO (this is Errett Bishop's terminology).

3. The law of excluded middle $A \vee \neg A$, for $\tilde{\exists}$-free formulas A, is benign. This form of excluded middle includes Π^0_1-LEM, i.e., $\forall n^0 A(n) \vee \neg\forall n^0 A(n)$ for A a first-order bounded formula. In view of the fact that $\mathsf{HA}^{\omega}_0 + \mathsf{bAC}^{\omega} + \mathsf{bIP}^{\omega}_{\tilde{\exists}\mathrm{free}} + \mathsf{MAJ}^{\omega}$ refutes Markov's principle, we are drawn to the conclusion that Π^0_1-LEM does not prove Markov's principle.

4. The choice principle $\forall x^{0/1}\exists y A(x, y) \to \exists f \forall x A(x, fx)$, where A is an arbitrary formula and y is of any type, is benign.

5. The following is a benign version of choice with no restrictions on the types: $\forall x \leq^* a\, \exists y A(x, y) \to \exists f \forall x \leq^* a\, A(x, fx)$, for arbitrary formulas A.

6. It is well-known that the FAN theorem is *true* (and intuitionistically acceptable) for quantifier-free matrices with parameters of type 0 or 1 *only.* A modification of its contrapositive, which is equivalent to *weak König's lemma* WKL, is rejected intuitionistically. Nevertheless, this modified contrapositive, namely $\forall n^0 \exists x \leq_1 1 \forall k \leq n\, A_{\mathrm{qf}}(x, k) \to \exists x \leq_1 1 \forall k\, A_{\mathrm{qf}}(x, k)$, is a benign principle. The familiar formulation of weak König's lemma states that every infinite subtree of the full binary tree has an infinite path. This principle is non-constructive in the following precise sense: There are infinite recursive subtrees of the full binary tree that have no recursive infinite path (this is a reformulation of Kleene's result that there are recursively inseparable r.e. sets).

7. The form of comprehension $\exists\Phi\forall y(\Phi y = 0 \leftrightarrow A(y))$, where A a $\tilde{\exists}$-free formula and y may be of any type, is benign. Comprehension for negated formulas is also a benign principle: $\exists\Phi\forall y(\Phi y = 0 \leftrightarrow \neg A(y))$, for arbitrary A.

8. Kohlenbach considered in [29] the principles F_{ρ}, a simplification of which are:

$$\forall\Phi^{\rho\to 0}\forall y^{\rho}\exists y_0 \leq_{\rho} y \forall z \leq_{\rho} y\, (\Phi(z) \leq \Phi(y_0)).$$

These principles are false for $\rho \neq 0$. They are, nevertheless, benign.

The proof that the above principles are benign relies on a careful study of the formulas which imply, or are implied by, their own bounded realizations.

For some years now, Kohlenbach and his co-workers have been showing the *practical* use of Proof Theory in obtaining numerical bounds from *classical* proofs of analysis. Kohlenbach's methods are *not* based on realizability because realizability notions (including bounded realizability) are *not* taylored for the analysis of *classical* proofs. In effect, even though a classical proof may be translated into an intuitionistic proof via (e.g.) the Gödel-Gentzen negative translation, the translation destroys existential statements — replacing them by negated universal statements — with the consequence that realizers yield no computational information. Of course, this shortcoming is related with the fact that Markov's principle is *not* benign. That notwithstanding, bounded modified realizability (and Kohlenbach's monotone modified realizability) supports many classical principles that go beyond intuitionistic logic.

3.7. Suggested reading and historical notes. The notion of (numerical) realizability was introduced by Stephen Kleene in [21]. Modified realizability is due to Kreisel in [40] and [41]. Proofs of the theorems of Section 3.1 can be found in [52]. An alternative is the very recent [35] (which has, nevertheless, circulated in preliminary form as a manuscript for quite some years). A survey on realizability until the mid-nineties can be found in [54] (where one can find the story of the truth variants of realizability). The FAN rule at the end of Section 3.2 appeared in [35] but is already essentially treated in [25]. Bounded modified realizability was introduced by Ferreira and Nunes in [11]. This paper includes full proofs of the results concerning this form of realizability, including the discussion of the benign principles. For the sake of space, we did not discuss the monotone version of realizability introduced by Kohlenbach in [29]. The result that $\Pi^0_1-\mathsf{LEM}$ does not prove Markov's principle first appeared in [1] using this version. Monotone modified realizability is also explained in [35].

§4. The *dialectica* interpretation.

4.1. Gödel's *dialectica* interpretation. In 1958, Gödel published an article in an issue of the journal *dialectica* dedicated to the seventieth anniversary of Paul Bernays. The article presents an interpretation of first-order Heyting arithmetic into a quantifier-free theory with finite-type functionals (Gödel's theory T). This theory is, essentially, the quantifier-free part of HA^ω_0. Gödel's *dialectica* interpretation appeared as a contribution to an extended Hilbert's program by means of the notion of computable functional of finite type. In the sequel, we present Gödel's interpretation extended to finite-type arithmetic HA^ω_0.

DEFINITION 4.1. To each formula A of the language $\mathcal{L}^\omega_0$ we assign formulas $(A)^{\mathrm{D}}$ and A_{D} so that $(A)^{\mathrm{D}}$ is of the form $\exists \underline{x} \forall \underline{y} A_{\mathrm{D}}(\underline{x}, \underline{y})$ with $A_{\mathrm{D}}(\underline{x}, \underline{y})$ a quantifier-free formula, according to the following clauses:

1. $(A)^{\mathrm{D}}$ and A_{D} are simply A, for atomic formulas A.

If we have already interpretations for A and B given by $\exists \underline{x} \forall \underline{y} A_{\mathrm{D}}(\underline{x}, \underline{y})$ and $\exists \underline{z} \forall \underline{w} B_{\mathrm{D}}(\underline{z}, \underline{w})$ (respectively) then we define:

2. $(A \wedge B)^{\mathrm{D}}$ is $\exists \underline{x}, \underline{z} \forall \underline{y}, \underline{w}(A_{\mathrm{D}}(\underline{x}, \underline{y}) \wedge B_{\mathrm{D}}(\underline{z}, \underline{w}))$,
3. $(A \vee B)^{\mathrm{D}}$ is $\exists n^0, \underline{x}, \underline{z} \forall \underline{y}, \underline{w}((n = 0 \rightarrow A_{\mathrm{D}}(\underline{x}, \underline{y})) \wedge (n \neq 0 \rightarrow B_{\mathrm{D}}(\underline{z}, \underline{w})))$,
4. $(A \rightarrow B)^{\mathrm{D}}$ is $\exists \underline{f}, \underline{g} \forall \underline{x}, \underline{w}(A_{\mathrm{D}}(\underline{x}, \underline{g}(\underline{x}, \underline{w})) \rightarrow B_{\mathrm{D}}(\underline{f}(\underline{x}), \underline{w}))$.
5. $(\forall z A(z))^{\mathrm{D}}$ is $\exists \underline{f} \forall z \forall \underline{y} A_{\mathrm{D}}(\underline{f}(z), \underline{y}, z)$.
6. $(\exists z A(z))^{\mathrm{D}}$ is $\exists z, \underline{x} \forall \underline{y} A_{\mathrm{D}}(\underline{x}, \underline{y}, z)$.

The definition of implication is the hardest to understand. Gödel motivates it as follows. First consider $(A)^{\mathrm{D}} \rightarrow (B)^{\mathrm{D}}$, that is, the implication $\exists \underline{x} \forall \underline{y} A_{\mathrm{D}}(\underline{x}, \underline{y}) \rightarrow \exists \underline{z} \forall \underline{w} B_{\mathrm{D}}(\underline{z}, \underline{w})$. In other words, a witness $\underline{x}$ to $\forall \underline{y} A_{\mathrm{D}}(\underline{x}, \underline{y})$ gives rise to a witness $\underline{z}$ to $\forall \underline{w} B_{\mathrm{D}}(\underline{z}, \underline{w})$. If this is done by a rule of computation, there should be finite-type computable functionals $\underline{f}$ such that $\forall \underline{x}(\forall \underline{y} A_{\mathrm{D}}(\underline{x}, \underline{y}) \rightarrow \forall \underline{w} B_{\mathrm{D}}(\underline{f}(\underline{x}), \underline{w}))$. Gödel invites us to interprete the inner implication in the following way: Whenever a counter-example $\underline{w}$ is given to $B_{\mathrm{D}}(\underline{f}(\underline{x}), \underline{w})$ then a counter-example $\underline{y}$ is given to $A_{\mathrm{D}}(\underline{x}, \underline{y})$. If the latter counter-example is given by a rule of computation in terms of the former, then there should be finite-type computable functionals $\underline{g}$ such that $\neg B_{\mathrm{D}}(\underline{f}(\underline{x}), \underline{w}) \rightarrow \neg A_{\mathrm{D}}(\underline{x}, \underline{g}(\underline{x}, \underline{w}))$. Using the decidability of quantifier-free formulas, we get the implication $A_{\mathrm{D}}(\underline{x}, \underline{g}(\underline{x}, \underline{w})) \rightarrow B_{\mathrm{D}}(\underline{f}(\underline{x}), \underline{w})$, as in the definition. The attentive reader will notice that several of the passages above are not intuitionistically justifiable. However, Gödel's definitions do sustain a soundness theorem. There are three important principles in connection with this theorem:

I. The axiom of choice AC^{ω}.

II. The principle of independence of premises stated for universal antecedents, denoted by $\mathsf{IP}^{\omega}_{\forall}$ (a formula is *universal* if it is of the form $\forall \underline{x} A_{\mathrm{qf}}(\underline{x})$, for A_{qf} quantifier-free).

III. *Markov's Principle* MP^{ω}: $\neg \forall \underline{z} A_{\mathrm{qf}}(\underline{z}) \rightarrow \exists \underline{z} \neg A_{\mathrm{qf}}(\underline{z})$, where A_{qf} is a quantifier-free formula and the z's are of *any* types.

Markov's principle *strictu sensu* applies only for z of type 0. The acceptance of this principle differentiates the school of Russian constructivism from the other constructivist schools. Its acceptance is based on the intuition that if $\forall z^0 A_{\mathrm{qf}}(z)$ leads to a contradiction then, if we test in succession each natural number z^0 for the (decidable) truth of $A_{\mathrm{qf}}(z)$ then we will eventually find a z^0

such that $\neg A_{\mathrm{qf}}(z)$. Note that this intuition does not apply for z of non-zero type. The name "Markov's Principle" for non-zero types is a misnomer.[5]

Using the decidability of quantifier-free formulas, we can see that MP^{ω} implies $(\forall \underline{z} A_{\mathrm{qf}}(\underline{z}) \to B_{\mathrm{qf}}) \to \exists \underline{z}(A_{\mathrm{qf}}(\underline{z}) \to B_{\mathrm{qf}})$, for A_{qf} and B_{qf} quantifier-free formulas. Note that MP^{ω} is the particular case when A_{qf} is $\perp$.

THEOREM 4.2 (Soundness of the *dialectica* interpretation). *Suppose that*

$$\mathsf{HA}_0^{\omega} + \mathsf{AC}^{\omega} + \mathsf{IP}_{\forall}^{\omega} + \mathsf{MP}^{\omega} + \Delta \vdash A(\underline{z}),$$

where Δ is a set of universal sentences and A is an arbitrary formula (with the free variables as shown). Then there are closed terms $\underline{t}$ of appropriate types such that

$$\mathsf{HA}_0^{\omega} + \Delta \vdash \forall \underline{z} \forall \underline{y} A_{\mathrm{D}}(\underline{t}(\underline{z}), \underline{y}, \underline{z}).$$

As usual, the proof is by induction on the length of formal derivations, and the terms are effectively constructed from the formal derivations. The discussion of the seemingly innocuous contraction axiom $A \to A \wedge A$ is subtle in three respects (with the advent of *Linear Logic* in the late eighties, we learned that contraction is not that innocuous). Firstly, the choice of witnessing functionals is not canonical at this point. Secondly, it involves a definition by cases functional. Finally, it uses the decidabilility of quantifier-free formulas. The interpretation of the hypothesis is $\exists x \forall y A_{\mathrm{D}}(x, y)$ and that of the conclusion is

$$\exists x_1 \exists x_2 \forall y_1 \forall y_2 (A_{\mathrm{D}}(x_1, y_1) \wedge A_{\mathrm{D}}(x_2, y_2)).$$

We must define functionals f_1, f_2 and g such that

$$A_{\mathrm{D}}(x, g(x, y_1, y_2)) \to A_{\mathrm{D}}(f_1(x), y_1) \wedge A_{\mathrm{D}}(f_2(x), y_2).$$

We take f_1 and f_2 as $\lambda x.x$. Suppose that y_1 and y_2 are of type σ. In order to define g we need a closed term C_σ of type $0 \to \sigma \to \sigma \to \sigma$ which satisfies in HA_0^{ω} the following "definition by cases" requirements:

$$B[C_\sigma(0, u, v)/w] \leftrightarrow B[u/w] \text{ and } B[C_\sigma(Sz^0, u, v)/w] \leftrightarrow B[v/w],$$

where w is a distinguished variable of type σ and B is a quantifier-free formula. The functional C_σ can be defined with the aid of the recursor R_σ by $R_\sigma(z, u, \lambda w \lambda s.v)$. Note that the checking of the contraction axiom needs a recursor, not merely the combinators. Since A_{D} is quantifier-free, there is a closed term t of $\mathcal{L}_0^{\omega}$ such that $\mathsf{HA}_0^{\omega} \vdash t(x, y) = 0 \leftrightarrow \neg A_{\mathrm{D}}(x, y)$. It is easy to see that g can be taken to be $\lambda x \lambda y_1 \lambda y_2 . C_\sigma(t(x, y_1), y_1, y_2)$. But, as remarked, the choice is not canonical at this point: we can also take the term $\lambda x \lambda y_1 \lambda y_2 . C_\sigma(t(x, y_2), y_2, y_1)$.

[5]A referee of this paper pointed out that in the model of continuous functionals ECF^{ω}, where there is a dense subset of finitary objects at any type, it makes some sense to retain the name "Markov's principle."

The following proposition is an immediate consequence of the soundness theorem:

PROPOSITION 4.3 (Extraction and conservation, *dialectica* case). *Suppose that*

$$\mathsf{HA}_0^\omega + \mathsf{AC}^\omega + \mathsf{IP}_\forall^\omega + \mathsf{MP}^\omega + \Delta \vdash \forall x \exists y A_{\mathrm{qf}}(x, y),$$

where Δ is a set of universal sentences and A_{qf} is a quantifier-free formula with free variables among x and y. Then there is a closed term t of appropriate type such that

$$\mathsf{HA}_0^\omega + \Delta \vdash \forall x A_{\mathrm{qf}}(x, tx).$$

THEOREM 4.4 (Characterization). *For any formula A,*

$$\mathsf{HA}_0^\omega + \mathsf{AC}^\omega + \mathsf{IP}_\forall^\omega + \mathsf{MP}^\omega \vdash A \leftrightarrow (A)^{\mathrm{D}}.$$

With the aid of the characterization theorem, we can prove the following sound extraction result for *arbitrary* formulas A:

PROPOSITION 4.5. *Let A be an* arbitrary *formula with free variables among x and y. Suppose that*

$$\mathsf{HA}_0^\omega + \mathsf{AC}^\omega + \mathsf{IP}_\forall^\omega + \mathsf{MP}^\omega \vdash \forall x \exists y A(x, y),$$

Then there is a closed term t of appropriate type such that

$$\mathsf{HA}_0^\omega + \mathsf{AC}^\omega + \mathsf{IP}_\forall^\omega + \mathsf{MP}^\omega \vdash \forall x A(x, tx).$$

PROOF. Let $(A)^{\mathrm{D}}(x, y)$ be $\exists \underline{z} \forall \underline{w} A_{\mathrm{D}}(y, \underline{z}, \underline{w}, x)$. By the soundness theorem, there are closed terms t and $\underline{q}$ of appropriate types such that

$$\mathsf{HA}_0^\omega + \mathsf{AC}^\omega + \mathsf{IP}_\forall^\omega + \mathsf{MP}^\omega \vdash \forall x \forall \underline{w} A_{\mathrm{D}}(tx, \underline{q}x, \underline{w}, x).$$

The result follows by the characterization theorem. ⊣

4.2. On the monotone functional interpretation. The reason why we could include universal sentences Δ in the statement of the soundness theorem for the *dialectica* interpretation is because the *dialectica* interpretation of a universal sentence is (essentially) itself. By weakening the conclusion of the soundness theorem it is possible to deal with a wider classe of sentences.

THEOREM 4.6 (Soundness of the monotone functional interpretation). *Suppose that*

$$\mathsf{HA}_0^\omega + \mathsf{AC}^\omega + \mathsf{IP}_\forall^\omega + \mathsf{MP}^\omega + \Delta \vdash A(\underline{z}),$$

with Δ is a set of sentences *of the form $\exists \underline{x} \leq \underline{r} \forall \underline{y} B_{\mathrm{qf}}(\underline{x}, \underline{y})$, where B_{qf} is quantifier-free, $\underline{r}$ is tuple of closed terms, and A is an arbitrary formula (with the free variables as shown). Then there are closed monotone terms $\underline{t}$ of appropriate types such that*

$$\mathsf{HA}_0^\omega + \Delta \vdash \exists \underline{x} \leq^* \underline{t} \forall \underline{z} \forall \underline{y} \, A_{\mathrm{D}}(\underline{x}(\underline{z}), \underline{y}, \underline{z}).$$

The proof is by induction on the length of formal derivations, and the terms are effectively constructed from the formal derivations. In the proof, when it comes to the contraction axiom, the choice of terms becomes canonical. However, both the decidability of quantifier-free formulas and the definition by cases functional are still required.

PROPOSITION 4.7 (Extraction and conservation, monotone case). *Suppose that*

$$\mathsf{HA}_0^\omega + \mathsf{AC}^\omega + \mathsf{IP}_\forall^\omega + \mathsf{MP}^\omega + \Delta \vdash \forall x \exists y A_{\mathrm{qf}}(x, y),$$

with Δ *is a set of* sentences *of the form* $\exists \underline{x} \leq \underline{r} \forall \underline{y} B_{\mathrm{qf}}(\underline{x}, \underline{y})$, *where* B_{qf} *is quantifier-free,* $\underline{r}$ *is tuple of closed terms, and* A_{qf} *is a quantifier-free formula with free variables among* x *and* y. *Then there is a closed monotone term* t *of appropriate type such that*

$$\mathsf{HA}_0^\omega + \Delta \vdash \forall x, u\, (x \leq^* u \to \exists y \leq^* tu\, A_{\mathrm{qf}}(x, y)).$$

When x is of type 1 we can put u as x^{M} and get $\forall x \exists y \leq^* tx A_{\mathrm{qf}}(x, y)$. When furthermore y is of type 0, even $\forall x A_{\mathrm{qf}}(x, tx)$ is in order.

With the aid of the characterization theorem of the *dialectica* interpretation (notice that the assignment of formulas for the *dialectica* and monotone interpretations are the same), the following sound extraction result for *arbitrary* formulas A can be proved in the manner of Proposition 4.5:

PROPOSITION 4.8. *Let* A *be an* arbitrary *formula with free variables among* x *and* y. *Suppose that*

$$\mathsf{HA}_0^\omega + \mathsf{AC}^\omega + \mathsf{IP}_\forall^\omega + \mathsf{MP}^\omega + \Delta \vdash \forall x \exists y A(x, y),$$

with Δ *as in the proposition above. Then there is a closed monotone term* t *of appropriate type such that*

$$\mathsf{HA}_0^\omega + \mathsf{AC}^\omega + \mathsf{IP}_\forall^\omega + \mathsf{MP}^\omega + \Delta \vdash \forall x, u\, (x \leq^* u \to \exists y \leq^* tu\, A(x, y)).$$

In order to compare the *dialectica* with the monotone interpretation, let us consider a toy, but illuminating, example: the lesser limited principle of omniscience LLPO (see Section 3.6). This principle is not intuitionistically acceptable. Furthermore, it does not have a *dialectica* interpretation. Let us see why not. Fix two recursively enumerable, recursively inseparable sets X and Y. Consider quantifier-free formulas $A_{\mathrm{qf}}(w, k)$ and $B_{\mathrm{qf}}(w, r)$ such that $X = \{w \in \mathbb{N} : \exists k \neg A_{\mathrm{qf}}(w, k)\}$ and $Y = \{w \in \mathbb{N} : \exists r \neg B_{\mathrm{qf}}(w, r)\}$. The *dialectica* assignment of the following instance of LLPO (with numerical parameter w)

$$\forall k, r (A_{\mathrm{qf}}(w, k) \vee B_{\mathrm{qf}}(w, k)) \to \forall k A_{\mathrm{qf}}(w, k) \vee \forall r B_{\mathrm{qf}}(w, r)$$

is essentially

$$\exists n^1 \exists f, g \forall w \forall k, r \big(A_{\mathrm{qf}}(w, fkrw) \vee B_{\mathrm{qf}}(w, gkrw) \to (nw = 0 \to A_{\mathrm{qf}}(w, k)) \wedge (nw \neq 0 \to B_{\mathrm{qf}}(w, r))\big).$$

A *dialectica* interpretation would provide a closed term t of type 1 such that

$$\forall w \forall k, r((tw = 0 \to A_{\mathrm{qf}}(w,k)) \wedge (tw \neq 0 \to B_{\mathrm{qf}}(w,r))),$$

is true (at this juncture, we are using the fact that X and Y are disjoint). But this would entail that the set $\{w \in \mathbb{N} : t^{S^\omega} w = 0\}$ is recursive and separates X from Y.

Nonetheless, it has a monotone functional interpretation (in a suitable verifying theory). One must find closed terms t and q of types $0 \to 0$ and $0 \to 0 \to 0 \to 0$, respectively, such that

$$(\star)\ \exists n \leq t \exists f, g \leq q \forall w \forall k, r(A_{\mathrm{qf}}(w, fwkr) \vee B_{\mathrm{qf}}(w, gwkr) \to$$
$$(nw = 0 \to A_{\mathrm{qf}}(w,k)) \wedge (nw \neq 0 \to B_{\mathrm{qf}}(w,r))).$$

It turns out that $t := \lambda w.1$ and $q := \lambda w \lambda k, r.\max(k,r)$ do the job. The proof is simple, though not completely obvious. One considers the modified predicates

$$\overline{A}(w,k) := \forall u \leq k A_{\mathrm{qf}}(w,u) \vee \exists u, v \leq k(\neg A_{\mathrm{qf}}(w,u) \wedge \neg B_{\mathrm{qf}}(w,v)) \text{ and}$$

$$\overline{B}(w,r) := \forall v \leq r B_{\mathrm{qf}}(w,v) \vee \exists u, v \leq r(\neg A_{\mathrm{qf}}(w,u) \wedge \neg B_{\mathrm{qf}}(w,v)),$$

and verifies that, for each w, $\forall k, r(\overline{A}(w,k) \vee \overline{B}(w,r))$. By LLPO, one gets, for each w, $\forall k \overline{A}(w,k) \vee \forall r \overline{B}(w,r)$. The function n is chosen to be 0 or 1 according to whether the first or second leg of the disjunction holds. At this juncture, we draw attention to the fact that a bit of choice is used. The functions f and g are, respectively, $\lambda w, k, r.(\mu m \leq \max(k,r) \neg A_{\mathrm{qf}}(w,m))$ and $\lambda w, k, r.(\mu m \leq \max(k,r) \neg B_{\mathrm{qf}}(w,m))$ (where $\mu m \leq t$ is the bounded minimization operation; we take $t+1$ as the default value if no pertinent value satisfies the matrices).

Let us take stock. Suppose that the theory $\mathsf{HA}_0^\omega + \mathsf{AC}^\omega + \mathsf{IP}_\forall^\omega + \mathsf{MP}^\omega + \mathsf{LLPO}$ proves $A(\underline{z})$. It is easy to see that $\mathsf{HA}_0^\omega + (\star) \vdash \mathsf{LLPO}$. Therefore, $\mathsf{HA}_0^\omega + \mathsf{AC}^\omega + \mathsf{IP}_\forall^\omega + \mathsf{MP}^\omega + (\star) \vdash A(\underline{z})$. Since $(\star)$ has the right syntactic form, by Theorem 4.6 there are closed terms $\underline{t}$ such that $\mathsf{HA}_0^\omega + (\star) \vdash \exists \underline{x} \leq^* \underline{t} \forall \underline{z} \forall \underline{y} A_{\mathrm{D}}(\underline{x}(\underline{z}), \underline{y}, \underline{z})$. In the paragraphs above we have shown that $(\star)$ is a consequence of HA_0^ω, LLPO, but a bit of choice is also needed (AC^ω is, of course, sufficient). So, in order to verify $\exists \underline{x} \leq^* \underline{t} \forall \underline{z} \forall \underline{y} A_{\mathrm{D}}(\underline{x}(\underline{z}), \underline{y}, \underline{z})$ we need more than just $\mathsf{HA}_0^\omega + \mathsf{LLPO}$.

The above is a typical phenomenom. Usually, one deals with principles of the form

$$(\star\star)\ \exists \underline{x} \leq \underline{r} \underline{w} \forall \underline{y} B_{\mathrm{qf}}(\underline{x}, \underline{y}, \underline{w}),$$

with *parameters* $\underline{w}$ (as axioms, one should of course take the universal closures of these principles). The effect of having parameters is that these principles are no longer covered by Theorem 4.6. One must instead use the *uniformization* of these principles, namely the *sentences* $\exists \underline{X} \leq \underline{r} \forall \underline{w} \forall \underline{y} B_{\mathrm{qf}}(\underline{X}(\underline{w}), \underline{y}, \underline{w})$. Note that these uniformizations have the right syntactical form for the application of Theorem 4.6. In other words, if one wants to state a monotone soundness theorem which includes in Δ principles of the form ($\star\star$), as Kohlenbach does, then in the verifying theory one must *strengthen* these principles by their uniformizations. As long as one's aim is just to extract true computational information (in the form of bounding terms $\underline{t}$) from proofs, the uniformization procedure is acceptable because if a certain *true* principle of the form ($\star\star$) is used in the proof then its uniformization is also true. Therefore, the verification of the role of the extracted terms takes place in a true theory and, hence, *correct* computational information is indeed extracted. The pre-eminent example is weak König's lemma (see Section 3.6). In its tree formulation, WKL is the statement

$$\forall T^1(Tree_\infty(T) \to \exists \alpha \leq_1 1 \forall n^0\, T(\overline{\alpha}n) = 0),$$

where we are using the sequence notation of Section 2.4, and $Tree_\infty(T)$ abbreviates the conjunction of

$$\forall s^0(Ts = 0 \to Seq_2(s)) \wedge \forall s, u(Tu = 0 \wedge s \preceq u \to Ts = 0)$$

with the infinity clause $\forall n^0 \exists s(Ts = 0 \wedge |s| = n)$. Here, $Seq_2(s)$ expresses that s is the number-code of a binary sequence, $s \preceq r$ means that the binary sequence given by s is an initial segment of r, and $|s|$ is the length of the binary sequence given by s. Within HA_0^ω, this principle can be put in the form ($\star\star$). NB the verification of this takes some work and was done in [26]. Therefore, by the above discussion, one has a monotone soundness theorem with WKL, although with a strengthened version of it in the verifying theory of the soundness theorem. In fact, this strengthened version can be taken to be the so-called *uniform* weak König's lemma:

$$\exists \Phi^{1\to 1} \forall T^1(Tree_\infty(T) \to \Phi(T) \leq_1 1 \wedge \forall n^0\, T(\overline{\Phi(T)}\, n) = 0).$$

Even though one needs in general to strengthen WKL to the uniform weak König's lemma in the verifying theory, in certain particular cases the *dialectica* interpretation together with some majorizability tricks also permits the *elimination* of WKL from the verifying theory: this happens when $A(\underline{z})$ is an existencial numerical statement with parameters $\underline{z}$ of types 0 or 1.

4.3. Negative translation. Let us consider Gödel-Gentzen's negative translation in the framework of arithmetic.

DEFINITION 4.9. To each formula A of the language $\mathcal{L}_0^\omega$ we associate its (Gödel-Gentzen) *negative translation* A^{g} according to the following clauses:

i. A^{g} is A, for atomic formulas A.
ii. $(A \wedge B)^{\mathrm{g}}$ is $A^{\mathrm{g}} \wedge B^{\mathrm{g}}$.
iii. $(A \vee B)^{\mathrm{g}}$ is $\neg(\neg A^{\mathrm{g}} \wedge \neg B^{\mathrm{g}})$.
iv. $(A \to B)^{\mathrm{g}}$ is $A^{\mathrm{g}} \to B^{\mathrm{g}}$.

v. $(\forall x A)^g$ is $\forall x A^g$.
vi. $(\exists x A)^g$ is $\neg\forall x \neg A^g$.

In the framework of pure logic we must define A^g as $\neg\neg A$ for atomic A, but this is not needed in the current framework because atomic formulas are decidable. The result of Gödel-Gentzen states that if a formula A is classically provable from Γ then A^g is intuitionistically provable from Γ^g, where $\Gamma^g = \{B^g : B \in \Gamma\}$. This result extends to HA_0^ω because the negative translation of an induction axiom is still an induction axiom and because the remainder axioms are universal formulas. It does *not* extend to $\mathsf{HA}_0^\omega + \mathsf{AC}^\omega + \mathsf{IP}_\forall^\omega + \mathsf{MP}^\omega$ because of the axiom of choice (the other principles pose no problem since they are laws of classical logic). However, the restriction of the axiom of choice for quantifier-free matrices $\mathsf{AC}_{\mathrm{qf}}^\omega$ behaves well under the negative translation *provided that* Markov's principle MP^ω is present in the verifying theory (as well as $\mathsf{AC}_{\mathrm{qf}}^\omega$ itself). In fact:

THEOREM 4.10 (Negative translation). *Suppose that*

$$\mathsf{PA}_0^\omega + \mathsf{AC}_{\mathrm{qf}}^\omega + \Delta \vdash A,$$

where Δ is a set of sentences and A is an arbitrary sentence. Then

$$\mathsf{HA}_0^\omega + \mathsf{AC}_{\mathrm{qf}}^\omega + \mathsf{MP}^\omega + \Delta^g \vdash A^g.$$

Let us analyze the negative translation of $\mathsf{AC}_{\mathrm{qf}}^\omega$. It is,

$$\forall x \neg\forall y \neg B_{\mathrm{qf}}(x, y) \to \neg\forall f \neg\forall x B_{\mathrm{qf}}(x, fx),$$

where B_{qf} is a quantifier-free formula. It is easy to check that this translation is provable in $\mathsf{HA}_0^\omega + \mathsf{AC}_{\mathrm{qf}}^\omega + \mathsf{MP}^\omega$. To see this, assume $\forall x \neg\forall y \neg B_{\mathrm{qf}}(x, y)$. By MP^ω, we get $\forall x \exists y B_{\mathrm{qf}}(x, y)$. By $\mathsf{AC}_{\mathrm{qf}}^\omega$, $\exists f \forall x B_{\mathrm{qf}}(x, fx)$ and, *a fortiori*, $\neg\forall f \neg\forall x B_{\mathrm{qf}}(x, fx)$.

4.4. Extraction and all that (III) & (IV). We may now state an extraction and conservation result which is applicable to a *classical* theory:

PROPOSITION 4.11 (Extraction and conservation, classical *dialectica* case). *Suppose that*

$$\mathsf{PA}_0^\omega + \mathsf{AC}_{\mathrm{qf}}^\omega + \Delta \vdash \forall x \exists y A_{\mathrm{qf}}(x, y),$$

where Δ is a set of universal sentences and A_{qf} is a quantifier-free formula with free variables among x and y. Then there is a closed term t of appropriate type such that

$$\mathsf{HA}_0^\omega + \Delta \vdash A_{\mathrm{qf}}(x, tx).$$

This is an easy corollary of the properties of the negative translation and the soundness of the *dialectica* interpretation. If $\forall x \exists y A_{\mathrm{qf}}(x, y)$ is provable in $\mathsf{PA}_0^\omega + \mathsf{AC}_{\mathrm{qf}}^\omega + \Delta$ then, by Theorem 4.10, $\forall x \neg\forall y \neg A_{\mathrm{qf}}(x, y)$ is provable in

$\mathsf{HA}_0^\omega + \mathsf{AC}_{\mathrm{qf}}^\omega + \mathsf{MP}^\omega + \Delta$. Then so is $\forall x \exists y A_{\mathrm{qf}}(x, y)$ (because MP^ω is available!). At this point, we use Proposition 4.3.

In a similar vein, the combination of the negative translation and the soundness of the monotone functional interpretation yields,

PROPOSITION 4.12 (Extraction and conservation, classical monotone case). *Suppose that*

$$\mathsf{PA}_0^\omega + \mathsf{AC}_{\mathrm{qf}}^\omega + \Delta \vdash \forall x \exists y A_{\mathrm{qf}}(x, y),$$

where Δ *is a set of sentences of the form* $\exists \underline{x} \leq \underline{r} \forall \underline{y} B_{\mathrm{qf}}(\underline{x}, \underline{y})$, *with* B_{qf} *quantifier-free,* $\underline{r}$ *a tuple of closed terms and* $\underline{y}$ *of any types, and* A_{qf} *is a quantifier-free formula with free variables among* x *and* y. *Then there is a closed monotone term* t *of appropriate type such that*

$$\mathsf{HA}_0^\omega + \Delta \vdash \forall x, u\, (x \leq^* u \to \exists y \leq^* tu\, A_{\mathrm{qf}}(x, y)).$$

Observe that each sentence in Δ proves intuitionistically its own Gödel-Gentzen translation. As in Proposition 4.7, when x is of type 1 we may conclude $\forall x \exists y \leq^* tx A_{\mathrm{qf}}(x, y)$. When furthermore y is of type 0, even $\forall x A_{\mathrm{qf}}(x, tx)$ is in order.

THEOREM 4.13 (Characterization). *For any formula* A,

$$\mathsf{PA}_0^\omega + \mathsf{AC}_{\mathrm{qf}}^\omega \vdash A \leftrightarrow (A^{\mathrm{g}})^{\mathrm{D}}.$$

4.5. The no-counterexample interpretation. It is a basic observation in pure logic that a first-order formula in prenex normal form

$$\exists x_0 \forall y_0 \exists x_1 \forall y_1 \ldots \exists x_k \forall y_k\, A_{\mathrm{qf}}(x_0, y_0, x_1, y_1, \ldots, x_k, y_k),$$

is classically valid if, and only if, its *Herbrandization*

$$\exists x_0 \exists x_1 \ldots \exists x_k\, A_{\mathrm{qf}}(x_0, f_0(x_0), x_1, f_1(x_0, x_1), \ldots, x_k, f_k(x_0, \ldots, x_k))$$

is classically valid, where $f_0, f_1, \ldots, f_k$ are *new* function symbols of appropriate arities (known as *index functions*). Perhaps the most intuitive way of seeing this is to show that the negation of the former formula is satisfiable if, and only if, the negation of the latter one is.

THEOREM 4.14 (No-counterexample interpretation). *Suppose that the sentence*

$$\exists x_0 \forall y_0 \ldots \exists x_k \forall y_k\, A_{\mathrm{qf}}(x_0, y_0, \ldots, x_k, y_k),$$

of the language of first-order arithmetic is provable in PA (*first-order Peano arithmetic*). *Then there are closed terms* $t_0, \ldots, t_k$ *of such that*

$$\mathsf{HA}_0^\omega \vdash \forall f_0 \ldots \forall f_k\, A_{\mathrm{qf}}(t_0(\underline{f}), f_0(t_0(\underline{f})), \ldots, t_k(\underline{f}), f_k(t_0(\underline{f}), \ldots, t_k(\underline{f}))),$$

where $\underline{f}$ *abbreviates a tuple of variables* $f_0, f_1, \ldots, f_k$ *of* (*essentially*) *type* 1.

Observation. A type of the form $0 \to 0 \to \cdots \to 0$ is essentially of type 1 via a pairing code.

PROOF. Note that $\exists x_0 \ldots \exists x_k A_{\rm qf}(x_0, f_0(x_0), \ldots, x_k, f_k(x_0, \ldots, x_k))$ follows from $\exists x_0 \forall y_0 \ldots \exists x_k \forall y_k A_{\rm qf}(x_0, y_0, \ldots, x_k, y_k)$ by logic alone. Therefore, the former formula is provable in PA_0^ω. We now use the extraction theorem of the previous section. ⊣

To make sense of the name of the theorem, we may think of $f_0, \ldots, f_k$ as denoting functions in (essentially) S_1 that attempt to provide a counterexample to the truth of $\exists x_0 \forall y_0 \ldots \exists x_k \forall y_k A_{\rm qf}(x_0, y_0, \ldots, x_k, y_k)$, by making $A(n_0, f_0(n_0), \ldots, n_k, f_k(n_0, \ldots, n_k))$ false for any given numerical values $n_0, \ldots, n_k$. However, such counterexample must fail for the values $n_0 = t_0^{S^\omega}(\underline{f}), \ldots, n_k = t_k^{S^\omega}(\underline{f})$. As we have argued at the end of Section 2.4, the functions $t_0^{S^\omega}, \ldots, t_k^{S^\omega}$ are computable (note that we can view the closed terms $t_0, \ldots, t_k$ as having type 2). Hence, the above theorem says that values which defeat a purported counterexample may be effectively constructed from the attempted counterexample *and* that the effective computations are specified by closed terms of $\mathcal{L}_0^\omega$.

The no-counterexample interpretations coincides with the *dialectica* interpretation (after a double negation translation) for $\forall\exists\forall$ statements. This is the case, for instance, with the modified infinite convergence theorem discussed in the introduction. However, the two interpretations already differ for $\exists\forall\exists$ statements. Kohlenbach discusses in several places the shortcomings of the no-counterexample interpretation vis-à-vis the *dialectica* interpretation. One such discussion can be found in his recent book [35].

4.6. Digression on provably total functions. The following result can be proved formalizing Tait's normalization argument:

PROPOSITION 4.15. *Let $t[x_1, \ldots, x_k]$ be a term with its (free) variables as shown, all of which are of type* 0. *The theory* HA_0^ω *proves the Π_2^0-sentence saying that for all natural numbers $n_1, \ldots, n_k$ the closed term $t[\overline{n_1}/x_1, \ldots, \overline{n_k}/x_k]$ normalizes. As a consequence, so does the theory* HA.

A word of caution: PA_0^ω does not prove the *sentence* that says that every closed term of $\mathcal{L}_0^\omega$ has a normal form because this would imply that PA_0^ω proves its own consistency, an impossibility by Gödel's second incompleteness theorem. To see this, suppose that PA_0^ω proves '0=1'. By the *proof* of the soundness of the *dialectica* interpretation there would be a sequence of closed terms $\underline{t}_n$ of $\mathcal{L}_0^\omega$ and a sequence of quantifier-free formulas $A_n(\underline{x}, \underline{y})$ ending in '0=1' such that $A_n(\underline{t}_n, \underline{q})$ holds for each list of closed terms $\underline{q}$ of appropriate types. Well, a truth predicate for these (universal) statements can be defined within PA_0^ω *provided that* each closed term normalizes (reducing the verifications of the quantifier-free matrices to checking whether pairs of numerals are, or are not, the same), and *this* would yield a consistency proof.

The following corollary is immediate, but worth an explicit formulation:

COROLLARY 4.16. *Let us fix a closed term t of type* 1. *The theory* HA_0^ω *proves the Π_2^0-sentence saying that, for each numeral $\overline{n}$, the term $t\overline{n}$, has a normal form (necessarily a numeral). As a consequence, so does the theory* HA.

The previous corollary and Proposition 4.11 imply that the provably total Σ_1^0-functions of PA are given by the closed terms of type 1 of $\mathcal{L}_0^\omega$. This provides an alternative characterization to the one based on Gentzen's work on the ordinal analysis of PA ($< \varepsilon_0$-recursion).

Let us now briefly consider a very natural subsystem of HA_0^ω. The theory iPRA^ω differs from HA_0^ω by only having the recursor R_0 and, correspondingly, induction in the following restricted form:

$$A_{\mathrm{qf}}(0) \wedge \forall x^0(A_{\mathrm{qf}}(x) \rightarrow A_{\mathrm{qf}}(Sx)) \rightarrow \forall x A_{\mathrm{qf}}(x),$$

where A_{qf} is quantifier-free. Due to the absence of higher-order recursors, in order to interprete the contraction axioms we need *primitive* constants C_σ in the language satisfying the "decision by cases" requirements discussed in Section 4.1.

The combination of Gödel's *dialectica* interpretation and the Gödel-Gentzen negative translation yields,

PROPOSITION 4.17 (Extraction and conservation, classical p.r. case). *Suppose that*

$$\mathsf{PRA}^\omega + \mathsf{AC}_{\mathrm{qf}}^\omega + \Delta \vdash \forall x \exists y A_{\mathrm{qf}}(x, y),$$

where Δ is a set of universal sentences and A_{qf} is a quantifier-free formula with free variables among x and y (PRA^ω is the classical *theory associated with iPRA^ω). Then there is a closed term t of appropriate type such that*

$$\mathsf{iPRA}^\omega \vdash A_{\mathrm{qf}}(x, tx).$$

In the present setting, there is a also a notion of contraction of terms and a corresponding strong normalization theorem, with the following consequence:

PROPOSITION 4.18. *For every closed term t of type* 1 *of the language of* iPRA^ω, *t^{S^ω} is a primitive recursive function.*

It is easy to see that the theory $\mathsf{PRA}^\omega + \mathsf{AC}_{\mathrm{qf}}^{0,0}$ and, *a fortiori* $\mathsf{PRA}^\omega + \mathsf{AC}_{\mathrm{qf}}^\omega$, proves the following two schemes:

(1) Σ_1^0-*induction*: $A(0) \wedge \forall x^0(A(x) \rightarrow A(Sx)) \rightarrow \forall x A(x)$, for Σ_1^0-formulas A, i.e., formulas of the form $\exists n^0 A_{\mathrm{qf}}(n, x)$ with A_{qf} quantifier-free.

(2) Δ_1^0-*comprehension*: $\forall x^0(A(x) \leftrightarrow \neg B(x)) \rightarrow \exists \alpha^1 \forall x^0(\alpha x = 0 \leftrightarrow A(x))$, where A and B are Σ_1^0-formulas.

The previous schemes are the distinguished axioms of the second-order theory RCA_0, which plays an important role in the studies of *Reverse Mathematics*. The above results show that the witnesses of the Π_2^0-consequences of RCA_0 are the primitive recursive functions. *A fortiori*, they are also the witnesses of the Π^0_2-consequences of the subsystem PA^1 of PA, characterized by having the scheme of induction restricted to Σ^0_1-formulas (this result is, essentially, due to Charles Parsons and, independently, to Grigori Mints and Gaisi Takeuti).

4.7. Suggested reading and historical notes. Gödel's functional interpretation was published in German in [17]. An English translation of this paper can be found in Gödel's collected works [19]. The soundness theorem, generalized to finite-type Heyting arithmetic, with the characteristic principles, is implicit in [40]. The characterization theorem is due to Mariko Yasugi [56]. The books [52] and [35] present proofs of these theorems. [2] is a good survey of the functional interpretations until the mid-nineties. The monotone functional interpretation is due to Kohlenbach and appeared in [27]. A good reference for this interpretation is [35]. The treatment of LLPO in Section 4.2 is related to, but not quite the same as in [31]. The elimination of WKL mentioned at the end of Section 4.2 appears in [24]. Similar eliminations are also explained in [2] and [35]. An alternative elimination of WKL via the elimination of LLPO is worked out in [31].

The negative translation is due, independently, to Gödel and Gerhard Gentzen (cf. [16]). The characterization theorem of Section 4.4 is essentially due to Kreisel in [40]. Instead of dealing with classical Peano arithmetic via the negative translation followed by the *dialectica* interpretation, a direct and elegant path is provided by Joseph Shoenfield in [45] (note, however, that Shoenfield's interpretation is the same as the combination of an appropriate negative translation followed by the *dialectica* interpretation: this was recently shown by Kohlenbach and Thomas Streicher in [38]). The no-counterexample interpretation is due to Kreisel in [39]; [30] discusses in detail the shortcomings of the no-counterexample interpretation. Proofs of the results in Section 4.6 can be found in [52]. The result of Charles Parsons appeared in [44]. For information regarding the program of Reverse Mathematics one should consult [46].

§5. Injecting uniformities.

5.1. Intensional majorizability. Bounded formulas are treated as computationally empty by the bounded modified realizability interpretation, in the sense that their realizers are trivial. This hinges on the fact that the majorizability relations of Howard-Bezem are given by $\tilde{\exists}$-free formulas and that these formulas have empty realizers. However, the functional interpretation acts non-trivially on these formulas. We solve this problem with the notion of *intensional* majorizability.

We introduce a modification of the language $\mathcal{L}^\omega_{\leq}$, dubbed $\mathcal{L}^\omega_{\trianglelefteq}$. The new language is an extension of $\mathcal{L}^\omega_0$ with the primitive binary relation symbol $\leq$ (as in $\mathcal{L}^\omega_{\leq}$), but also with *primitive* binary relation symbols $\trianglelefteq_\sigma$, for each type σ.

Note that the terms of $\mathcal{L}^{\omega}_{\trianglelefteq}$ and $\mathcal{L}^{\omega}_{0}$ are the same. The symbols $\trianglelefteq_\sigma$ are the intensional counterparts of $\leq^*_\sigma$. There are now new atomic formulas, and the syntactic device of bounded quantification is modified in such a way that it now concerns the intensional symbols instead of the extensional ones. I.e., we have quantifications of the form $\forall x \trianglelefteq t A(x)$ and $\exists x \trianglelefteq t A(x)$, for terms t not containing x. In the current framework, we use the terminology of the bounded formulas and monotone functionals *with respect to the intensional symbols*.

DEFINITION 5.1. The theory $\mathsf{HA}^{\omega}_{\trianglelefteq}$ is an altered version of $\mathsf{HA}^{\omega}_{\leq}$ with the axiom schemes:

$$\mathsf{B}_\forall : \forall x \trianglelefteq t A(x) \leftrightarrow \forall x (x \trianglelefteq t \to A(x)),$$
$$\mathsf{B}_\exists : \exists x \trianglelefteq t A(x) \leftrightarrow \exists x (x \trianglelefteq t \wedge A(x))$$

instead of the corresponding extensional ones, and with the further axioms

$$\mathsf{M}_1 : x \trianglelefteq_0 y \leftrightarrow x \leq y,$$
$$\mathsf{M}_2 : x \trianglelefteq_{\rho\to\sigma} y \to \forall u \trianglelefteq_\rho v (xu \trianglelefteq_\sigma yv \wedge yu \trianglelefteq_\sigma yv)$$

and a *rule* $\mathsf{RL}_{\trianglelefteq}$

$$\frac{A_{\mathrm{bd}} \wedge u \trianglelefteq v \to su \trianglelefteq tv \wedge tu \trianglelefteq tv}{A_{\mathrm{bd}} \to s \trianglelefteq t},$$

where s and t are terms of $\mathcal{L}^{\omega}_{\trianglelefteq}$, A_{bd} is a bounded formula and u and v are variables that do not occur free in the conclusion. Moreover, the induction scheme is now extended to all formulas of the language $\mathcal{L}^{\omega}_{\trianglelefteq}$.

We observe that it is not needed to state the axioms for combinators, the axiom of equality and the axioms for recursors for the new atomic formulas of the language, since these already follow from the more restricted statements.

The crucial feature of the above theory is that we have *rules* instead of the purported axioms: $\forall u \trianglelefteq_\rho v (xu \trianglelefteq_\sigma yv \wedge yu \trianglelefteq_\sigma yv) \to x \trianglelefteq_{\rho\to\sigma} y$. Note that the previous formula would pose a problem for the functional interpretation. The rules, however, do *not* pose such a problem! We dubbed the new majorizability symbols $\trianglelefteq_\sigma$ *intensional* because they are (partially) governed by *rules*. The presence of rules entails the failure of the deduction theorem in $\mathsf{HA}^{\omega}_{\trianglelefteq}$, a feature that many do not find attractive. However, one should keep in mind that the rules are introduced for *mathematical reasons* and that "mathematical attraction" is partly a question of familiarity. In the end, systems where rules play an essential role must be judged by their own mathematical merits.

Even though the rules are *weaker* than the axioms we still have:

LEMMA 5.2. *The theory* $\mathsf{HA}^{\omega}_{\trianglelefteq}$ *proves that the relations* $\trianglelefteq_\sigma$ *are transitive and that* $x \trianglelefteq y \to y \trianglelefteq y$. *For type* 1, *we have* $x \trianglelefteq_1 y \to x \leq^*_1 y$, $x \trianglelefteq_1 x^{\mathrm{M}}$ *and* $\min_1(x, y) \trianglelefteq_1 y^{\mathrm{M}}$ (*where the* $\min_1$ *function is the minimum function defined pointwise*). *Howard's majorizability theorem holds*: *For each closed term* t, *there is a closed term q such that $\mathsf{HA}^{\omega}_{\trianglelefteq} \vdash t \trianglelefteq q$. Furthermore, it holds of the very same term constructed for the extensional case.*

5.2. Bounded functional interpretation. Let us now define the bounded functional interpretation:

DEFINITION 5.3. To each formula A of the language $\mathcal{L}^{\omega}_{\trianglelefteq}$ we assign formulas $(A)^{\mathrm{B}}$ and A_{B} so that $(A)^{\mathrm{B}}$ is of the form $\tilde{\exists}\underline{b}\tilde{\forall}\underline{c}A_{\mathrm{B}}(\underline{b},\underline{c})$, with $A_{\mathrm{B}}(\underline{b},\underline{c})$ a bounded formula, according to the following clauses:

1. $(A)^{\mathrm{B}}$ and A_{B} are simply A, for atomic formulas A.

If we have already interpretations for A and B given by $\tilde{\exists}\underline{b}\tilde{\forall}\underline{c}A_{\mathrm{B}}(\underline{b},\underline{c})$ and $\tilde{\exists}\underline{d}\tilde{\forall}\underline{e}B_{\mathrm{B}}(\underline{d},\underline{e})$ (respectively) then we define:

2. $(A \wedge B)^{\mathrm{B}}$ is $\tilde{\exists}\underline{b},\underline{d}\tilde{\forall}\underline{c},\underline{e}(A_{\mathrm{B}}(\underline{b},\underline{c}) \wedge B_{\mathrm{B}}(\underline{d},\underline{e}))$,
3. $(A \vee B)^{\mathrm{B}}$ is $\tilde{\exists}\underline{b},\underline{d}\tilde{\forall}\underline{c},\underline{e}(\tilde{\forall}\underline{c}' \trianglelefteq \underline{c}A_{\mathrm{B}}(\underline{b},\underline{c}') \vee \tilde{\forall}\underline{e}' \trianglelefteq \underline{e}B_{\mathrm{B}}(\underline{d},\underline{e}'))$,
4. $(A \to B)^{\mathrm{B}}$ is $\tilde{\exists}\underline{f},\underline{g}\tilde{\forall}\underline{b},\underline{e}(\tilde{\forall}\underline{c} \trianglelefteq \underline{gbe}A_{\mathrm{B}}(\underline{b},\underline{c}) \to B_{\mathrm{B}}(\underline{fb},\underline{e}))$.

For bounded quantifiers we have:

5. $(\forall x \trianglelefteq tA(x))^{\mathrm{B}}$ is $\tilde{\exists}\underline{b}\tilde{\forall}\underline{c}\forall x \trianglelefteq t\, A_{\mathrm{B}}(\underline{b},\underline{c},x)$,
6. $(\exists x \trianglelefteq tA(x))^{\mathrm{B}}$ is $\tilde{\exists}\underline{b}\tilde{\forall}\underline{c}\exists x \trianglelefteq t\tilde{\forall}\underline{c}' \trianglelefteq \underline{c}\, A_{\mathrm{B}}(\underline{b},\underline{c}',x)$.

And for unbounded quantifiers we define:

7. $(\forall x A(x))^{\mathrm{B}}$ is $\tilde{\exists}\underline{f}\tilde{\forall}a,\underline{c}\forall x \trianglelefteq aA_{\mathrm{B}}(\underline{f}a,\underline{c},x)$.
8. $(\exists x A(x))^{\mathrm{B}}$ is $\tilde{\exists}a,\underline{b}\tilde{\forall}\underline{c}\exists x \trianglelefteq a\tilde{\forall}\underline{c}' \trianglelefteq \underline{c}A_{\mathrm{B}}(\underline{b},\underline{c}',x)$.

There are five important principles in connection with this interpretation:

I. *Intensional Bounded Choice* $\mathsf{bAC}^{\omega}_{\trianglelefteq}$:

$$\forall \underline{x}\exists \underline{y}A(\underline{x},\underline{y}) \to \tilde{\exists}\underline{f}\tilde{\forall}\underline{b}\forall \underline{x} \trianglelefteq \underline{b}\exists \underline{y} \trianglelefteq \underline{f}\,\underline{b}\, A(\underline{x},\underline{y}),$$

where A is an arbitrary formula of $\mathcal{L}^{\omega}_{\trianglelefteq}$. It is clear that the forms of choice $\mathsf{bAC}^{i,j}_{\mathrm{qf}}$, $\forall x^i \exists y^j B_{\mathrm{qf}}(x,y) \to \exists f \forall x \exists y \leq_j fx\, B_{\mathrm{qf}}(x,y)$, for $i,j \in \{0,1\}$ and B_{qf} quantifier-free, follow from $\mathsf{bAC}^{\omega}_{\trianglelefteq}$ in $\mathsf{HA}^{\omega}_{\trianglelefteq}$ (even if $\mathsf{bAC}^{\omega}_{\trianglelefteq}$ were restricted to bounded matrices only, a principle which we denote by $\mathsf{bBAC}^{\omega}_{\trianglelefteq}$). Due to the availability of minimization for quantifier-free formulas, note that $\mathsf{bAC}^{i,0}_{\mathrm{qf}} \Rightarrow \mathsf{AC}^{i,0}_{\mathrm{qf}}$, for $i \in \{0,1\}$. Here, $\mathsf{AC}^{i,j}_{\mathrm{qf}}$, $i,j \in \{0,1\}$, is the usual form of choice $\forall x^i \exists y^j B_{\mathrm{qf}}(x,y) \to \exists f \forall x B_{\mathrm{qf}}(x,fx)$, with B_{qf} quantifier-free. Note that appropriate tuple versions of these principles also follow from $\mathsf{bBAC}^{\omega}_{\trianglelefteq}$.

II. *Intensional Bounded Independence of Premises* $\mathsf{bIP}^{\omega}_{\trianglelefteq}$:

$$(A \to \exists \underline{y}B(\underline{y})) \to \tilde{\exists}\underline{b}(A \to \exists \underline{y} \trianglelefteq \underline{b}\, B(\underline{y})),$$

where A is a universal formula (with bounded matrix) and B is an arbitrary formula. In the present setting, by "universal with bounded matrix" we mean a formula of the form $\forall \underline{x}A_{\mathrm{bd}}(\underline{x})$, with A_{bd} a *bounded (intensional) matrix.*

III. *Intensional Bounded Markov's Principle* $\mathsf{MP}^{\omega}_{\trianglelefteq}$:

$$(\forall \underline{y} A_{\text{bd}}(\underline{y}) \to B_{\text{bd}}) \to \tilde{\exists} \underline{b} (\forall \underline{y} \trianglelefteq \underline{b} A_{\text{bd}}(\underline{y}) \to B_{\text{bd}}),$$

where A_{bd} is a bounded matrix and B_{bd} is a bounded formula. When B is the formula $0 = 1$, the above principle specializes to $\neg \forall \underline{y} A_{\text{bd}}(\underline{y}) \to \tilde{\exists} \underline{b} \neg \forall \underline{y} \trianglelefteq \underline{b} A_{\text{bd}}(\underline{y})$. If y is of type 0 and the matrix is quantifier-free, the principle further specializes to the familiar Markov's principle of type 0: $\neg \forall y A_{\text{qf}}(y) \to \exists y \neg A_{\text{qf}}(y)$. In this, we are using bounded numerical search.

IV. *Intensional Bounded Contra-Collection Principle* $\mathsf{bBCC}^{\omega}_{\trianglelefteq}$:

$$\tilde{\forall} \underline{b} \exists \underline{z} \trianglelefteq \underline{c} \forall \underline{y} \trianglelefteq \underline{b} A_{\text{bd}}(\underline{y}, \underline{z}) \to \exists \underline{z} \trianglelefteq \underline{c} \forall \underline{y} A_{\text{bd}}(\underline{y}, \underline{z}),$$

where $\underline{c}$ is a tuple of monotone functionals and A_{bd} is a bounded formula. Note that this is *classically* equivalent to collection restricted to bounded matrices (see Proposition 5.4 below). This principle allows the conclusion of certain existentially bounded statements from the assumption of weakenings thereof. Let us see that it implies weak König's lemma WKL (the statement of WKL is in the end of Section 4.2). Suppose that $Tree_{\infty}(T^1)$. The infinity clause says that $\forall n^0 \exists s (Ts = 0 \wedge |s| = n)$. We introduce some notation. Given s a binary sequence, let $\hat{s}$ be the infinite binary path which extends s by appending an infinite string of zeros. With the aid of the rule $\mathsf{RL}_{\trianglelefteq}$ one can show that $\mathsf{HA}^{\omega}_{0} \vdash \forall s (Seq_2(s) \to \hat{s} \trianglelefteq_1 1)$. Therefore, from the infinity clause one gets $\forall n^0 \exists \alpha \trianglelefteq_1 1 \, (T(\overline{\alpha} n) = 0)$. Of course, by the definition of tree,

$$\forall n^0 \exists \alpha \trianglelefteq_1 1 \forall k \leq n \, (T(\overline{\alpha} k) = 0).$$

Applying $\mathsf{bBCC}^{\omega}_{\trianglelefteq}$, we may infer $\exists \alpha \trianglelefteq_1 \forall n (T(\overline{\alpha} n) = 0)$. Since $\alpha \trianglelefteq_1 1 \to \alpha \leq 1$, we get our infinite path through T.

The following principle, dubbed *Intensional Bounded Disjunction Property*, also follows from $\mathsf{bBCC}^{\omega}_{\trianglelefteq}$:

$$\tilde{\forall} \underline{b} \tilde{\forall} \underline{c} (\forall \underline{x} \trianglelefteq \underline{b} A_{\text{bd}}(\underline{x}) \vee \forall \underline{y} \trianglelefteq \underline{c} B_{\text{bd}}(\underline{y})) \to \forall \underline{x} A_{\text{bd}}(\underline{x}) \vee \forall \underline{y} B_{\text{bd}}(\underline{y}),$$

where A_{bd} and B_{bd} are bounded formulas. This property clearly implies (within $\mathsf{HA}^{\omega}_{\trianglelefteq}$) the lesser limited principle of omniscience LLPO, mentioned in Section 3.6.

V. *Intensional Majorizability Axioms* $\mathsf{MAJ}^{\omega}_{\trianglelefteq}$: $\forall x \exists y (x \trianglelefteq y)$.

The following result is similar to Proposition 3.13:

PROPOSITION 5.4. *The theory* $\mathsf{HA}^{\omega}_{\trianglelefteq} + \mathsf{bAC}^{\omega}_{\trianglelefteq} + \mathsf{bIP}^{\omega}_{\trianglelefteq}$ *proves the* Intensional Collection Principle $\mathsf{bC}^{\omega}_{\trianglelefteq}$:

$$\forall \underline{z} \trianglelefteq \underline{c} \exists \underline{y} A(\underline{y}, \underline{z}) \to \tilde{\exists} \underline{b} \forall \underline{z} \trianglelefteq \underline{c} \exists \underline{y} \trianglelefteq \underline{b} \, A(\underline{x}, \underline{y}),$$

where A is an arbitrary formula and $\underline{c}$ is a tuple of monotone functionals.

Obviously, the restricted theory $\mathsf{HA}^{\omega}_{\trianglelefteq}+\mathsf{bBAC}^{\omega}_{\trianglelefteq}+\mathsf{bIP}^{\omega}_{\trianglelefteq}$ proves the Intensional *Bounded* Collection Scheme $\mathsf{bBC}^{\omega}_{\trianglelefteq}$, that is, the above scheme restricted to bounded formulas A.

The theory $\mathsf{HA}^{\omega}_{\trianglelefteq} + \mathsf{bAC}^{\omega}_{\trianglelefteq} + \mathsf{bIP}^{\omega}_{\trianglelefteq}$ is *classically inconsistent*. E.g., it refutes the *limited principle of omniscience* LPO (in Errett Bishop's terminology):

$$\forall x^1(\forall n^0(xn = 1) \vee \exists n^0(xn \neq 1)).$$

To see this, assume the above. Hence $\forall x^1 \exists n^0(\forall k^0(xk = 1) \vee xn \neq 1)$, by intuitionistic logic. *A fortiori*, $\forall x \trianglelefteq_1 1 \exists n^0(\forall k(xk = 1) \vee xn \neq 1)$. By the intensional collection principle $\mathsf{bC}^{\omega}_{\trianglelefteq}$, we get $\exists m^0 \forall x \trianglelefteq_1 1 \exists n \leq m(\forall k(xk = 1) \vee xn \neq 1)$. Take such m and consider the sequence s of length $m + 1$ with constant value 1. Clearly, $\hat{s} \trianglelefteq_1 1$ but it is not the case that $\exists n \leq m(\forall k(\hat{s}k = 1) \vee \hat{s}n \neq 1)$.

Also, the restricted theory $\mathsf{HA}^{\omega}_{\trianglelefteq} + \mathsf{bBAC}^{\omega}_{\trianglelefteq} + \mathsf{bIP}^{\omega}_{\trianglelefteq} + \mathsf{MP}^{\omega}_{\trianglelefteq}$ already refutes a basic form of extensionality. In fact, it proves the *negation* of

$$\forall \Phi^2 \forall \alpha^1, \beta^1 \, (\forall k^0(\alpha k = \beta k) \rightarrow \Phi\alpha = \Phi\beta).$$

Towards a contradiction, assume the above. In particular, one has

$$\forall \Phi \trianglelefteq_2 1^2 \, \forall \alpha, \beta \trianglelefteq_1 1^1 \exists k \, (\alpha k = \beta k \rightarrow \Phi\alpha = \Phi\beta),$$

where $1^1 := \lambda k^0.1^0$ and $1^2 := \lambda y^1.1^0$ (we used here $\mathsf{MP}^{\omega}_{\trianglelefteq}$). By $\mathsf{bBC}^{\omega}_{\trianglelefteq}$, one may infer

$$\exists n \forall \Phi \trianglelefteq_2 1 \forall \alpha, \beta \trianglelefteq_1 1 \, (\forall k < n(\alpha k = \beta k) \rightarrow \Phi\alpha = \Phi\beta).$$

Take one such $n = n_0$. Define Φ according to:

$$\gamma^1 \leadsto_{\Phi} \begin{cases} 0 & \text{if } \forall k \leq n_0 \, (\gamma k = 0) \\ 1 & \text{otherwise} \end{cases}$$

It is clear that for $\alpha := \lambda k.0$ and $\beta := \lambda k.\delta_{n_0,k}$ (Kronecker's delta) one has $\forall k < n_0 \, (\alpha k = \beta k)$ but $\Phi\alpha \neq \Phi\beta$. Since it is easy to show that $\Phi \trianglelefteq 1^2$ and $\alpha, \beta \trianglelefteq 1^1$, we are faced with a contradiction.

The above two examples are not set-theoretically sound. The bounded functional interpretation *injects uniformities* which are absent in the universe of sets and which are incompatible with it. Given this state of affairs, it is pressing to assure that the theory $\mathsf{HA}^{\omega}_{\trianglelefteq}+\mathsf{bAC}^{\omega}_{\trianglelefteq}+\mathsf{bIP}^{\omega}_{\trianglelefteq}+\mathsf{MP}^{\omega}_{\trianglelefteq}+\mathsf{bBCC}^{\omega}_{\trianglelefteq}+\mathsf{MAJ}^{\omega}_{\trianglelefteq}$ is consistent.

THEOREM 5.5 (Soundness of the bounded functional interpretation). *Suppose that*

$$\mathsf{HA}^{\omega}_{\trianglelefteq} + \mathsf{bAC}^{\omega}_{\trianglelefteq} + \mathsf{bIP}^{\omega}_{\trianglelefteq} + \mathsf{MP}^{\omega}_{\trianglelefteq} + \mathsf{bBCC}^{\omega}_{\trianglelefteq} + \mathsf{MAJ}^{\omega}_{\trianglelefteq} + \Delta \vdash A(\underline{z}),$$

where Δ is a set of universal sentences (with bounded matrices) and A is an arbitrary formula (with the free variables as shown). Then there are closed *monotone terms of appropriate types such that*

$$\mathsf{HA}^{\omega}_{\trianglelefteq} + \Delta \vdash \tilde{\forall} \underline{a} \forall \underline{z} \trianglelefteq \underline{a} \tilde{\forall} \underline{c}\, A_{\mathrm{B}}(\underline{t}(\underline{a}), \underline{c}, \underline{z}).$$

Given an appropriate restatement, we may include in Δ sentences of the form $\exists \underline{x} \leq \underline{r} \forall \underline{y} B_{\mathrm{qf}}(\underline{x}, \underline{y})$, as in the soundness theorem of the monotone functional interpretation (see Theorem 4.6). Let us see what we mean by this. Given a tuple of closed terms $\underline{r}$, by Lemma 5.2 there is a tuple of closed terms $\underline{t}$ such that $\mathsf{HA}^{\omega}_{\trianglelefteq} \vdash \underline{r} \trianglelefteq \underline{t}$. The sentence $\exists \underline{x} \trianglelefteq \underline{t}(\underline{x} \leq \underline{r} \wedge \forall \underline{y} B_{\mathrm{qf}}(\underline{x}, \underline{y}))$ obviously implies the original sentence $\exists \underline{x} \leq \underline{r} \forall \underline{y} B_{\mathrm{qf}}(\underline{x}, \underline{y})$. Furthermore, the modified sentence is of the form $\exists \underline{x} \trianglelefteq \underline{t}\, \forall \underline{w}\, C_{\mathrm{qf}}(\underline{x}, \underline{w})$ for quantifier-free C_{qf}, because the formula $\underline{x} \leq \underline{r}$ is universal. By $\mathsf{bBCC}^{\omega}_{\trianglelefteq}$, the modified sentence is implied by $\tilde{\forall} \underline{b} \exists \underline{x} \trianglelefteq \underline{t}\, \forall \underline{w} \trianglelefteq \underline{b}\, C_{\mathrm{qf}}(\underline{x}, \underline{w})$, and this latter sentence has the right form for applying the soundness theorem above. Of course, in the conclusion of the soundness theorem, the verifying theory must include this sentence. At this juncture, we draw attention to the fact that this sentence "flattens" (see Section 5.5) to a sentence which is implied by $\forall \underline{u} \exists \underline{x} \leq \underline{r} \forall \underline{y} \leq^{*} \underline{u}\, B_{\mathrm{qf}}(\underline{x}, \underline{y})$ (we are using here (iii) of Lemma 2.4). NB this is a *weaker* statement than the original one. This should be compared with the monotone functional interpretation (see the discussion in Section 4.2) where a strengthening of the original statement is needed.

As usual, the proof of the above soundness theorem is by induction on the length of formal derivations, and the terms are effectively constructed from the formal derivations. However, for the bounded functional interpretation the choice of the witnessing terms is *always canonical*, and there is no need for a definition by cases functional nor for the decidability of bounded formulas.

PROPOSITION 5.6. (Extraction and conservation, intuitionistic intensional case) *Suppose that*

$$\mathsf{HA}^{\omega}_{\trianglelefteq} + \mathsf{bAC}^{\omega}_{\trianglelefteq} + \mathsf{bIP}^{\omega}_{\trianglelefteq} + \mathsf{MP}^{\omega}_{\trianglelefteq} + \mathsf{bBCC}^{\omega}_{\trianglelefteq} + \mathsf{MAJ}^{\omega}_{\trianglelefteq} + \Delta \vdash \forall x \exists y A_{\mathrm{bd}}(x, y),$$

where Δ is a set of universal sentences (with bounded matrices) and A_{bd} is a bounded formula with free variables among x and y. Then there is a closed monotone term t of appropriate type such that

$$\mathsf{HA}^{\omega}_{\trianglelefteq} + \Delta \vdash \tilde{\forall} a \forall x \trianglelefteq a \exists y \trianglelefteq ta\, A_{\mathrm{bd}}(x, y).$$

THEOREM 5.7 (Characterization). *For any formula A,*

$$\mathsf{HA}^{\omega}_{\trianglelefteq} + \mathsf{bAC}^{\omega}_{\trianglelefteq} + \mathsf{bIP}^{\omega}_{\trianglelefteq} + \mathsf{MP}^{\omega}_{\trianglelefteq} + \mathsf{bBCC}^{\omega}_{\trianglelefteq} + \mathsf{MAJ}^{\omega}_{\trianglelefteq} \vdash A \leftrightarrow (A)^{\mathrm{B}}.$$

Using the above characterization theorem and an argument as in the proof of Proposition 4.5, we get the following fact:

PROPOSITION 5.8. *Let A be an* arbitrary *formula with free variables among x and y. Suppose that*

$$\mathsf{HA}^{\omega}_{\trianglelefteq} + \mathsf{bAC}^{\omega}_{\trianglelefteq} + \mathsf{bIP}^{\omega}_{\trianglelefteq} + \mathsf{MP}^{\omega}_{\trianglelefteq} + \mathsf{bBCC}^{\omega}_{\trianglelefteq} + \mathsf{MAJ}^{\omega}_{\trianglelefteq} \vdash \forall x \exists y A(x, y),$$

Then there is a closed monotone term t of appropriate type such that

$$\mathsf{HA}^{\omega}_{\trianglelefteq}+\mathsf{bAC}^{\omega}_{\trianglelefteq}+\mathsf{bIP}^{\omega}_{\trianglelefteq}+\mathsf{MP}^{\omega}_{\trianglelefteq}+\mathsf{bBCC}^{\omega}_{\trianglelefteq}+\mathsf{MAJ}^{\omega}_{\trianglelefteq} \vdash \tilde{\forall} a \forall x \trianglelefteq a \exists y \trianglelefteq ta\, A(x, y).$$

It is not clear what is accomplished by the above extraction result, given that the verifying theory is not sound. This is in sharp contrast with Proposition 4.8, where the extraction is indeed sound. We are here facing the same phenomenon as the one discussed at the end of Section 3.5. However, note that for restricted A, Proposition 5.6 yields a sound extraction result since the characteristic principles are absent from the verifying theory (the forthcoming Section 5.5 is also relevant for this discussion).

In the presence of classical logic, the situation concerning the monotone and the bounded interpretations changes. In the classical case, the matrix A must necessarily be narrowed down (it is well-known that computational extraction in the classical case may already fail for universal matrices). The reader should compare the extraction results of Proposition 4.12 and Proposition 5.15 below.

5.3. Digression on a new conservation result. In Section 4.6, we mentioned the classical second-order theory RCA_0, the ordinary base theory for Reverse Mathematics. By iRCA_0 we mean an intuitionistic version of this theory, whereby the logic used is intuitionistic, adjoined with the axiom $\forall X \forall x (x \in X \vee x \notin X)$. Let us introduce some further second-order principles. $\mathsf{bIP}_{\forall}$ is the following *bounded* version of the scheme of independence of premises:

$$(A \to \exists y B(y)) \to \exists y (A \to \exists w \leq y B(w)),$$

where A is a formula starting with a string of universal (first or second-order) quantifiers followed by a formula without unbounded first-order quantifications (it can have bounded first-order quantifications as well as second-order quantifications), and B is an arbitrary formula. MP is the usual (numerical) Markov's principle. $\mathsf{bAC}^{\mathbb{N}}$ is *bounded* version of the countable axiom of choice:

$$\forall x \exists y A(x, y) \to \exists X \big(\mathrm{Func}(X) \wedge \forall x, y(\langle x, y\rangle \in X \to \exists w \leq y A(x, w))\big),$$

where A is any formula, and $\mathrm{Func}(X)$ says that the set X is constituted by the codes of the pairs of a total function from $\mathbb{N}$ to $\mathbb{N}$. Finally, FAN is the scheme

$$\forall X \exists x A(x, X) \to \exists z \forall X \exists x \leq z A(x, X),$$

with arbitrary A. Weak König's lemma WKL has its usual formulation in terms of infinite binary trees.

THEOREM 5.9. *The second-order intuitionistic theory*

$$\mathsf{iRCA}_0 + \mathsf{WKL} + \mathsf{MP} + \mathsf{bIP}_{\forall} + \mathsf{bAC}^{\mathbb{N}} + \mathsf{FAN}$$

is conservative over PA^1 *with respect to* Π^0_2*-sentences.*[6]

[6]There is a careless misstep in our abstract [9], where the scheme of independence of premisses $\mathsf{IP}_{\forall}$ (i.e., $(A \to \exists y B(y)) \to \exists y(A \to B(y))$, for A and B as in $\mathsf{bIP}_{\forall}$) and the principle of countable choice $\mathsf{AC}^{\mathbb{N}}$ (i.e., $\forall x \exists y A(x, y) \to \exists X[\mathrm{Func}(X) \wedge \forall x, y(\langle x, y\rangle \in X \to A(x, y))]$)

PROOF. The second-order language of arithmetic can be embedded in $\mathcal{L}^{\omega}_{\trianglelefteq}$ by letting the first-order variables run over type 0 arguments, letting the second-order variables run over type 1 variables X such that $X \trianglelefteq_1 1$, and by interpreting $x \in X$ by $Xx = 0$. Under this embedding, the theory $\mathsf{iRCA}_0 + \mathsf{WKL} + \mathsf{MP} + \mathsf{bIP}_\forall + \mathsf{bAC}^{\mathbb{N}} + \mathsf{FAN}$ is clearly a sub-theory of $\mathsf{iPRA}^{\omega}_{\trianglelefteq} + \mathsf{bAC}^{\omega}_{\trianglelefteq} + \mathsf{bIP}^{\omega}_{\trianglelefteq} + \mathsf{MP}^{\omega}_{\trianglelefteq} + \mathsf{bBCC}^{\omega}_{\trianglelefteq} + \mathsf{MAJ}^{\omega}_{\trianglelefteq}$ (where $\mathsf{iPRA}^{\omega}_{\trianglelefteq}$ is the obvious intensional version of iPRA^{ω}). Therefore, if the former theory proves a Π^0_2-sentence then, by (an adaptation of) Proposition 5.6, then $\mathsf{PRA}^{\omega}_{\trianglelefteq}$ already proves it. By *flattening* (see the forthcoming Section 5.5), so does the theory PRA^{ω}. By an internalization argument in the manner of Proposition 2.3, it can be argued that PRA^{ω} is (first-order) conservative over PA^1. We are done. ⊣

As a consequence, the provably total Σ^0_1-functions of $\mathsf{iRCA}_0 + \mathsf{WKL} + \mathsf{MP} + \mathsf{bIP}_\forall + \mathsf{bAC}^{\mathbb{N}} + \mathsf{FAN}$ are the primitive recursive functions (even though this theory is *classically* inconsistent; see the argument after Proposition 5.4). In Reverse Mathematics, WKL_0 is the classical theory RCA_0 together with weak König's lemma WKL. The importance of the theory WKL_0 is well documented from the work in Reverse Mathematics. The following elimination result is originally due to Harvey Friedman.

COROLLARY 5.10. *The theory* WKL_0 *is* Π^0_2*-conservative over* RCA_0.

PROOF. Suppose that WKL_0 proves a certain Π^0_2-sentence. Note that $(\mathsf{WKL}_0)^g$ is a subtheory of $\mathsf{iRCA}_0 + \mathsf{WKL} + \mathsf{MP}$. Therefore, by the negative translation, $\mathsf{iRCA}_0 + \mathsf{WKL} + \mathsf{MP}$ also proves the given Π^0_2-sentence (notice the presence of MP). The result follows from the previous theorem. ⊣

5.4. Extraction and all that (V). The negative translation of Gödel-Gentzen can be easily extended to the language $\mathcal{L}^{\omega}_{\trianglelefteq}$ according to the extra clauses:

vii. $(\forall x \trianglelefteq t\, A)^g$ *is* $\forall x \trianglelefteq t\, A^g$.
viii. $(\exists x \trianglelefteq t\, A)^g$ *is* $\neg\forall x \trianglelefteq t\, \neg A^g$.

It is clear that the negative translation of a bounded formula is still bounded, and it can be shown by induction on the type that the intensional majorizability relations are *stable*, i.e., $\mathsf{HA}^{\omega}_{\trianglelefteq} \vdash \neg\neg(x \trianglelefteq y) \to x \trianglelefteq y$. Now, in analogy with the *dialectica* interpretation, the next result is not difficult:

THEOREM 5.11 (Negative translation). *Suppose that*

$$\mathsf{PA}^{\omega}_{\trianglelefteq} + \mathsf{bBAC}^{\omega}_{\trianglelefteq} + \mathsf{MAJ}^{\omega}_{\trianglelefteq} + \Delta \vdash A,$$

where Δ *is a set of sentences,* A *is an arbitrary sentence and* $\mathsf{PA}^{\omega}_{\trianglelefteq}$ *is the classical version of* $\mathsf{HA}^{\omega}_{\trianglelefteq}$. *Then,*

$$\mathsf{HA}^{\omega}_{\trianglelefteq} + \mathsf{bBAC}^{\omega}_{\trianglelefteq} + \mathsf{MAJ}^{\omega}_{\trianglelefteq} + \mathsf{MP}^{\omega}_{\trianglelefteq} + \Delta^g \vdash A^g.$$

---

appear instead of the weaker principles $\mathsf{bIP}_\forall$ and $\mathsf{bAC}^{\mathbb{N}}$, respectively. We do not know if the theorem still holds with $\mathsf{IP}_\forall$ and $\mathsf{AC}^{\mathbb{N}}$.

Remember that $\mathsf{PA}^{\omega}_{\trianglelefteq} + \mathsf{bBAC}^{\omega}_{\trianglelefteq}$ proves the intensional collection principle restricted to bounded formulas $\mathsf{bBC}^{\omega}_{\trianglelefteq}$ and, by *classical logic*, the contra-collection scheme $\mathsf{bBCC}^{\omega}_{\trianglelefteq}$ follows. The next proposition is a consequence of the combination of the negative translation with the soundness theorem:

THEOREM 5.12 (Extraction and conservation, classical intensional case). *Suppose that*

$$\mathsf{PA}^{\omega}_{\trianglelefteq} + \mathsf{bBAC}^{\omega}_{\trianglelefteq} + \mathsf{MAJ}^{\omega}_{\trianglelefteq} + \Delta \vdash \forall x \exists y A_{\mathrm{bd}}(x, y),$$

where Δ is a set of universal sentences (with bounded matrices) and A_{bd} is a bounded formula with free variables among x and y. Then there is a closed monotone term t of appropriate type such that

$$\mathsf{PA}^{\omega}_{\trianglelefteq} + \Delta \vdash \tilde{\forall} a \forall x \trianglelefteq a \exists y \trianglelefteq ta\, A_{\mathrm{bd}}(x, y).$$

In contrast with the analogous result of Section 4.3, the verification theory is a *classical* theory because intensional bounded formulas are not decidable in general. However, if y is of type 0 and A_{bd} is quantifier-free, the verification can be done in $\mathsf{HA}^{\omega}_{\trianglelefteq} + \Delta^{\mathrm{g}}$.

THEOREM 5.13 (Characterization). *For any formula A,*

$$\mathsf{PA}^{\omega}_{\trianglelefteq} + \mathsf{bBAC}^{\omega}_{\trianglelefteq} + \mathsf{MAJ}^{\omega}_{\trianglelefteq} \vdash A \leftrightarrow (A^{\mathrm{g}})^{\mathrm{B}}.$$

5.5. Flattening. We want to use this *intensional* technology in "real world" applications. The following lemma is the passageway from the intensional theory to plain $\mathsf{PA}^{\omega}_{\leq}$:

LEMMA 5.14 (Flattening). *Suppose $\mathsf{PA}^{\omega}_{\trianglelefteq} + \Gamma \vdash A$, where A is a sentence and Γ is a set of sentences, formulated in the intensional language $\mathcal{L}^{\omega}_{\trianglelefteq}$. Then $\mathsf{PA}^{\omega}_{\leq} + \Gamma^* \vdash A^*$, where B^* is the sentence of $\mathcal{L}^{\omega}_{\leq}$ obtained from B by replacing throughout the binary symbols $\trianglelefteq_\sigma$ by the formulas $\leq^*_\sigma$* (mutatis mutandis *for sets of sentences).*

We call B^* the *flattening* of B (*mutatis mutandis* for sets of sentences).

PROPOSITION 5.15 (Extraction and conservation, classical flattened case). *Suppose that*

$$\mathsf{PA}^{\omega}_{\trianglelefteq} + \mathsf{bBAC}^{\omega}_{\trianglelefteq} + \mathsf{MAJ}^{\omega}_{\trianglelefteq} + \Delta \vdash \forall x \exists y A_{\mathrm{bd}}(x, y),$$

where Δ is a set of universal sentences (with bounded intensional matrices) and A_{bd} is a bounded (intensional) formula with free variables among x and y. Then there is a closed monotone term t of appropriate type such that

$$\mathsf{PA}^{\omega}_{\leq} + \Delta^* \vdash \forall a \forall x \leq^* a \exists y \leq^* ta\, A^*_{\mathrm{bd}}(x, y).$$

In particular, the theory $\mathsf{PA}^{\omega}_{\trianglelefteq} + \mathsf{bBAC}^{\omega}_{\trianglelefteq} + \mathsf{MAJ}^{\omega}_{\trianglelefteq}$ is conservative over $\mathsf{PA}^{\omega}_{\leq}$ with respect to Π^0_2-sentences.

It is interesting to inquire what happens if one flattens the theory $\mathsf{PA}^{\omega}_{\trianglelefteq}$ together with the characteristic principles of Theorem 5.13. It so happens that $\mathsf{PA}^{\omega}_{\leq} + \mathsf{bBAC}^{\omega}$ is *inconsistent* (here bBAC^{ω} is the flattening of $\mathsf{bBAC}^{\omega}_{\trianglelefteq}$). Let us see why. Well, $\mathsf{PA}^{\omega}_{\trianglelefteq}$ proves

$$\forall x^1(x \leq_1^* 0 \to x \trianglelefteq_1 0) \to \forall x \trianglelefteq_1 1\,\exists n^0\,(\neg x \trianglelefteq_1 0 \to xn \neq 0).$$

If we now apply the intensional collection principle restricted to bounded formulas to the consequent above, we get

$$\forall x^1(x \leq_1^* 0 \to x \trianglelefteq_1 0) \to \exists m^0 \forall x \trianglelefteq_1 1 \exists n \leq m(\neg x \trianglelefteq_1 0 \to xn \neq 0).$$

We remind the reader that the intensional collection principle restricted to bounded formulas is provable in $\mathsf{PA}^{\omega}_{\trianglelefteq} + \mathsf{bBAC}^{\omega}_{\trianglelefteq}$. Hence, by flattening, $\mathsf{PA}^{\omega}_{\leq} + \mathsf{bBAC}^{\omega}$ proves

$$\forall x^1(x \leq_1^* 0 \to x \leq_1^* 0) \to \exists m^0 \forall x \leq_1^* 1 \exists n \leq m(\neg x \leq_1^* 0 \to xn \neq 0).$$

Since the antecedent is logically true, we infer

$$\exists m^0 \forall x \leq 1(\exists n(xn \neq 0) \to \exists n \leq m\,(xn \neq 0)),$$

a statement which is obviously refuted in $\mathsf{PA}^{\omega}_{\leq}$. Therefore, the theory $\mathsf{PA}^{\omega}_{\leq} + \mathsf{bBAC}^{\omega}$ is inconsistent.

5.6. On extensionality and uniform boundedness. We say that a formula $A(x^1)$ is *extensional* in the type 1 variable x if $\forall x^1, y^1\,(x =_1 y \wedge A(x) \to A(y))$, and we write $Ext_x[A]$.

PROPOSITION 5.16. *Let $A_{\mathrm{bd}}(x^1, k^0)$ be a bounded (intensional) formula. Then the theory $\mathsf{PA}^{\omega}_{\trianglelefteq} + \mathsf{bBAC}^{\omega}_{\trianglelefteq} + \mathsf{MAJ}^{\omega}_{\trianglelefteq}$ proves the implication whose antecedent is $Ext_x[A_{\mathrm{bd}}]$ and whose consequent is*

$$\forall x \leq_1 z \exists k A_{\mathrm{bd}}(x,k) \to \exists n \forall x \leq_1 z \exists k \leq n A_{\mathrm{bd}}(x,k).$$

PROOF. Assume $\forall x \leq_1 z \exists k A_{\mathrm{bd}}(x,k)$. By the extensionality of A_{bd} with respect to x, we have $\forall x \exists k A_{\mathrm{bd}}(\min_1(x,z),k)$. *A fortiori*, we get $\forall x \trianglelefteq z^{\mathrm{M}}$ $\exists k A_{\mathrm{bd}}(\min_1(x,z),k)$. Hence, by the intensional bounded collection scheme $\mathsf{bBC}^{\omega}_{\trianglelefteq}$, we may infer that there is n such that $\forall x \trianglelefteq z^{\mathrm{M}} \exists k \leq n A_{\mathrm{bd}}(\min_1(x,z),k)$. Since $\min_1(x,z) \trianglelefteq z^{\mathrm{M}}$, we have $\forall x \exists k \leq n A_{\mathrm{bd}}(\min_1(\min_1(x,z),z),k)$. By extensionality again, we conclude $\forall x \leq_1 z \exists k \leq n A_{\mathrm{bd}}(x,k)$. ⊣

The principle of the previous proposition is a uniform boundedness principle. As they were stated originally by Kohlenbach, they also incorporate a bit of choice. The following corollary can be proved similarly to the above result with the aid of $\mathsf{bBAC}^{\omega}_{\trianglelefteq}$:

COROLLARY 5.17. *Let $A_{\mathrm{bd}}(x^1, z^{0\to 1}, m^0, k^0)$ be a bounded (intensional) formula. Then the theory $\mathsf{PA}^{\omega}_{\trianglelefteq} + \mathsf{bBAC}^{\omega}_{\trianglelefteq} + \mathsf{MAJ}^{\omega}_{\trianglelefteq}$ proves the implication whose antecedent is $Ext_x[A_{\mathrm{bd}}]$ and whose consequent is*

$$\forall m \forall x \leq_1 zm \exists k A_{\mathrm{bd}}(x,z,m,k) \to \exists f^1 \forall m \forall x \leq zm \exists k \leq fm A_{\mathrm{bd}}(x,z,m,k).$$

When the formula A_{bd} is a Σ_1^0-formula (possibly with higher order parameters), the consequent above is dubbed Σ_1^0-UB. Note that the previous corollary includes this case because one can collapse two existential numerical quantifiers into a single one. Observe also that Σ_1^0-UB is a *false* principle. For instance, it entails that *all* type 2 functionals are bounded on compact sets:

$$\forall \Phi^2 \forall z^1 \exists k^0 \forall x \leq_1 z\, (\Phi x \leq k).$$

Full extensionality (see the definition at the end of Section 2.2) cannot be added to the theory $\mathsf{PA}^\omega_\trianglelefteq + \mathsf{bBAC}^\omega_\trianglelefteq + \mathsf{MAJ}^\omega_\trianglelefteq$: as we saw in Section 5.2, this would entail a contradiction. Instead, we consider the theory $\mathsf{PA}^\omega_\leq + \mathsf{AC}^{1,0}_{\mathrm{qf}} + \mathsf{AC}^{0,1}_{\mathrm{qf}} + \Sigma_1^0\text{-}\mathsf{UB}$ (stated in the language $\mathcal{L}^\omega_\leq$). We will see that adding full extensionality to this theory is consistent (and more). In order to accomplish this, we briefly review a (streamlined version of the) method of elimination of extensionality due to Horst Luckhardt:

Definition 5.18. We define, by recursion on the type:
a) $x \approx_0 y \equiv x = y$,
b) $x \approx_{\rho\to\tau} y \equiv \forall u^\rho, v^\rho (u \approx_\rho v \to xu \approx_\tau yv)$.

Lemma 5.19. *For each finite type σ, the theory HA^ω_0 proves:*
(i) $x \approx_\sigma y \to y \approx_\sigma x$;
(ii) $x \approx_\sigma y \to y \approx_\sigma y$;
(iii) $x \approx_\sigma y \wedge y \approx_\sigma z \to x \approx_\sigma z$;
(iv) $x =_\sigma y \wedge y \approx_\sigma z \to x \approx z$.

Proof. All the claims can be proved by induction on the complexity of the type σ, but (ii) and (iii) should be proved simultaneously. ⊣

Let the expression $E(x)$ abbreviate $x \approx x$ and, given a formula A of $\mathcal{L}^\omega_\leq$, let A^E be the *relativization* of A to the predicate E. Note that $\forall x^1 E(x)$.

Theorem 5.20 (Elimination of extensionality). *Suppose that*

$$\mathsf{E\text{-}PA}^\omega_\leq + \mathsf{AC}^{1,0}_{\mathrm{qf}} + \Delta \vdash A(\underline{z}),$$

where Δ is a set of universal closures of formulas with bound variables of type 0 *or* 1 *only, and A is an arbitrary formula with its free variables as shown (all this is stated in the language $\mathcal{L}^\omega_\leq$). Then,*

$$\mathsf{PA}^\omega_\leq + \mathsf{AC}^{1,0}_{\mathrm{qf}} + \Delta \vdash E(\underline{z}) \to A^E(\underline{z}).$$

Usually, this theorem is also stated with the inclusion of $\mathsf{AC}^{0,1}_{\mathrm{qf}}$. There is however no restriction because this form of choice can be included in Δ (the type $0 \to (0 \to 0)$ is essentially of type 1). Even though $\mathsf{AC}^{1,0}_{\mathrm{qf}}$ is not of the form Δ, the elimination of extensionality still goes through because the type 2 witness functional of $\mathsf{AC}^{1,0}_{\mathrm{qf}}$ can be taken as giving the *least* numerical witness satisfying the matrix of choice, and this forces the functional to satisfy the predicate E (if the parameters of the matrix also do). The following result is due to Kohlenbach:

THEOREM 5.21 (Extraction and conservation, uniform boundedness). *Suppose that*

$$\mathsf{E\text{-}PA}^{\omega}_{\leq} + \mathsf{AC}^{1,0}_{\mathrm{qf}} + \mathsf{AC}^{0,1}_{\mathrm{qf}} + \Sigma^0_1\text{-}\mathsf{UB} + \Delta \vdash \forall x^{0/1} \exists y A_{\mathrm{qf}}(x, y),$$

where A_{qf} is a quantifier-free formula with free variables among x and y, and Δ is a set of universal *sentences (all this is stated in the language $\mathcal{L}^{\omega}_{\leq}$). Then there is a closed monotone term t of appropriate type such that*

$$\mathsf{PA}^{\omega}_{\leq} + \Delta \vdash \forall x^{0/1} \exists y \leq^* tx\, A_{\mathrm{qf}}(x, y).$$

PROOF. For this proof, let us introduce the scheme $(\Sigma^0_1\text{-}\mathsf{UB})^-$ constituted by (universal closures of) implications of the following form: the antecedent is $Ext_x[F]$ and the consequent is $\forall m \forall x \leq_1 zm \exists k F(x, z, m, k) \to \exists f^1 \forall m \forall x \leq zm \exists k \leq fmF(x, z, m, k)$, where $F(x^1, z^{0\to 1}, m^0, k^0)$ is a Σ^0_1-formula (possibly with higher order parameters). Observe that this is a scheme of formulas whose bound variables are of type 0 and 1 only. By hypothesis (notice the presence of full extensionality),

$$\mathsf{E\text{-}PA}^{\omega}_{\leq} + \mathsf{AC}^{1,0}_{\mathrm{qf}} + \mathsf{AC}^{0,1}_{\mathrm{qf}} + (\Sigma^0_1\text{-}\mathsf{UB})^- + \Delta \vdash \forall x^{0/1} \exists y A_{\mathrm{qf}}(x, y).$$

Hence, by elimination of extensionality (previous theorem),

$$\mathsf{PA}^{\omega}_{\leq} + \mathsf{AC}^{1,0}_{\mathrm{qf}} + \mathsf{AC}^{0,1}_{\mathrm{qf}} + (\Sigma^0_1\text{-}\mathsf{UB})^- + \Delta \vdash \forall x^{0/1} \exists y A_{\mathrm{qf}}(x, y).$$

By the comments in Section 5.2 apropos intensional bounded choice and by Corollary 5.17, we get $\mathsf{PA}^{\omega}_{\trianglelefteq} + \mathsf{bBAC}^{\omega}_{\trianglelefteq} + \mathsf{MAJ}^{\omega}_{\trianglelefteq} + \Delta \vdash \forall x^{0/1} \exists y A_{\mathrm{qf}}(x, y)$. Hence, according to Theorem 5.12, there is a closed term t of appropriate type such that

$$\mathsf{PA}^{\omega}_{\trianglelefteq} + \Delta \vdash \forall x^{0/1} \exists y \trianglelefteq tx\, A_{\mathrm{qf}}(x, y).$$

The desired result follows by flattening. ⊣

The above result can be refined in two ways. On the one hand, Δ may be constituted by sentences of the form $\exists \underline{x} \leq \underline{r} \forall \underline{y} B_{\mathrm{qf}}(\underline{x}, \underline{y})$, with B_{qf} quantifier-free, $\underline{r}$ a tuple of closed terms of type 0 or 1 (the same types of $\underline{x}$) and $\underline{y}$ a tuple of any types. One can apply Luckhardt's elimination of extensionality technique to such sentences and, afterwards, use the techniques discussed after Theorem 5.5. As observed in that discussion, the verification can even be done with the weaker (corresponding) sentences $\forall \underline{z} \exists \underline{x} \leq \underline{r} \forall \underline{y} \leq^* \underline{z} B_{\mathrm{qf}}(\underline{x}, \underline{y})$. On the other hand, the above result can also be stated with the form of choice $\mathsf{AC}^{1,1}_{\mathrm{qf}}$. We discuss in detail the latter improvement. This improvement follows from

PROPOSITION 5.22. *The theory* $\mathsf{E\text{-}PA}^{\omega}_{0} + \mathsf{AC}^{1,0}_{\mathrm{qf}} + \Sigma^0_1\text{-}\mathsf{UB}$ *proves* $\mathsf{AC}^{1,1}_{\mathrm{qf}}$.

PROOF. We need a preliminary lemma:

LEMMA 5.23. *Let $A_{\mathrm{qf}}(y^1)$ be a quantifier-free formula. Then*

$$\text{E-PA}_0^\omega + \mathrm{AC}_{\mathrm{qf}}^{1,0} + \Sigma_1^0\text{-UB} \vdash \forall y^1(A_{\mathrm{qf}}(y) \to \exists m^0 A_{\mathrm{qf}}(\widehat{\overline{y}m})).$$

PROOF OF THE LEMMA. By extensionality, we have $\forall y, z\,(y =_1 z \wedge A_{\mathrm{qf}}(y) \to A_{\mathrm{qf}}(z))$. Therefore, $\forall y, z \exists n\,(yn = zn \wedge A_{\mathrm{qf}}(y) \to A_{\mathrm{qf}}(z))$. It is easy to see that $\mathrm{AC}_{\mathrm{qf}}^{1,0}$ entails the existence of a functional Σ of type $1 \to (1 \to 0)$ such that

$$\forall y, z(y(\Sigma yz) = z(\Sigma yz) \wedge A_{\mathrm{qf}}(y) \to A_{\mathrm{qf}}(z)).$$

Fix y. By Σ_1^0-UB, there is m^0 such that $\forall z \leq y(\Sigma yz < m)$. The lemma follows. ⊣(of lemma)

We now prove that $\mathrm{AC}_{\mathrm{qf}}^{1,1}$ follows from $\text{E-PA}_0^\omega + \mathrm{AC}_{\mathrm{qf}}^{1,0} + \Sigma_1^0\text{-UB}$. Let $A_{\mathrm{qf}}(x^1, y^1)$ be a quantifier-free formula and suppose that $\forall x \exists y A_{\mathrm{qf}}(x, y)$. By the previous lemma, we may infer that $\forall x \exists s^0(Seq(s) \wedge A_{\mathrm{qf}}(x, \hat{s}))$, where $Seq(s)$ means that s is a finite sequence of natural numbers. Therefore, by $\mathrm{AC}_{\mathrm{qf}}^{1,0}$, there is a functional Φ of type 2 such that $\forall x(Seq(\Phi x) \wedge A_{\mathrm{qf}}(x, \widehat{\Phi x}))$. Clearly, $\Psi^{1\to 1} := \lambda x.\widehat{\Phi x}$ is a choice function. ⊣(of proposition)

As a consequence of the results above, we obtain an extraction result for the fully extensional, *true* theory, $\text{E-PA}_0^\omega + \mathrm{AC}_{\mathrm{qf}}^{1,1}$. However, this result is obtained in a very roundabout way, via a false extension. Is there a more direct route?

5.7. Suggested reading and historical notes. The bounded functional interpretation appeared in [12], where the proofs of the theorems in Sections 5.1 and 5.2 can be found. The intuitionistic application in Section 5.3 appears here for the first time. The negative translation within the setting of the bounded functional interpretation is also discussed in [12]. A direct interpretation of Peano arithmetic — in the style of Shoenfield — was recently defined in [10]. In this paper, the characterization theorem is formulated with different, but equivalent, characteristic principles. Notwithstanding, Theorem 5.13 regards an indirect interpretation, via a negative translation. However, one could (for instance) use the factorization of Jaime Gaspar [14] to get the result in the text (albeit for the so-called Krivine's negative translation; it is easy to see, though, that the result also holds for the Gödel-Gentzen translation using the fact that these negative translations are intuitionistically equivalent). Flattening is already introduced in [12], but only in [13] it is given its name. The latter article includes a study of the elimination of weak König's lemma in the feasible setting (the elimination technique mentioned at the end of Section 4.2 does not apply to the feasible setting since it uses bounded search in an essential manner). The discussion on extensionality and the uniform boundedness principles in Section 5.6 are based on [12], where stronger results are proved.

Luckhardt's result on the elimination of extensionality is from [42] but, as noticed, it is here simplified. The extraction and conservation result on uniform boundedness is due to Kohlenbach and appears (essentially) in [35] under a treatment which is a combination of results in [28] and [32]. Our treatment is rather different, in that it is based on the bounded functional interpretation.

§6. Coda. The emphasis of this paper is on the theoretical aspects of proof mining in the style of Kohlenbach and his co-workers. This means studying proof interpretations where the concept of majorizability plays a central role. Even though the applied work has been revolving around *functional* interpretations, for the sake of rounding up we also included in our discussions the realizability interpretations. The main theoretical tool of the applied work has been the monotone functional interpretation but, from a theoretical point of view, we believe that the bounded functional interpretations provide a fresh perspective. We also took the chance of making some comparisons between the monotone and bounded interpretations. It remains to be seen whether the latter interpretations prove to be useful in the applied work. Proof mining itself was left out. I suggest the surveys [37] and [34] for a first reading (see also [36] for a systematic list of statements of the results obtained until 2006). Kohlenbach's book [35] is recommended for a detailed treatment. In [8] the reader can find some *general* discussions on the theoretical and practical benefits of (functional) majorizability interpretations.

We left out two important theoretical topics, one old, the other quite recent. The old one is Clifford Spector's *deep* generalization of Gödel's interpretation to second-order classical arithmetic using bar-recursive functionals (see [47]). The systems which Kohlenbach and his co-workers use in proof mining include, as a matter of course, full second-order comprehension. Kohlenbach's book [35] is a good source for an exposition of Spector's interpretation. Very recently, it was shown in [7] that the bar-recursive functionals of Spector can also be used to obtain a bounded functional interpretation of second-order classical artithmetic. In the words of Part 5, it is possible to inject uniformities into systems containing full second-order comprehension (see [8] for a brief discussion on how far one can go on in doing this). The other topic is the generalization of the monotone functional interpretation to new base types, typically metric or normed spaces. This generalization was introduced in [33] and is also treated in [35]. It has been proved very useful in the applied work and it is rather illuminating from a theoretical point of view.

REFERENCES

[1] Y. Akama, S. Berardi, S. Hayashi, and U. Kohlenbach, *An arithmetical hierarchy of the law of excluded middle and related principles*, ***Proceedings of the 19th Annual IEEE Symposium on Logic and Computer Science***, vol. 117, IEEE Press, 2004, pp. 192–201.

[2] J. Avigad and S. Feferman, *Gödel's functional ("Dialectica") interpretation*, ***Handbook of Proof Theory*** (S. R. Buss, editor), Studies in Logic and the Foundations of Mathematics, vol. 137, North Holland, Amsterdam, 1998, pp. 337–405.

[3] J. Avigad and H. Towsner, *Functional interpretation and inductive definition*, ***The Journal of Symbolic Logic***, vol. 74 (2009), pp. 1100–1120.

[4] M. Bezem, *Strongly majorizable functionals of finite type: a model for bar recursion containing discontinuous functionals*, ***The Journal of Symbolic Logic***, vol. 50 (1985), pp. 652–660.

[5] W. Burr, *A Diller-Nahm-style functional interpretation of* $\mathsf{KP}\omega$, ***Archive for Mathematical Logic***, vol. 39 (2000), pp. 599–604.

[6] J. Diller and W. Nahm, *Eine Variante zur Dialectica-Interpretation der Heyting-Arithmetik endlicher Typen*, ***Archive für mathematische Logik und Grundlagenforschung***, vol. 16 (1974), pp. 49–66.

[7] P. Engrácia and F. Ferreira, *The bounded functional interpretation of the double negation shift*, to appear in ***The Journal of Symbolic Logic***.

[8] F. Ferreira, *A most artistic package of a jumble of ideas*, ***dialectica***, vol. 62 (2008), pp. 205–222, Special Issue: Gödel's *dialectica* Interpretation. Guest editor: Thomas Strahm.

[9] ———, *On a new functional interpretation (abstract)*, ***The Bulletin of Symbolic Logic***, vol. 14 (2008), pp. 128–129.

[10] ———, *Injecting uniformities into Peano arithmetic*, ***Annals of Pure and Applied Logic***, vol. 157 (2009), pp. 122–129, Special Issue: Kurt Gödel Centenary Research Prize Fellowships. Editors: Sergei Artemov, Matthias Baaz and Harvey Friedman.

[11] F. Ferreira and A. Nunes, *Bounded modified realizability*, ***The Journal of Symbolic Logic***, vol. 71 (2006), pp. 329–346.

[12] F. Ferreira and P. Oliva, *Bounded functional interpretation*, ***Annals of Pure and Applied Logic***, vol. 135 (2005), pp. 73–112.

[13] ———, *Bounded functional interpretation and feasible analysis*, ***Annals of Pure and Applied Logic***, vol. 145 (2005), pp. 115–129.

[14] J. Gaspar, *Factorization of the Shoenfield-like bounded functional interpretation*, ***Notre Dame Journal of Formal Logic***, vol. 50 (2009), pp. 53–60.

[15] P. Gerhardy and U. Kohlenbach, *General logical metatheorems for functional analysis*, ***Transactions of the American Mathematical Society***, vol. 360 (2008), pp. 2615–2660.

[16] K. Gödel, *Zur intuitionistischen Arithmetik und Zahlentheorie*, ***Ergebnisse eines Mathematischen Kolloquiums***, vol. 4 (1933), pp. 34–38.

[17] ———, *Über eine bisher noch nicht benützte Erweiterung des finiten Standpunktes*, ***dialectica***, vol. 12 (1958), pp. 280–287, Reprinted with an English translation in [19].

[18] ———, *On an extension of finitary mathematics which has not yet been used*, 1972, Published in [19], pages 271–280. Revised version of [17].

[19] ———, ***Collected Works, Vol. II***, (S. Feferman, editor), Oxford University Press, Oxford, 1990.

[20] W. A. Howard, *Hereditarily majorizable functionals of finite type*, ***Metamathematical investigation of intuitionistic Arithmetic and Analysis*** (A. S. Troelstra, editor), Lecture Notes in Mathematics, vol. 344, Springer, Berlin, 1973, pp. 454–461.

[21] S. C. Kleene, *On the interpretation of intuitionistic number theory*, ***The Journal of Symbolic Logic***, vol. 10 (1945), pp. 109–124.

[22] ———, *Countable functionals*, ***Constructivity in Mathematics*** (A. Heyting, editor), North Holland, Amsterdam, 1959, pp. 81–100.

[23] U. Kohlenbach, ***Theorie der majorisierbaren und stetigen Funktionale und ihre Anwendung bei der Extraktion von Schranken aus inkonstruktiven Beweisen: Effektive Eindeutigkeitsmodule bei besten Approximationen aus ineffektiven Beweisen***, Ph.D. thesis, Frankfurt, pp. xxii+278, 1990.

[24] ———, *Effective bounds from ineffective proofs in analysis: an application of functional interpretation and majorization*, ***The Journal of Symbolic Logic***, vol. 57 (1992), pp. 1239–1273.

[25] ———, *Pointwise hereditary majorization and some applications*, ***Archive for Mathematical Logic***, vol. 31 (1992), pp. 227–241.

[26] ———, *Effective moduli from ineffective uniqueness proofs. An unwinding of de La Vallée Poussin's proof for Chebycheff approximation*, ***Annals of Pure and Applied Logic***, vol. 64 (1993), pp. 27–94.

[27] ———, *Analysing proofs in analysis*, ***Logic: from Foundations to Applications*** (W. Hodges, M. Hyland, C. Steinhorn, and J. Truss, editors), Oxford University Press, 1996, pp. 225–260.

[28] ———, *Mathematically strong subsystems of analysis with low rate of growth of provably recursive functionals*, ***Archive for Mathematical Logic***, vol. 36 (1996), pp. 31–71.

[29] ———, *Relative constructivity*, ***The Journal of Symbolic Logic***, vol. 63 (1998), pp. 1218–1238.

[30] ———, *On the no-counterexample interpretation*, ***The Journal of Symbolic Logic***, vol. 64 (1999), pp. 1491–1511.

[31] ———, *Intuitionistic choice and restricted classical logic*, ***Mathematical Logic Quarterly***, vol. 47 (2001), pp. 455–460.

[32] ———, *Foundational and mathematical uses of higher types*, ***Reflections on the Foundations of Mathematics: Essay in Honor of Solomon Feferman*** (W. Sieg et al., editors), Lecture Notes in Logic, vol. 15, A. K. Peters, 2002, pp. 92–116.

[33] ———, *Some logical metatheorems with applications in functional analysis*, ***Transactions of the American Mathematical Society***, vol. 357 (2005), pp. 89–128.

[34] ———, *Proof interpretations and the computational content of proofs in mathematics*, ***Bulletin of the EATCS***, vol. 93 (2007), pp. 143–173.

[35] ———, ***Applied Proof Theory: Proof Interpretations and Their Use in Mathematics***, Springer Monographs in Mathematics, Springer, Berlin, 2008.

[36] ———, *Effective bounds from proofs in abstract functional analysis*, ***New Computational Paradigms: Changing Conceptions of What is Computable*** (S. B. Cooper, B. Löwe, and A. Sorbi, editors), Springer Verlag, to appear.

[37] U. Kohlenbach and P. Oliva, *Proof mining: a systematic way of analysing proofs in mathematics*, ***Proceedings of the Steklov Institute of Mathematics***, vol. 242 (2003), pp. 136–164.

[38] U. Kohlenbach and T. Streicher, *Shoenfield is Gödel after Krivine*, ***Mathematical Logic Quarterly***, vol. 53 (2007), pp. 176–179.

[39] G. Kreisel, *On the interpretation of non-finitist proofs, part I*, ***The Journal of Symbolic Logic***, vol. 16 (1951), pp. 241–267.

[40] ———, *Interpretation of analysis by means of constructive functionals of finite types*, ***Constructivity in Mathematics*** (A. Heyting, editor), North Holland, Amsterdam, 1959, pp. 101–128.

[41] ———, *On weak completeness of intuitionistic predicate logic*, ***The Journal of Symbolic Logic***, vol. 27 (1962), pp. 139–158.

[42] H. Luckhardt, ***Extensional Gödel Functional Interpretation: A Consistency Proof of Classical Analysis***, Lecture Notes in Mathematics, vol. 306, Springer, Berlin, 1973.

[43] P. Oliva, *Unifying functional interpretations*, ***Notre Dame Journal of Formal Logic***, vol. 47 (2006), pp. 263–290.

[44] C. Parsons, *On a number theoretic choice schema and its relation to induction*, ***Intuitionism and Proof Theory*** (J. Myhill, A. Kino, and R. E. Vesley, editors), North-Holland, 1970, pp. 459–473.

[45] J. R. Shoenfield, ***Mathematical Logic***, Addison-Wesley Publishing Company, 1967, Republished in 2001 by AK Peters.

[46] S. G. Simpson, ***Subsystems of Second Order Arithmetic***, Perspectives in Mathematical Logic, Springer, Berlin, 1999.

[47] C. Spector, *Provably recursive functionals of analysis: a consistency proof of analysis by an extension of principles in current intuitionistic mathematics*, ***Recursive Function Theory:***

Proceedings of Symposia in Pure Mathematics (F. D. E. Dekker, editor), vol. 5, AMS, Providence, Rhode Island, 1962, pp. 1–27.

[48] T. Strahm (editor), *Gödel's* dialectica *Interpretation*, ***dialectica***, vol. 62 (2008), pp. 145–290.

[49] W. Tait, *Intentional interpretations of functionals of finite type I*, ***The Journal of Symbolic Logic***, vol. 32 (1967), pp. 198–212.

[50] T. Tao, *The correspondence principle and finitary ergodic theory*, Essay posted August 30, 2008. Available at: http://terrytao.wordpress.com/2008/08/30/the-correspondence-principle-and-finitary-ergodic-theory/.

[51] ———, *Soft analysis, hard analysis, and the finite convergence principle*, ***Structure and Randomness: Pages from Year One of a Mathematical Blog***, AMS, 2008, pp. 17–29.

[52] A. S. Troelstra (editor), ***Metamathematical Investigation of Intuitionistic Arithmetic and Analysis***, Lecture Notes in Mathematics, vol. 344, Springer, Berlin, 1973.

[53] ———, *Introductory note to* [17] *and* [18], 1990, Published in [19], pages 214–241.

[54] ———, *Realizability*, ***Handbook of Proof Theory*** (S. R. Buss, editor), Studies in Logic and the Foundations of Mathematics, vol. 137, North Holland, Amsterdam, 1998, pp. 408–473.

[55] A. S. Troelstra and D. van Dalen, ***Constructivism in Mathematics. An Introduction***, Studies in Logic and the Foundations of Mathematics, vol. 121, North Holland, Amsterdam, 1988.

[56] M. Yasugi, *Intuitionistic analysis and Gödel's interpretation*, ***Journal of the Mathematical Society of Japan***, vol. 15 (1963), pp. 101–112.

DEPARTAMENTO DE MATEMÁTICA
UNIVERSIDADE DE LISBOA
CAMPO GRANDE, EDIFíCIO C6
P-1749-016 LISBOA, PORTUGAL
E-mail: ferferr@cii.fc.ul.pt

PROOF MINING IN PRACTICE

PHILIPP GERHARDY

Abstract. In this paper, we present some aspects of a recent application of proof mining by J. Avigad, H. Towsner and the author. In this case study, we analysed a proof of the Mean Ergodic Theorem and obtained a computable rate of convergence for the ergodic averages. Proof mining generally falls into two main categories: Establishing general metatheorems that classify theorems and proofs from which additional information may be extracted and carrying out case studies. The aim of presenting aspects of a proof analysis in detail in this paper is to illustrate how the general logical results and the techniques they rely on translate into a proof analysis in practice.

§1. Introduction. 'Proof mining' is the subfield of mathematical logic concerned with extracting additional information from proofs in mathematics and computer sciences. This activity has its roots in Kreisel's so-called 'unwinding' program and is motivated by the following quote by G. Kreisel:

"What more do we know if we have proved a theorem by restricted means than if we merely know the theorem is true."

Kreisel proposed to use techniques developed in proof theory (e.g. to settle questions of consistency) to analyse proofs and unwind the extra information hidden in them. This additional information can both be of qualitative nature, such as computable realizers and bounds, as well as of quantitative nature, such as uniformities or weakenings of premises. An example of the former is extracting a computable rate of convergence even from an ineffective proof using full classical logic that a certain kind of iteration sequence converges. An example of the latter is establishing that the rate of convergence is uniform in the starting point of the iteration or that the convergence not only holds in compact metric spaces, but already in bounded metric spaces.

Proof mining falls into two major categories: On the one hand, one establishes general metatheorems that classify theorems and proofs from which such additional information can be extracted and develops general techniques to carry out these extractions. On the other hand, one carries out case studies and analyses actual mathematical proofs. This paper will be concerned with the latter part.

Logic Colloquium '07
Edited by Françoise Delon, Ulrich Kohlenbach, Penelope Maddy, and Frank Stephan
Lecture Notes in Logic, 35

The main techniques in proof mining are proof interpretations. Proof interpretations are used to inductively transform given proofs into enriched proofs from which the desired additional information can be read off. While the soundness of these methods rests on results in mathematical logic, the resulting enriched proof itself is again an ordinary mathematical proof. The main examples of proof interpretations are so-called functional interpretations, where the most widely used variants are Gödel's Dialectica interpretation (sometimes itself referred to as functional interpretation) and Kreisel's modified realizability interpretation, as well as variants of these two interpretations.

The idea behind functional interpretation is to give a computational interpretation of constants, axioms and derivation rules of a given formal system by functionals of higher type. Thus, starting with the axioms, one may inductively transform the proof into an enriched proof and, in the end, read off a computable realizer for the conclusion of the proof. In monotone variants of functional interpretations, one seeks to obtain bounds rather than exact realizers. This has the advantage that the monotone interpretation of axioms and rules often is much simpler. Furthermore, certain non-constructive principles, e.g. weak König's lemma, do not allow for an exact realizer, but do allow for bounds, such that these principles can be safely added to a formal system, when one merely is interested in extracting bounds.

Naturally, there are certain limits to what can be achieved with proof mining: The halting problem can be formulated as a simple $\forall\exists\forall$-statement, where the $\exists$-quantifier tells us when a Turing machine stops, if it stops at all. This has a simple classical proof, but of course we can extract neither realizers nor bounds for the $\exists$-quantifier as this would contradict the undecidability of the halting problem. Still, proof interpretations allow one to prove very general logical metatheorems stating when and how theorems and proofs may be analysed and what information can be extracted from them. For some of the most recent results on such metatheorems, see [4, 2]. For a comprehensive overview of proof interpretations see [9, 5].

While metatheorems apply to formal systems and thus, in principle, to fully formal proofs, it is in general not necessary to completely formalize a proof in order to analyse it. In practice, case studies in proof mining often consist of preprocessing a "normal" (mathematical, but still relatively informal) proof by putting the statement and the main concepts involved into a suitable form and then identifying the key inferences in the proof that need to be given a computational interpretation. The metatheorems guarantee the extractability of additional information based on a formal, mechanical transformation of entire proofs. In practice, these transformations translate into a number of heuristics applied to the key inferences and concepts of the proof.

In this paper, we illustrate proof mining in practice by describing some details of a recent application of proof mining ([1], joint work by J. Avigad, H. Towsner and the author). In this application, we analysed a standard textbook proof of the Mean Ergodic Theorem and obtained a computable rate of convergence for the ergodic averages. In the next three sections, we will treat the following three aspects of this application: (1) Putting the theorem into a suitable form, checking that the proof in principle is formalizable in a formal system for which metatheorems exist and using this to make predictions about the bounds we can extract from the proof. (2) Giving a computational interpretation to an instance of the law of excluded middle that is used in the proof. More precisely, the proof uses that "a given real number r is either zero or not zero". (3) Weakening the assumption that a full rate of convergence for a certain monotone bounded sequence of real numbers exists to the assumption that there exists no counterexample to the convergence of that sequence. The latter always has a simple computational interpretation, while in general a full rate of convergence for bounded monotone sequences of real numbers need not exist at all, as illustrated by so-called Specker sequences.

§2. The Mean Ergodic Theorem. There are several ways to state the Mean Ergodic Theorem. Originally, it is stated in the context of ergodic theory, where it asserts that the ergodic averages for measure preserving maps on a measure space converge. We state a more general version, asserting the convergence of ergodic averages for Hilbert spaces and nonexpansive mappings. Recall, that a Hilbert space is a normed linear space with an inner product. The inner product provides a notion of orthogonality, where $x \perp y \Leftrightarrow \langle x, y \rangle = 0$, and induces a norm $\|x\| :\equiv \sqrt{\langle x, x \rangle}$ on X.

DEFINITION 1. Let $(X, \langle \cdot, \cdot \rangle)$ be a Hilbert space. A function $T : X \to X$ is nonexpansive (short: f n.e.), if $\|Tf\| \leq \|f\|$ for all $f \in X$.

MEAN ERGODIC THEOREM. *Let $(X, \langle \cdot, \cdot \rangle)$ be a Hilbert space, let $T : X \to X$ be a linear, nonexpansive selfmapping of X. For $f \in X$ define $A_n f := \frac{1}{n+1} \sum_{i=0}^{n} T^i f$, then $A_n f$ converges to a limit.*

The first reformulation we will undertake is to rewrite "$A_n f$ converges to a limit" as

$$\forall \varepsilon > 0 \exists n \forall m > n (\|A_m f - A_n f\| < \varepsilon),$$

where $\| \cdot \|$ is the norm induced by the inner product.

The theorem has the logical form '$\forall T \forall f \forall \varepsilon > 0 \exists n \forall m > n(\dots)$', so the computational challenge of the theorem is to find a realizer or bound for $\exists n$. If we furthermore restrict the $\varepsilon > 0$ to those that can be written as 2^{-k} for $k \in \mathbb{N}$, we see that the theorem is almost purely arithmetical — except for parameters f and T which involve the types X. In a moment, we shall see how to cast even those parameters in a such a way that they only contribute to the computational challenge of the theorem by natural numbers.

This theorem can be formalized in the formal system $\mathcal{A}^\omega[X, \langle \cdot, \cdot \rangle]$. The theory $\mathcal{A}^\omega$ is Peano arithmetic extended to all finite types and extended with the axiom schema of dependent choice. Most of classical analysis can be formalized in this theory. The theory $\mathcal{A}^\omega[X, \langle\cdot,\cdot\rangle]$ denotes the extension of $\mathcal{A}^\omega$ with an abstract Hilbert space in the following way:

- We add a new ground type X for the elements of the Hilbert space and extend $\mathcal{A}^\omega$ to all finite types $\boldsymbol{T}^X$ over $\mathbb{N}$ and X.
- We add new constants representing the inner product and the vector space operations on X.
- We add new axioms describing the algebraic properties of the inner product and the vector space operations.

Using a monotone variant of Gödel's Dialectica interpretation, one proves general metatheorems about the extraction of bounds from proofs of $\forall\exists$-statements in $\mathcal{A}^\omega[X, \langle\cdot,\cdot\rangle]$ and similar theories. For details, see [2].

The Mean Ergodic Theorem is a $\forall\exists\forall$-statement. As mentioned above, such theorems do not always allow to extract realizers or bounds for the existential quantifier. This is also the case with the Mean Ergodic Theorem, where we can construct a Hilbert space and a mapping T such that a full rate of convergence for the ergodic averages $A_n f$ would solve the halting problem for Turing machines. Even in the measure theoretic setting, one may construct counterexamples. Only when the measure space is ergodic, one obtains a full rate of convergence. For details, see [1].

Instead, we will extract bounds for the classically equivalent but constructively weaker no-counterexample version of the Mean Ergodic Theorem:

Mean Ergodic Theorem (no-counterexample version). *Let $(X, \langle\cdot,\cdot\rangle)$ be a Hilbert space, let $T : X \to X$ be a linear, nonexpansive selfmapping of X. For $f \in X$ define $A_n f := \frac{1}{n+1}\sum_{i=0}^{n} T^i f$, then for every $\varepsilon > 0$ and and every number-theoretic function $M : \mathbb{N} \to \mathbb{N}$ there exists an $n \in \mathbb{N}$ s.t. $\|A_{M(n)} f - A_n f\| < \varepsilon$.*

The function M claims the existence of a counterexample to the convergence of $A_n f$, and the existence of n shows that any such supposed counterexample can be refuted. This no-counterexample version is classically equivalent to full convergence. Constructively, this establishes the local stability (up to an $\varepsilon > 0$) of the ergodic averages $A_n f$ for arbitrarily long periods of time.

The theorem is now in a suitable form for the metatheorems mentioned above to be applicable. From the most general form of the metatheorems in [2] we derive the following special case:

Definition 2. The finite types $\boldsymbol{T}^X$ are defined as follows:

$$(i)\ \mathbb{N}, X \in \boldsymbol{T}^X, \quad (ii)\ \rho, \tau \in \boldsymbol{T}^X \Rightarrow \rho \to \tau \in \boldsymbol{T}^X.$$

We often write 0 for the type $\mathbb{N}$ and 1 for the type $\mathbb{N} \to \mathbb{N}$.

Definition 3. A formula F is called a $\forall$-formula, resp. $\exists$-formula, if it is of the form $\forall \underline{a}^{\underline{\sigma}} F_{qf}$, resp. $\exists \underline{a}^{\underline{\sigma}} F_{qf}$, where F_{qf} is quantifier-free and the types of

σ are *small*[1], which includes amongst others the types $\mathbb{N}$, X, $\mathbb{N} \to \mathbb{N}$, $\mathbb{N} \to X$ and $X \to X$. We write $F_\forall$ and $F_\exists$ for $\forall$-formulas and $\exists$-formulas respectively.

COROLLARY 4. *Let $(X, \langle \cdot, \cdot \rangle)$ be an abstract Hilbert space. If*

$$\forall f^X, T^{X \to X}, k^0, M^1(T \text{ n.e.} \wedge \forall u^0 B_\forall(f, T, k, M, u) \to \exists v^0 C_\exists(f, T, k, M, v))$$

is provable in $\mathcal{A}^\omega[X, \langle \cdot, \cdot \rangle]$, then there is a computable $\phi : \mathbb{N} \times \mathbb{N} \times \mathbb{N}^\mathbb{N} \to \mathbb{N}$ such that

$$\forall f^X, T^{X \to X}, k^0, M^1(\|f\| \leq b \wedge T \text{ n.e.} \wedge \\ \forall u^0 \leq \phi(b, k, M) B_\forall \to \exists v^0 \leq \phi(b, k, M) C_\exists)$$

holds in any Hilbert space $(X, \langle \cdot, \cdot \rangle)$.

This is exactly the logical form of the no-counterexample version of the Mean Ergodic Theorem, and it is easy to see that the standard proof of the (no-counterexample version) Mean Ergodic Theorem can be formalized in $\mathcal{A}^\omega[X, \langle \cdot, \cdot \rangle]$, as the proof only uses the basic algebraic properties of normed spaces and inner products. Thus, this corollary predicts a computable bound on $\exists n$ in the Mean Ergodic Theorem in terms of (an integer bound b on) $\|f\|$, ε and the counterexample function M, but independent of the particular f (as long as its norm is bounded by b), the particular T or the particular space $(X, \langle \cdot, \cdot \rangle)$.

In [1], the following bounds were obtained:

$$i_0 = 0, \qquad n_k = \left\lceil \frac{b}{\varepsilon^2} \sum_{l=0}^{i_k} M\left(\frac{2lb}{\varepsilon}\right) \right\rceil$$

$$i_k + 1 = i_k + \left\lceil \frac{2^{15} M(n_k)^4 b^4}{\varepsilon^4} \right\rceil$$

Let $d = \frac{512b^2}{\varepsilon^2}$, then for some $n \leq N(b, \varepsilon, M) = \frac{2n_d b}{\varepsilon}$, we have that $\|A_{M(n)} f - A_n f\| < \varepsilon$.

In the next sections, we discuss details of the proof analysis carried out in [1] which resulted in these bounds.

§3. Interpreting the law of the excluded middle. The main argument of the proof of the Mean Ergodic Theorem goes as follows: Any element $f \in X$ can be written as a sum $f = h + g$, where h is T-invariant (i.e. $Th = h$) and g can be approximated arbitrarily well by an element of the form $u - Tu$ with $u \in X$. Since T is linear, we may thus study $A_n h$ and $A_n g$ seperately. Obviously, $A_n h = h$ for all $n \in \mathbb{N}$. Also, $\|A_n(u - Tu)\| \leq \frac{\|u\|}{2n}$, which quickly converges to 0. In other words, the main argument is: Either $\|f - Tf\| = 0$, f is T-invariant and $A_n f$ is constant, or $\|f - Tf\| > 0$ and then $A_n f$ converges, because the part that is not T-invariant tends to 0 quickly. To estimate how

[1]For the exact definition, see [2].

fast $\|A_n(u - Tu)\|$ converges to 0, we need to know the norm $\|u\|$, or at least an upper bound on $\|u\|$. As the decomposition of f into g and h corresponds to a decompostion of the whole space X into orthogonal subspaces U (the closure of the space of elements of the form $u - Tu$) and V (the subspace of T-invariant elements), this is equivalent to approximating the projection of f onto U by some $u - Tu$ and obtaining an upper bound on $\|u\|$.

As the averages $A_n f$ lie in the subspace generated by $\{f, Tf, T^2 f, \dots\}$, it suffices to consider the projection of f onto $U_f = \{T^i f - T^{i+1} f | i = 0, 1, 2, \dots\}$. For this subspace $U_f \subseteq U$, we can explicitly define a sequence u_n such that $g_n = u_n - Tu_n$ converges to the projection of f onto U_f, and it is from this sequence that we derive an upper bound on the norms $\|u_n\|$. The elements u_n are defined as follows:

$$u_0 = \frac{\langle f, f - Tf\rangle}{\|f - Tf\|^2} f,$$

$$u_{i+1} = \frac{\langle f - (u_i - Tu_i), T^{i+1} f - T^{i+2} f\rangle}{\|T^{i+1} f - T^{i+2} f\|^2} T^{i+1} f.$$

The numerator of the fraction has an easy upper bound by $\langle f, f - Tf\rangle \leq \|f\|\|f - Tf\|$, resp. $\langle f - (u_i - Tu_i), T^{i+1} f - T^{i+2} f\rangle \leq \|f\|\|T^{i+1} f - T^{i+2} f\|$. Thus we only need to find a *lower bound* for $\|f - Tf\|$, resp. $\|T^{i+1} f - T^{i+2} f\|$ in the denominator — but the norm of elements $T^i f - T^{i+1} f$ can be arbitrarily small.

The solution lies in the original argument: If the norm of $T^i f - T^{i+1} f$ is zero for some i, then $T^i f$ is T-invariant and we are done. If not, it must be larger than some $\delta > 0$. This is close to the solution, but not quite there yet. As we are only looking for local stability for $A_n f$, not full convergence, it suffices that $\|T^i f - T^{i+1} f\|$ is small enough, as then repeated use of the triangle inequality yields local stability. Given $\varepsilon > 0$ and the counterexample function M, we can say how small $\|T^i f - T^{i+1} f\|$ needs to be to directly imply local stability, i.e. we can produce an explicit $\delta > 0$ such that $\|T^i f - T^{i+1} f\| < \delta$ yields the result. Otherwise, we have a lower bound on $\|T^i f - T^{i+1} f\|$ and therefore an upper bound for u_i!

In [1], we prove the following lemma (Lemma 2.15 in [1]), reflecting the above discussion:

LEMMA 5. *For any $i \geq 0$ and $\varepsilon > 0$, either*

1. *there is an $n \leq 2i\lceil\frac{\|f\|}{\varepsilon}\rceil$ such that $\|A_{M(n)} f - A_n f\| \leq \varepsilon$, or*
2. $\|u_i\| \leq \frac{\|f\|^2}{2\varepsilon}\sum_{j=0}^{i} M(2j\lceil\frac{\|f\|}{\varepsilon}\rceil)$.

In summary, we use the fact that for any $i \in \mathbb{N}$ either $\|T^i f - T^{i+1} f\|$ is zero or it is not zero. We give a computational interpretation of this instance of the law of the excluded middle by producing, for each $i \in \mathbb{N}$, a $\delta > 0$ such that both $\|T^i f - T^{i+1} f\| \leq \delta$ and $\|T^i f - T^{i+1} f\| > \delta$ yield an $n \in \mathbb{N}$ such that $\|A_{M(n)}f - A_n f\| < \varepsilon$. The maximum of the two is then an upper bound on an $n \in \mathbb{N}$ such that the result holds.

§4. The principle of convergence for monotone sequences. Another important aspect of the proof is the use of the convergence of the sequence $g_n = u_n - Tu_n$ to the projection of f onto U_f. The convergence of g_n is established in the following way: For every $n \in \mathbb{N}$, the norm $\|g_n\| \leq \|f\|$. Also, $\|g_n\| \leq \|g_{n+1}\|$, so the sequence $a_n = \|g_n\|$ is a bounded monotone sequence of real numbers, and hence it is a Cauchy sequence. A simple argument shows that for every $\delta > 0$ there is a $\gamma > 0$ such that if $|a_i - a_{i+d}| < \gamma$ then $\|g_i - g_{i+d}\| < \delta$, so that we can obtain a rate of convergence for g_n from a rate of convergence for a_n.

The convergence of g_n is used to give an estimate of $\|A_{M(n)}f - A_n f\|$. For any $i, n \in \mathbb{N}$:

$$\|A_{M(n)}f - A_n f\| \leq \|A_{M(n)}(f - g_i) - A_n(f - g_i)\| + \|A_{M(n)}g_i\| + \|A_n g_i\|.$$

One then shows that there is $\delta > 0$ such that if $\|g_i - g_{i+d}\| < \delta$ for all $d > 0$ then $\|A_{M(n)}(f - g_i) - A_n(f - g_i)\| < \varepsilon/2$ for all n. With an upper bound on $\|u_i\|$ (such as the one we sketched in the previous section), we can find an n large enough so that $\|A_{M(n)}g_i\|, \|A_n g_i\| < \varepsilon/4$. Combined, this yields $\|A_{M(n)}f - A_n f\| < \varepsilon$.

Classically, the full convergence of the sequence g_n is established using the principle of convergence for monotone sequences, but constructively, we cannot obtain a full rate of convergence for the g_n, as the convergence of g_n and the $\forall\exists\forall$ version of the Mean Ergodic Theorem are equivalent constructively. However, we may still give a computational interpretation of this particular appeal to the principle of convergence for monotone bounded sequences in the proof of the Mean Ergodic Theorem. As this may sound confusing, here is another explanation: We cannot obtain a computable rate of convergence, but we can give a computational interpretation to the use of its existence as a lemma. This is because a proof of a $\forall\exists$-theorem cannot fully exploit the computational strength of a $\forall\exists\forall$-lemma.

The principle of convergence for monotone bounded sequences is an instance of arithmetical comprehension. The computational interpretation of arithmetical comprehension in general requires bar recursion (see [6]). Even though comprehension is applied to a sequence $a_n = \|g_n\|$ which is explicitly given in the parameters of the proof, a computational interpretation of this instance of comprehension — although simpler — would still vastly increase the complexity of the extracted bounds. Instead we opt for a different approach.

In [3], Kohlenbach describes a technique to eliminate certain instances of arithmetical comprehension without increasing the growth rate (relative to the Grzegorczyk classes of computational complexity) of the bounds extracted by monotone functional interpretation. In this particular case, the informal idea amounts to this: Although we obtain the Mean Ergodic Theorem from the full convergence of g_n, we only need to establish $\|g_i - g_{i+d}\| < \delta$ for a particular d which is expressed in the other parameters and a hypothetical i. The exact d can be read off of the proof — as made explicit in the lemmas stated below — and from this we may form a sequence of non-overlapping intervals $[i_k, i_k + d_k]$ where $i_0 = 0$ and $i_{k+1} = i_k + d_k$. Now recall that there was a $\gamma > 0$ such that $|a_i - a_{i+d}| < \gamma$ implies $\|g_i - g_{i+d}\| < \delta$. Since the sequence a_n is monotone and bounded $|a_{i_k} - a_{j_k}|$ cannot exceed $\gamma > 0$ infinitely often. Hence, $|a_{i_k} - a_{j_k}| < \gamma$ for some $k \in \mathbb{N}$, and the result follows.

In total, this amounts to replacing the use of full convergence with the statement that there is no counterexample to convergence, and the latter has, as sketched, a simple primitive recursive interpretation. Interestingly, the usefulness of this computationally weaker, no-counterexample version of convergence was recently (re-)discovered independently by Terrence Tao [7] under the name "Finite convergence principle". Tao uses this principle to establish finitary, combinatorial proofs of some convergence results in ergodic theory, but without proving explicit bounds. See [8] for details.

In [1], this argument yields the following sequence of lemmas, concerned with establishing $\|A_{M(n)}(f - g_i) - A_n(f - g_i)\| \leq \varepsilon/2$ by making $\|g_i - g_{i+d}\|$ small enough for some i and d:

LEMMA 6. *Let $\varepsilon > 0$, let $d = d(\varepsilon) = \lceil \frac{32\|f\|^4}{\varepsilon^4} \rceil$. Then for every i there is a j in the interval $[i, i + d)$ such that $\|T(f - g_j) - (f - g_j)\| \leq \varepsilon$.*

LEMMA 7. *Let $\varepsilon > 0$, let $n \geq 1$, let $d' = d'(n, \varepsilon) = d(2\varepsilon/n) = \lceil \frac{2n^4\|f\|^4}{\varepsilon^4} \rceil$. Then for every i there is a j in the interval $[i, i + d')$ such that $\|A_n(f - g_j) - (f - g_j)\| \leq \varepsilon$.*

LEMMA 8. *Let $\varepsilon > 0$, let $m \geq 1$, let $d'' = d''(m, \varepsilon) = d'(m, \varepsilon/2) = \lceil \frac{32m^4\|f\|^4}{\varepsilon^4} \rceil$. Furthermore suppose $\|g_i - g_{i+d''}\| \leq \frac{\varepsilon}{4}$. Then for any $n \leq m$, $\|A_n(f - g_i) - (f - g_i)\| \leq \varepsilon$.*

LEMMA 9. *Let $\varepsilon > 0$, let $m \geq 1$, let $d''' = d'''(m, \varepsilon) = d''(m, \varepsilon/2) = \lceil \frac{2^9 m^4\|f\|^4}{\varepsilon^4} \rceil$. Furthermore suppose $\|g_i - g_{i+d''}\| \leq \frac{\varepsilon}{8}$. Then for any $n \leq m$, $\|A_n(f - g_i) - A_m(f - g_i)\| \leq \varepsilon$.*

The last three lemmas follow fairly easy from the first one, repeatedly appealing to the triangle inequality. The first is proved using general properties of the inner product in Hilbert spaces; see [1] for details. The last lemma shows exactly, how the convergence of g_n allows us to bound $\|A_n(f - g_i) - A_m(f - g_i)\|$, where eventually m is replaced by $M(n)$.

In other words, this insight allows us to prove the above lemmas and later the no-counterexample convergence of $A_n f$ already from the no-counterexample convergence of g_n: The expression defining d_k in terms of i_k and the other parameters is nothing but a counterexample function. In this case, there is a nested appeal to the no-counterexample convergence of g_n, resp. a_n, as d_k needs to be big enough enough to ensure that $|a_{d_k} - a_{d_k+1}| < \gamma_k$ where γ_k is given in terms of i_k and the other parameters. Thus s_k depends on i_k, while i_k depends on previous values d_{k-1} and thereby on i_{k-1}. The required $\delta > 0$ such that $\|g_{i_k} - g_{i_k+d_k}\| < \delta$ implies $\|A_{M(n)}f - A_n f\| \leq \varepsilon$, and thus the *number* of intervals $[i_k, i_k + d_k]$ we need to consider, is independent of k though. Thus we consider longer and longer intervals $[i_k, i_k + d_k]$, but only a fixed number of those. For the details, see [1].

§5. Conclusions. In the previous three sections, we illustrated three aspects of a recent application of proof mining to the Mean Ergodic Theorem. At the heart of this proof analysis are proof theoretic ideas that originate from the general logical metatheorems in [2]. On the surface, this analysis is based on ideas not dependent of particular aspects of mathematical logic: (1) making the computational meaning or challenge of a theorem explicit, (2) giving a computational interpretation of an instance of the law of the excluded middle, and (3) refining the appeal, within the proof, to the principle of convergence for monotone bounded sequences of real numbers. In this way, the author hopes that the above examples illustrate that proof mining yields analyses and refinements of mathematical proofs that ought to be considered mathematical even by those who do not consider logic and proof theory proper mathematics. The proof theoretic perspective on mining proofs merely provides a systematic tool to carry out these analyses and to guide intuition where the result may not be obvious otherwise.

Acknowledgements. The author would like to thank the reviewers for making helpful suggestions for improving the presentation in this paper.

REFERENCES

[1] J. Avigad, P. Gerhardy, and H. Towsner, *Local stability of ergodic averages*, to appear in ***Transactions of the American Mathematical Society***.

[2] P. Gerhardy and U. Kohlenbach, *General logical metatheorems for functional analysis*, ***Transactions of the American Mathematical Society***, vol. 360 (2008), no. 5, pp. 2615–2660.

[3] U. Kohlenbach, *Elimination of Skolem functions for monotone formulas in analysis*, ***Archive for Mathematical Logic***, vol. 37 (1998), no. 5-6, pp. 363–390, Logic Colloquium '95 (Haifa).

[4] ———, *Some logical metatheorems with applications in functional analysis*, ***Transactions of the American Mathematical Society***, vol. 357 (2005), no. 1, pp. 89–128.

[5] ———, ***Applied Proof Theory: Proof Interpretations and Their Use in Mathematics***, Springer Monographs in Mathematics, Springer-Verlag, Berlin, 2008.

[6] C. Spector, *Provably recursive functionals of analysis: a consistency proof of analysis by an extension of principles formulated in current intuitionistic mathematics*, ***Proceedings of symposia in pure mathematics***, AMS, Providence, 1962, pp. 1–27.

[7] T. Tao, *Soft analysis, hard analysis, and the finite convergence principle*, On T. Tao's blog: `terrytao.wordpress.com`.

[8] ———, *Norm convergence of multiple ergodic averages for commuting transformations*, ***Ergodic Theory and Dynamical Systems***, vol. 28 (2008), no. 2, pp. 657–688.

[9] A. S. Troelstra (editor), ***Metamathematical Investigation of Intuitionistic Arithmetic and Analysis***, Lecture Notes in Mathematics, vol. 344, Springer-Verlag, Berlin, 1973.

DEPARTMENT OF MATHEMATICS
UNIVERSITY OF OSLO
BLINDERN, N-0316 OSLO, NORWAY
E-mail: Philipp.Gerhardy@gmail.com

CARDINAL STRUCTURE UNDER AD

STEVE JACKSON

Our aim in this paper is to survey the theory of cardinal structure assuming AD, the axiom of determinacy (defined below). We work throughout in the theory ZF + DC + AD.

The axiom of determinacy was introduced by Mycielski and Steinhaus in the early 60's and quickly developed into a powerful tool for extending the ZF results of the classical descriptive set theorists about $\boldsymbol{\Sigma}^1_1$ (analytic) and $\boldsymbol{\Pi}^1_1$ (co-analytic) sets to higher levels of the projective hierarchy. The axiom contradicts the axiom of choice AC, but it was understood that it should be applied in a restricted universe such as $L(\mathbb{R})$ (the smallest model of set theory containing the reals) where it seemed like a reasonable axiom. It wasn't until much later, through the work of Martin, Steel, and Woodin (see [12], [20]), that it was shown that ZFC plus large cardinal assumptions actually imply that AD holds in $L(\mathbb{R})$. The work of the Cabal from the late 60's through the 80's developed an extensive theory of pointclasses and associated properties from this axiom. A good reference for those developments is Moschovakis' book [13] and the Cabal volumes themselves ([8], [9], [18], and [4]).

This theory of the projective sets (and beyond) was largely developed in terms of certain ordinals called the *projective ordinals*, the $\boldsymbol{\delta}^1_n$ (Definition 1.12 below). It was also realized that AD had much to say about the properties of cardinals in general. An example would be Solovay's early result that $\aleph_n$ is singular with cofinality $\aleph_2$ for $n \geq 3$ (see Theorem 8.2 of [8]). The theory at the time, however, was not sufficient to calculate the values of the $\boldsymbol{\delta}^1_n$ for $n \geq 5$ nor to develop the theory of the cardinals past $\aleph_\omega$. Kunen and Martin independently discovered the idea of a *homogeneous tree*. The precise concept and definition was formulated by Kechris. Using this concept, Kunen initiated a program for computing the $\boldsymbol{\delta}^1_n$. The program stalled, however. Kechris' article [8] gives an exposition of what was known at this time. In the early 80's Martin proved a result which in current terminology showed the existence of the *Martin tree*. This generalized the earlier construction of the *Kunen tree* which played an important role in the AD theory of the $\aleph_n$. Building on this and some joint work with Martin, the author computed the $\boldsymbol{\delta}^1_n$ and developed

2000 *Mathematics Subject Classification.* 03E60, 03E55, 03E05.

Logic Colloquium '07
Edited by Françoise Delon, Ulrich Kohlenbach, Penelope Maddy, and Frank Stephan
Lecture Notes in Logic, 35

a theory for analyzing the cardinal structure through the supremum of the $\boldsymbol{\delta}^1_n$ and a ways beyond. In particular, this theory of the projective ordinals generalizes in a straightforward manner to determine the cardinal structure through $\aleph_{\omega_1}$ (the supremum of the $\boldsymbol{\delta}^1_n$ is $\aleph_{\varepsilon_0}$, where $\varepsilon_0 = \sup_n \omega(n)$ where we set $\omega(0) = 1$ and $\omega(n+1) = \omega^{\omega(n)}$).

In hindsight, a key concept that was missing in the earlier theory was the notion of a *description*. These are finitary objects which, roughly speaking, describe how to generate ordinals through the iterated ultrapowers of certain canonical measures. The earlier theory can in fact be viewed as a very simple case of the general theory, using only "trivial descriptions" (which are just integers). This point is explained in more detail in [3]. It turn out that the descriptions in fact completely describe the cardinal structure. One of the main goals of this paper is to introduce the descriptions in as simple a manner as possible, and to show how they determine the cardinal structure. In the complete analysis one must, in addition to defining the descriptions, analyze the measures (countably additive ultrafilters) on the cardinals and establish certain partition properties (the strong partition property on the $\boldsymbol{\delta}^1_{2n+1}$) as well as verify certain inductive hypotheses. These additional arguments, however, use the same notion of description. In this paper, we focus on the descriptions themselves and the cardinal structure they generate. The reader wishing to see the details of the complete inductive analysis can consult [5] for the complete first step of the inductive analysis (including the computation of $\boldsymbol{\delta}^1_5$) and [4] for the general step. The reader can also consult [3] for a more complete exposition of how descriptions are used in other ways (e.g., in analyzing the measures and establishing the partition properties).

Knowledge of these other papers is not necessary for this paper, however. Rather than trying to explain in detail how descriptions are used in the complete inductive analysis, we introduce them here through two "exemplary problems." These two problems can be stated entirely down at the level of the $\aleph_n$, but require the development of the same descriptions needed to compute $\boldsymbol{\delta}^1_5$. As is turns out, $\boldsymbol{\delta}^1_5 = \aleph_{\omega^{\omega^\omega}+1}$, and the descriptions we introduce through these two problems suffice to analyze the cardinal structure below this point. Through this approach, the reader can see clearly the underlying combinatorics in a much simpler setting. Our aim is that reader with a knowledge of basic determinacy theory at the level of [13] and perhaps [8] can follow the main thread of this paper. The reader could consult [10] or [13] for more general background on descriptive set theory.

In §1 we give some background and sketch some of the "global" theory of AD. By this we mean the more general results that hold for all pointclasses and are independent of the more detailed analysis using the descriptions. This section is definitely of a survey nature and is included for the sake of completeness. In §2 we give a brief sketch of the overall plan of how the inductive analysis of the projective ordinals goes. This section is also included for the sake of completeness and to give the reader a sense of how descriptions fit into the bigger picture. The later sections do not specifically depend on these sections, and the reader wishing to quickly see the notion of description can skip them. In §3 we recall some basic definitions and facts about partition relations. Partition relations play a central role in how descriptions are used to generate ordinals. Our approach in this paper is to assume the strong partition property on the odd projective ordinals, the $\boldsymbol{\delta}^1_{2n+1}$, and proceed straight to the descriptions. In the actual full analysis, the descriptions are also used to prove the partition relations. In §4 we introduce the two canonical families of measure W_1^m and S_1^m and the first exemplary problem: compute the ultrapower $j_{S_1^m}(\omega_n)$ of the cardinal ω_n by the measure S_1^m. Solving this problem will only need the introduction of the "trivial" descriptions (actually slight generalizations of these). In §5 we consider the second exemplary problem: compute the iterated ultrapower $j_{S_1^{m_1}} \circ j_{S_1^{m_2}} \circ \cdots \circ j_{S_1^{m_t}}(\omega_n)$. The answer is actually immediate from the solution to the first problem, however we can describe the cardinal structure in the iterated ultrapower by introducing the next level of description. These are the same objects one uses in the full analysis to compute $\boldsymbol{\delta}^1_5$, prove the strong partition relation on $\boldsymbol{\delta}^1_3$, and the weak partition relation on $\boldsymbol{\delta}^1_5$ (which is the first step of the inductive analysis). We carry out an example in computational detail. In §6 we sketch how these descriptions actually are used to generate the cardinal structure between $\boldsymbol{\delta}^1_3 = \aleph_{\omega+1}$ and $\boldsymbol{\delta}^1_5 = \aleph_{\omega^{\omega^\omega}}$. We also give a brief hint at what goes into the more general definition of description. In §7 we give an entirely different mechanism which can be used to present the cardinal structure below the projective ordinals. This is done via a certain algebra which is defined directly and does not need descriptions. Descriptions are needed, however, to prove that this alternate formulation works (these proofs are not given here). This notational framework involves joint work with B. Löwe [7] and extends earlier joint work of the author and F. Khafizov [6]. Thus, although the proofs that this method works are quite involved, this gives a direct and self-contained mechanism for understanding the cardinal structure. This should make the techniques of this area accessible to a wider audience.

Finally, in §8 we present as an application of the theory of cardinal structure under AD a result concerning the collapse of cardinals from $L(\mathbb{R})$ to V. Namely, we sketch a proof of the fact that assuming large cardinals in V plus the saturation of the non-stationary ideal on ω_1, some regular cardinal in $L(\mathbb{R})$ below $(\aleph_{\omega_2})^{L(\mathbb{R})}$ is collapsed in V. The proof we present here is only a sketch, as complete details will be given elsewhere. This section is not intended to be self-contained, and is included to give an example of how the theory of the cardinal structure under AD can be used to get results in the ZFC world.

§1. Basic concepts. Though our intention is to eventually specialize to results of a "local" nature, that is, pertaining to the smaller cardinals for which the inductive combinatorial methods work, we present as background here AD results of a more general nature. There results are "global" in nature in that they hold throughout the Wadge hierarchy under AD. The reader already comfortable with the basics of AD theory can skip this section, or skim it to see our (mostly standard) notation.

Let ω^ω denote the Baire space with the usual topology, i.e., the product of the discrete topology on ω. ω^ω is homeomorphic to the set of irrationals (as a subspace of $\mathbb{R}$), and as is customary, we often call ω^ω the "reals."

We let $\boldsymbol{\Sigma}^0_1$ denote the collection of open sets, and $\boldsymbol{\Pi}^0_1$ the collection of closed sets in ω^ω. The levels of the Borel hierarchy are defined as usual by $A \in \boldsymbol{\Sigma}^0_\alpha$ (for $\alpha < \omega_1$) iff $A = \bigcup_{n \in \omega} A_n$ where $A_n \in \boldsymbol{\Pi}^0_{\alpha_n}$ for some $\alpha_n < \alpha$. Also, $A \in \boldsymbol{\Pi}^0_\alpha$ iff $A^c \in \boldsymbol{\Sigma}^0_\alpha$ (we use A^c to denote $\omega^\omega - A$). The collection of Borel sets is $\bigcup_{\alpha < \omega_1} \boldsymbol{\Sigma}^0_\alpha$. The *analytic* or $\boldsymbol{\Sigma}^1_1$ sets are the continuous images of the closed sets (from ω^ω). Equivalently, they are the sets of the form $A(x) \leftrightarrow \exists y\, F(x, y)$, where $F \subseteq \omega^\omega \times \omega^\omega$ is closed. We abbreviate this by writing $\boldsymbol{\Sigma}^1_1 = \exists^{\omega^\omega} \boldsymbol{\Pi}^0_1$. The *co-analytic* or $\boldsymbol{\Pi}^1_1$ sets are the complements of the analytic sets, which we abbreviate by $\boldsymbol{\Pi}^1_1 = (\boldsymbol{\Sigma}^1_1)^c$. A set is $\boldsymbol{\Delta}^1_1$ if it is both $\boldsymbol{\Sigma}^1_1$ and $\boldsymbol{\Pi}^1_1$, which we abbreviate $\boldsymbol{\Delta}^1_1 = \boldsymbol{\Sigma}^1_1 \cap \boldsymbol{\Pi}^1_1$. Suslin's theorem says that $\boldsymbol{\Delta}^1_1$ coincides with the collection of Borel sets. We extend these sets to form the *projective hierarchy* as follows. We set $\boldsymbol{\Sigma}^1_{n+1} = \exists^{\omega^\omega} \boldsymbol{\Pi}^1_n$, $\boldsymbol{\Pi}^1_{n+1} = (\boldsymbol{\Sigma}^1_{n+1})^c$, and $\boldsymbol{\Delta}^1_{n+1} = \boldsymbol{\Sigma}^1_{n+1} \cap \boldsymbol{\Pi}^1_{n+1}$.

The classical descriptive set theorists of the 20's through the 40's developed a theory of the $\boldsymbol{\Sigma}^1_1$ and $\boldsymbol{\Pi}^1_1$ sets in ZF. To extend the theory to higher levels requires stronger set theoretic assumptions. The axiom we consider is the axiom of determinacy, AD.

By AD we mean the axiom that every two player integer game is determined. More precisely, For $A \subseteq \omega^\omega$, we have the game G_A:

I	$x(0)$		$x(2)$		$x(4)$		$\dots$
II		$x(1)$		$x(3)$		$x(5)$	$\dots$

We say I wins the run iff $x \in A$, where

$$x = (x(0), x(1), x(2), \dots) \in \omega^\omega.$$

A *strategy* (for an integer game) for I (or II) is a function σ from the sequences $s \in \omega^{<\omega}$ of even (odd) length to ω. We say σ is a *winning strategy* for I (or II) if for all runs $x \in \omega^\omega$ of the game where I (or II) has followed σ, we have $x \in A$ (respectively $x \notin A$).

If σ is a strategy for I (or II), and $x = (x(1), x(3), \dots) \in \omega^\omega$, let $\sigma(x) \in \omega^\omega$ be the result of following σ against II's play of x. Note that $\sigma[\omega^\omega]$ is a $\mathbf{\Sigma}^1_1$ subset of ω^ω.

We employ variations of this notation, e.g., we might describe a game by saying I plays out x, II plays out y. Here we might use $\sigma(y)$ to denote just I's response x following σ against II's play of y. The meaning should be clear from the context.

These basic concepts generalize in natural ways to games on sets X other than ω. If X cannot be wellorderded in ZF then we usually consider *quai-strategies*, which are functions which assign a non-empty set $\sigma(s) \subseteq X$ of possible moves to each $s \in X^{<\omega}$ of appropriate parity length. Let AD_X denote the axiom that all games on the set X are determined. Quasi-strategies are used, for example, in considering AD_R, which is a strong form of determinacy. It implies, for example, that all sets of reals are Suslin (see below). For the purposes of this paper, however, all games considered will be integer games, so this concept will not be used.

If $\mathbf{\Gamma}$ is a collection of subsets of ω^ω, we write $\mathbf{\Gamma}$-determinacy to denote that G_A is determined for all $A \in \mathbf{\Gamma}$. Although the determinacy of all projective games suffices to establish much of the general theory of the projective sets, it does not suffice for the methods of the inductive analysis (using descriptions). For this we must assume full AD.

Although AD contradicts AC, it is consistent with weaker forms of choice. Recall DC (Dependent Choice) is the following axiom: For any set X, If $R \subseteq X^{<\omega}$ and

$$\forall (x_0, \dots x_{n-1}) \in R \, \exists x_n \, (x_1, \dots, x_{n-1}, x_n) \in R,$$

then $\exists \vec{x} \in X^\omega \, \forall n \, (\vec{x} \restriction n \in R)$. This is equivalent to asserting that every ill-founded relation has an infinite descending sequence. DC adds no consistency strength to the theory ZF + AD since Kechris has shown that AD implies that DC holds in $L(\mathbb{R})$. DC is implicitly used in many arguments, and we assume it in our background theory.

We next list some background facts about AD. Points (1), (3) and (5) are probably folklore. For (2) see 6A.7 and 6A.8 of [13]. A proof of (6) can be found in [15]. For (7) see for example 6A.16 and 6A.17 of [13]. A proof of (8) is given in Theorem 8.2 of [8].

FACT 1.1. (1) AD contradicts ZFC but is consistent with restricted forms of determinacy, e.g., projective determinacy PD or $\mathsf{AD}^{L(\mathbb{R})}$ (the determinacy of games in $L(\mathbb{R})$, see below for the definition).

(2) AD is equivalent to AD_X, where X is any countable set with at least two elements.

(3) AD_{ω_1} is inconsistent.

(4) $\mathsf{AD}_{\mathbb{R}}$ is a (presumably) consistent strengthening of AD. It is equivalent (Martin-Woodin) to AD+ every set has a scale.
(5) $\mathsf{AD} \Rightarrow \mathsf{AD}^{L(\mathbb{R})}$.
(6) DC is independent of even $\mathsf{AD}_{\mathbb{R}}$ (Solovay), but on the other hand $\mathsf{AD} \Rightarrow \mathsf{DC}^{L(\mathbb{R})}$ (Kechris).
(7) AD implies regularity properties for sets of reals, e.g., every set of reals has the perfect set property, the Baire property, is measurable, is Ramsey. The determinacy needed is local, e.g., $\boldsymbol{\Pi}^1_1$-determinacy implies perfect set property for $\boldsymbol{\Sigma}^1_2$.
(8) Under AD, successor cardinals need not be regular (but $\mathrm{cof}(\kappa^+) > \omega$ assuming countable choice).

The results on cardinal structure we describe in this paper are all done just assuming AD, without reference to a particular model. Nevertheless, it is appropriate to mention the natural inner model $L(\mathbb{R})$ of AD.

Recall that $L(\mathbb{R})$ is the smallest inner-model (i.e., transitive, proper class model) containing the reals $\mathbb{R}$ (or ω^ω). It is defined through a hierarchy similar to L, except at the bottom we throw in all the reals (or equivalently $V_{\omega+1}$)

$$J_0(\mathbb{R}) = V_{\omega+1},$$
$$J_\alpha(\mathbb{R}) = \bigcup_{\beta<\alpha} J_\beta(\mathbb{R}) \text{ for } \alpha \text{ limit},$$
$$J_{\alpha+1}(\mathbb{R}) = \mathrm{rud}(J_\alpha(\mathbb{R}) \cup \{J_\alpha(\mathbb{R})\}).$$

Here $\mathrm{rud}(X) = \bigcup_n S^n(X)$, where $S(X)$ is the result of applying a certain finite list of "rudimentary" functions to X. The reader can consult [18] for further details.

Each level $J_\alpha(\mathbb{R})$, in fact each sub-level $S^n(J_\alpha(\mathbb{R}) \cup \{J_\alpha(\mathbb{R})\})$, is transitive. It follows easily from the definition that every set in $L(\mathbb{R})$ is ordinal definable from a real. In fact, there is uniformly in α a $\Sigma_1(J_\alpha(\mathbb{R}))$ map from $\omega\alpha^{<\omega} \times \mathbb{R}$ onto $J_\alpha(\mathbb{R})$. Steel [18] has developed the "global" scale theory of $L(\mathbb{R})$ assuming AD, using among other things a generalization of Jensen's fine structure theory of L. In fact, most of this scale theory can be obtained from just AD, as shown in [3]. For the purposes of this paper, we will not need to consider the model $L(\mathbb{R})$, although for some arguments it does become necessary to assume $V = L(\mathbb{R})$, or at least some strengthening of AD such as Woodin's AD^+ axiom.

1.1. Global results: separation and prewellordering. We first make a "first pass" at the global theory of AD, sketching the theory of pointclasses it gives. This is largely given in terms of the separation, reduction, and most importantly, the prewellordering property. After this, we make a second pass, describing the theory of scales which AD gives (the scale property is a strengthening of the prewellordering property).

DEFINITION 1.2. For $A, B \subseteq \omega^\omega$, we say that A is *Wadge reducible* to B, $A \leq_w B$, if there is a continuous function $f: \omega^\omega \to \omega^\omega$ such that $A = f^{-1}(B)$, i.e., $x \in A$ iff $f(x) \in B$.

For $A, B \subseteq \omega^\omega$, we have the basic Wadge game $G_W(A, B)$: I plays out $x \in \omega^\omega$, II plays out $y \in \omega^\omega$, and II wins the run iff $(x \in A \leftrightarrow y \in B)$. It is not hard to see (Wadge's lemma) that if II has a winning strategy then $A \leq_W B$, and if I has a winning strategy then $B \leq_W A^c$. It is therefore natural to consider pairs $\{A, A^c\}$ of sets together with their complements. We say $\{A, A^c\} \leq_W \{B, B^c\}$ if $A \leq_W B$ or $A \leq_W B^c$. We let $[\{A, A^c\}]$ denote the equivalence class under the relation $\{A, A^c\} \equiv_W \{B, B^c\}$ iff $\{A, A^c\} \leq_W \{B, B^c\}$ and $\{B, B^c\} \leq_W \{A, A^c\}$.

A basic determinacy fact is the following.

THEOREM 1.3 (Martin-Monk). *$\leq_W$ is a wellordering on the equivalence classes* $[\{A, A^c\}]$.

We say $A \subseteq \omega^\omega$ is *selfdual* if $A \equiv_W A^c$. We write $\{A\}$ in the case of a selfdual degree.

In view of the Martin-Monk result, we can assign an ordinal rank to every set of reals as follows.

DEFINITION 1.4. If $A \subseteq \omega^\omega$, then $o(A)$ denotes the rank of $\{A, A^c\}$ in $\leq_W$.

Thus, assuming AD the sets of reals are stratified into a natural hierarchy, the Wadge hierarchy. This opens up the possibility of an inductive approach to analyzing the sets of reals,

DEFINITION 1.5. We define the ordinal Θ by $\Theta = \sup\{o(A): A \subseteq \omega^\omega\}$.

Thus, Θ is the length of the entire Wadge hierarchy of sets of reals. It can also be characterized as the supremum of the lengths of the prewellorderings of ω^ω.

The following result of [19] gives the picture of the Wadge degrees.

THEOREM 1.6. *The selfdual and non-selfdual Wadge degrees alternate. At limit ordinals of cofinality ω there is a selfdual degree and at limit ordinals of uncountable cofinality there is a non-selfdual degree.*

The basic objects of interest in descriptive set theory are the *pointclasses*.

DEFINITION 1.7. A pointclass is a collection $\boldsymbol{\Gamma} \subseteq \mathcal{P}(\omega^\omega)$ closed under Wadge reduction.

For $\boldsymbol{\Gamma}$ a pointclass, we let $o(\boldsymbol{\Gamma}) = \sup\{o(A): A \in \boldsymbol{\Gamma}\}$. We say a pointclass $\boldsymbol{\Gamma}$ is selfdual if $A \in \boldsymbol{\Gamma} \Rightarrow A^c \in \boldsymbol{\Gamma}$. Otherwise $\boldsymbol{\Gamma}$ is non-selfdual. We usually use $\boldsymbol{\Gamma}$ to denote a non-selfdual pointclass, and $\boldsymbol{\Delta}$ to denote a selfdual one.

The weakest of the structural properties for pointclasses is the *separation* property.

DEFINITION 1.8. $\mathbf{\Gamma}$ has the *separation* property, $\text{sep}(\mathbf{\Gamma})$, if whenever $A, B \in \mathbf{\Gamma}$ and $A \cap B = \emptyset$, then there is a $C \in \mathbf{\Delta} = \mathbf{\Gamma} \cap \check{\mathbf{\Gamma}}$ with $A \subseteq C$, $B \cap C = \emptyset$.

THEOREM 1.9 (Steel-Van Wesep, [19], [17]). *For any non-selfdual* $\mathbf{\Gamma}$, *exactly one of* $\text{sep}(\mathbf{\Gamma})$, $\text{sep}(\check{\mathbf{\Gamma}})$ *holds.*

Thus, the separation property can be used to distinguish $\mathbf{\Gamma}$ from $\check{\mathbf{\Gamma}}$ for a general non-selfdual pointclass.

The notions of norms and prewellorderings are fundamental in descriptive set theory. The concepts are interchangeable as they refer to essentially the same thing.

DEFINITION 1.10. A *prewellordering* $\preceq$ on a set A is a connected, reflexive, transitive binary relation on A, whose strict part $\prec$ is wellfounded ($\prec$ is defined by $x \prec y$ iff $x \preceq y$ and $\neg y \preceq x$). A *norm* φ on a set $A \subseteq$ is a map $\varphi \colon A \to \text{On}$. We say φ is *regular* if $\text{ran}(\varphi) \in \text{On}$.

A norm φ on A induces a prewellordering on A given by $x \preceq y$ iff $\varphi(x) \leq \varphi(y)$. Conversely, a prewellordering induces a wellordering on the equivalence classes given by:

$$[x] = \{y \colon x \preceq y \wedge y \preceq x\}.$$

The corresponding norm is $\varphi(x) = \text{rank of } [x]$.

The Wadge ordinal $o(\mathbf{\Gamma})$ defined above is one ordinal associated to a pointclass. The next definition associates another.

DEFINITION 1.11. $\delta(\mathbf{\Gamma}) =$ the supremum of the lengths of the $\mathbf{\Delta}$ prewellorderings of ω^ω, where $\mathbf{\Delta} = \mathbf{\Gamma} \cap \check{\mathbf{\Gamma}}$.

The two ordinals $o(\mathbf{\Gamma})$, $\delta(\mathbf{\Gamma})$ do not in general coincide. However, a result of [11] (see Lemma 2.3.1) says that if $\mathbf{\Delta}$ is selfdual and closed under $\wedge$ and $\exists^{\omega^\omega}$ (and hence also $\vee$ and $\forall^{\omega^\omega}$), then $o(\mathbf{\Delta}) = \delta(\mathbf{\Delta})$. An example of such a $\mathbf{\Delta}$ is the collection of all projective sets. Thus, the difference between the two ordinals is localized to a projective hierarchy.

The following special case of Definition 1.11 is important enough to warrant giving separately.

DEFINITION 1.12. The projective ordinal $\boldsymbol{\delta}^1_n$ is defined by $\boldsymbol{\delta}^1_n = \delta(\mathbf{\Sigma}^1_n) =$ the supremun of the lengths of the $\mathbf{\Delta}^1_n$ prewellorderings of ω^ω.

The projective ordinals are important because the theory of the projective sets is largely given in term of these ordinals. The reader can consult [13] or [8] for these results. We mention in Theorem 1.23 the earlier known results about the projective ordinals.

The next definition introduces definability into the notion of a norm,

DEFINITION 1.13. A $\mathbf{\Gamma}$-norm φ on $A \subseteq \omega^\omega$ is a norm such the relations

$$x <^* y \leftrightarrow x \in A \wedge (y \notin A \vee (y \in A \wedge \varphi(x) < \varphi(y))),$$
$$x \leq^* y \leftrightarrow x \in A \wedge (y \notin A \vee (y \in A \wedge \varphi(x) \leq \varphi(y)))$$

are both in $\mathbf{\Gamma}$.

If φ is a regular $\mathbf{\Gamma}$-norm on A of length λ, then the norm gives a representation of A as an increasing union $A = \bigcup_{\alpha<\lambda} A_{\leq\alpha}$, where $A_{\leq\alpha} = \{x \in A : \varphi(x) \leq \alpha\}$. It is easy to see from the definition that each $A_{\leq\alpha} \in \mathbf{\Delta} = \mathbf{\Gamma} \cap \check{\mathbf{\Gamma}}$ To see this, note that if $y \in A$ with $\varphi(y) = \alpha$, then $A_{\leq\alpha} = \{x : x \leq^* y\} = \{x : \neg(y <^* x)\}$.

The next definition is an important structural property of pointclasses. Though not as strong as the scale property (considered below), it is more general. For example, in $L(\mathbb{R})$ there is a largest scaled pointclass, namely $\mathbf{\Sigma}^2_1$, but the pointclasses with the prewellordering property are cofinal in the Wadge degrees.

DEFINITION 1.14. $\mathbf{\Gamma}$ has the *prewellordering property*, pwo($\mathbf{\Gamma}$), if every $A \in \mathbf{\Gamma}$ admits a $\mathbf{\Gamma}$-norm.

For φ a norm on A, let $A_\alpha = \{x \in A : \varphi(x) = \alpha\}$. If $\mathbf{\Delta}$ is closed under $\wedge$ (and hence also $\vee$), we clearly have that $A_\alpha \in \mathbf{\Delta}$.

The initial segment $\preceq_\alpha$ of the prewellordering $\prec$ associated to the $\mathbf{\Gamma}$-norm φ can also be computed as:

$$\begin{aligned} x \preceq_\alpha y &\leftrightarrow x, y \in A_{\leq\alpha} \wedge (x \leq^* y) \\ &\leftrightarrow x, y \in A_{\leq\alpha} \wedge \neg(y <^* x). \end{aligned}$$

So, $\preceq_\alpha \in \mathbf{\Delta}$ if $\mathbf{\Gamma}$ is closed under $\wedge$, $\vee$. In view of this, if $\mathbf{\Gamma}$ is closed under $\wedge$, $\vee$, and φ is a regular $\mathbf{\Gamma}$-norm then $|\varphi| \leq \delta(\mathbf{\Gamma})$, where $|\varphi|$ denotes the length of φ.

For classes resembling $\mathbf{\Pi}^1_1$, a $\mathbf{\Gamma}$-norm on a complete $\mathbf{\Gamma}$ set must have norm the maximum possible length according to the next fact (a proof can be found in 4C.14 of [13]).

FACT 1.15. If $\mathbf{\Gamma}$ is closed under $\forall^{\omega^\omega}$, $\wedge$, $\vee$, and φ is a $\mathbf{\Gamma}$ norm on a $\mathbf{\Gamma}$-complete set, then $|\varphi| = \delta(\mathbf{\Gamma})$.

The pointclasses of primary interest are those closed under either existential or universal real quantification (or both). These generalize the familiar *projective sets* defined below. We make this into a definition.

DEFINITION 1.16. $\mathbf{\Gamma}$ is a *Lévy class* if it is a non-selfdual pointclass closed under either $\exists^{\omega^\omega}$ or $\forall^{\omega^\omega}$ (or both).

We let $\mathbf{\Sigma}^1_\alpha$ enumerate the Lévy classes closed under $\exists^{\omega^\omega}$ and $\mathbf{\Pi}^1_\alpha$ those closed under $\forall^{\omega^\omega}$. So, $\mathbf{\Sigma}^1_0 =$ open, $\mathbf{\Sigma}^1_1$ is the collection of analytic sets, and $\mathbf{\Sigma}^1_n$ agrees with the definition given in Definition 1.12.

The next result from [16] shows that the prewellordering property is very general.

THEOREM 1.17 (Steel). *For every Lévy class* $\boldsymbol{\Gamma}$, *either* $\mathrm{pwo}(\boldsymbol{\Gamma})$ *or* $\mathrm{pwo}(\check{\boldsymbol{\Gamma}})$.

Steel's analysis gives more information. It shows that the Lévy classes fall into projective-like hierarchies. The first such hierarchy is the usual projective hierarchy. These hierarchies can be classified into four basic types.

Following Steel, let $C \subseteq \Theta$ be the c.u.b. set of limit α such that $\Lambda_\alpha \doteq \{A\colon o(A) < \alpha\}$ is closed under quantifiers. For $\boldsymbol{\Gamma}$ a Lévy class, let α be the largest element of C such that $\Lambda_\alpha \subseteq \boldsymbol{\Gamma}$. Then $\boldsymbol{\Gamma}$ is in a projective-like hierarchy over $\Lambda = \Lambda_\alpha$. We define this hierarchy as follows. If $\mathrm{cof}(\alpha) = \omega$, let $\boldsymbol{\Sigma}_0 = \bigcup_\omega \Lambda$ be the collection of countable unions of sets in Λ. Then define the $\boldsymbol{\Sigma}_n, \boldsymbol{\Pi}_n$ from $\boldsymbol{\Sigma}_0$ as usual by applying quantifiers. We call this case a type I hierarchy. If $\mathrm{cof}(\alpha) > \omega$, by [16] there is a non-selfdual pointclass $\boldsymbol{\Gamma}_0$ closed under $\forall^{\omega^\omega}$ with $o(\boldsymbol{\Gamma}_0) = \alpha$. From [16] we also have $\mathrm{sep}(\check{\boldsymbol{\Gamma}}_0)$. We call $\boldsymbol{\Gamma}_0$ a *Steel pointclass*. We assume for the current case that $\boldsymbol{\Gamma}_0$ is not closed under $\exists^{\omega^\omega}$. In this case we generate the hierarchy over $\boldsymbol{\Gamma}_0$ by applying quantifiers as usual. It is customary to split this case into two cases, called type II and type III hierarchies, according to whether $\boldsymbol{\Gamma}_0$ is closed under $\vee$ (it is automatically closed under $\wedge$ since it is closed under $\forall^{\omega^\omega}$). The last case, type IV, is when $\boldsymbol{\Gamma}_0$ is closed under quantifiers. In this case we let $\boldsymbol{\Sigma}_0 = \exists^{\omega^\omega}(\boldsymbol{\Gamma}_0 \wedge \check{\boldsymbol{\Gamma}}_0)$ (here $\boldsymbol{\Gamma}_0 \wedge \check{\boldsymbol{\Gamma}}_0$ is the pointclass of sets which can be written as an intersection of a $\boldsymbol{\Gamma}_0$ set and a $\check{\boldsymbol{\Gamma}}_0$ set). We then apply quantifiers to generate the hierarchy.

We summarize this discussion as well as the prewellordering analysis (from [16]) in the following list. The ordinal α in the following cases is the ordinal from the previous paragraph.

Types of Projective Hierarchies.

- Type I. $\mathrm{cof}(\alpha) = \omega$.
  Let $\boldsymbol{\Sigma}_0 = \bigcup_\omega \Lambda$. Then $\mathrm{pwo}(\boldsymbol{\Sigma}_0)$, $\mathrm{pwo}(\boldsymbol{\Pi}_{2n+1})$, $\mathrm{pwo}(\boldsymbol{\Sigma}_{2n+2})$.
- Type II. $\mathrm{cof}(\alpha) > \omega$ and $\boldsymbol{\Gamma}_0$ is not closed under $\vee$.
  Let $\boldsymbol{\Sigma}_0 = \exists^{\omega^\omega}\boldsymbol{\Gamma}_0$. Then $\mathrm{pwo}(\boldsymbol{\Gamma}_0)$, $\mathrm{pwo}(\boldsymbol{\Sigma}_0)$, $\mathrm{pwo}(\boldsymbol{\Pi}_{2n+1})$, $\mathrm{pwo}(\boldsymbol{\Sigma}_{2n+2})$.
- Type III. $\mathrm{cof}(\alpha) > \omega$, $\boldsymbol{\Gamma}_0$ is closed under $\vee$ but not $\exists^{\omega^\omega}$.
  Same conclusion as in type 2.
- Type IV. $\boldsymbol{\Gamma}_0$ is closed under real quantifiers.
  Then $\mathrm{pwo}(\boldsymbol{\Gamma}_0)$. Let $\boldsymbol{\Sigma}_0 = \exists^{\omega^\omega}(\boldsymbol{\Gamma}_0 \wedge \check{\boldsymbol{\Gamma}}_0)$. Then $\mathrm{pwo}(\boldsymbol{\Sigma}_0)$, $\mathrm{pwo}(\boldsymbol{\Pi}_{2n+1})$, $\mathrm{pwo}(\boldsymbol{\Sigma}_{2n+2})$.

1.2. Second pass: scales and Suslin cardinals. We now consider the notion of scales and the scale property. These are the basic structural notions in descriptive set theory. This notion was isolated by Moschovakis, and has its roots in the Novikov-Kondo proof of $\boldsymbol{\Pi}^1_1$ uniformization. We first introduce some standard notation.

A *tree* on a set X is a subset of $X^{<\omega}$ closed under initial segment. We frequently identify trees on $X \times Y$ with subsets of $X^{<\omega} \times Y^{<\omega}$. If T is a tree on X, then

$$[T] = \{f \in X^\omega : \forall n\, f \restriction n \in T\}.$$

If T is a tree on $X \times Y$ then we define its projection by:

$$\begin{aligned} p[T] &= \{f \in X^\omega : \exists g \in Y^\omega\ (f, g) \in [T]\} \\ &= \{f \in X^\omega : T_f \text{ is illfounded}\}, \end{aligned}$$

where $T_f = \{s \in Y^{<\omega} : (f \restriction \mathrm{lh}(s), s) \in T\}$ is the section of the tree at f.

Definition 1.18. $A \subseteq \omega^\omega$ is κ-Suslin if $A = p[T]$ for some tree T on $\omega \times \kappa$. $S(\kappa)$ denotes the pointclass of κ-Suslin sets. κ is a *Suslin cardinal* if $S(\kappa) - \bigcup_{\lambda<\kappa} S(\lambda) \neq \emptyset$.

We note that it makes sense to speak of κ-Suslin subsets of λ^ω for $\lambda \in \mathrm{On}$ as well. We have the following general closure properties of the κ-Suslin sets.

Fact 1.19 (Kechris [9]). For any Suslin cardinal κ, $S(\kappa)$ is closed under countable unions and intersection, $\exists^{\omega^\omega}$ and is non-selfdual.

Suslin representations are essentially the same thing as scales, which we introduce next.

Definition 1.20. A *semi-scale* on $A \subseteq \omega^\omega$ is a sequence of norms $\{\varphi_n\}_{n\in\omega}$ such that if $x_n \in A$, $x_n \to x$, and for each n, $\varphi_n(x_m)$ is eventually equal to some λ_n, then $x \in A$. If in addition we have $\varphi_n(x) \leq \lambda_n$ for all n, then the $\{\varphi_n\}$ is said to be a *scale*.

The last property is called the lower semi-continuity property of scales. As with norms and prewellorderings, we can add definability of the norms into the definition which gives the following.

Definition 1.21. A $\boldsymbol{\Gamma}$-scale on a set A is a scale $\{\varphi_n\}$ all of whose norms φ_n are $\boldsymbol{\Gamma}$-norms. A pointclass $\boldsymbol{\Gamma}$ has the *scale property* if every $A \in \boldsymbol{\Gamma}$ has a $\boldsymbol{\Gamma}$-scale.

The scale property implies both the prewellordering property as well as the existence of Suslin representations. The notion of semi-scale can also be thought of as generalizing the definition of a closed set. The following fact (see e.g. [3] for a proof) give the equivalence of semi-scales, scales, and Suslin representations.

Fact 1.22. For all cardinals κ, A is κ-Suslin iff A admits a semi-scale with norms into κ iff A admits a scale with norms into κ.

If $\vec{\varphi}$ is a semi-scale, the corresponding Suslin representation is given by the tree

$$T_{\vec{\varphi}} = \{((a_0, \dots, a_{n-1}), (\alpha_0, \dots, \alpha_{n-1})) \colon \exists x \in A\, (x \restriction n = (a_0, \dots, a_{n-1}) \\ \wedge \forall i < n\; \varphi_i(x) = \alpha_i\}.$$

Note that if $\{\varphi_n\}$ is a $\mathbf{\Gamma}$-scale on A, then the corresponding Suslin representation will be a tree on $\omega \times \delta(\mathbf{\Gamma})$, as each $\mathbf{\Gamma}$-norm has length $\leq \delta(\mathbf{\Gamma})$.

On the other hand, if $A = p[T]$, let for $x \in A$, $\varphi_n(x) = n^{\text{th}}$ coordinate of the leftmost branch of T_x. Then $\vec{\varphi}_T = \{\varphi_n\}$ is a semi-scale on A. Let $\psi_n = \langle \varphi_0(x), \dots \varphi_{n-1}(x) \rangle$ be the rank of $(\varphi_0(x), \dots \varphi_{n-1}(x))$ in the lexicographic ordering on κ^n. Then $\vec{\psi}$ is a scale on A. [with a little adjustment to T we can make the ψ_i map into κ.]

We note that if $\vec{\varphi}$ is a scale, then $\vec{\varphi}(T_{\vec{\varphi}}) = \vec{\varphi}$. But, not every tree is the tree of a scale, so we only have $T_{\vec{\varphi}(T)} \subseteq T$.

The Moschovakis periodicity theorems [13] show that under projective determinacy the pointclasses $\mathbf{\Pi}^1_{2n+1}$, $\mathbf{\Sigma}^1_{2n}$ have the scale property. In particular, all $\mathbf{\Pi}^1_{2n+1}$ sets are $\boldsymbol{\delta}^1_{2n+1}$-Suslin.

Using arguments just based upon the scale property one can establish some general properties about the projective ordinals and the Suslin cardinals below the projective ordinals. We summarize these earlier (pre-description theory) results in the next theorem.

THEOREM 1.23. *The projective ordinals $\boldsymbol{\delta}^1_n$ and Suslin cardinals below their supremum satisfy the following.*

(1) *All the $\boldsymbol{\delta}^1_n$ are regular, in fact measurable.*
(2) $\boldsymbol{\delta}^1_{2n+2} = (\boldsymbol{\delta}^1_{2n+1})^+$.
(3) *Each odd projective ordinal is of the form $\boldsymbol{\delta}^1_{2n+1} = (\lambda_{2n+1})^+$, where λ_{2n+1} is a cardinal of cofinality ω.*
(4) *The Suslin cardinals below the projective ordinals are the λ_{2n+1} and $\boldsymbol{\delta}^1_{2n+1}$. Also, $S(\lambda_{2n+1}) = \mathbf{\Sigma}^1_{2n+1}$ and $S(\boldsymbol{\delta}^1_{2n+1}) = \mathbf{\Sigma}^1_{2n+2}$.*

The projective ordinals were first defined by Moschovakis. Property (1) is due to Moschovakis for regularity (see Theorem 7D.11 of [13]). Solovay showed ω_1 and ω_2 are measurable and Kunen showed extended this to all of the $\boldsymbol{\delta}^1_n$. Property (2) is due to Martin and Kunen independently. Property (3) is due to Kechris (see [8] for proofs of (2)-(4)). It was also known that $\boldsymbol{\delta}^1_1 = \omega_1$, $\boldsymbol{\delta}^1_2 = \omega_2$ (classical), $\boldsymbol{\delta}^1_3 = \omega_{\omega+1}$ (Martin-Solovay), and $\boldsymbol{\delta}^1_4 = \omega_{\omega+2}$ (Martin-Kunen, see 8H.11 of [13] for proofs of the last two). Also Martin showed ω_1 has the strong partition property (defined in §3), and Kunen showed that $\boldsymbol{\delta}^1_3$ has the weak partition property. As we mentioned earlier, all of these results can be done from the description point of view, using only "trivial descriptions" which we discuss in §4 (for other reasons). The reader wishing to see this theory redone from this point of view can consult [3].

Past the projective hierarchy the same basic pattern of Theorem 1.23 continues. Before stating this, we first state the following important theorem of Steel and Woodin. Steel first showed the result under the assumption $V = L(\mathbb{R})$ (see [18]), and Woodin then gave the general result by a completely different argument.

THEOREM 1.24 (Steel-Woodin). *Assume* AD. *The Suslin cardinals are closed below their supremum.*

The Suslin classes $S(\kappa)$ are Lévy classes, and so we can attempt to classify them using the framework of the projective hierarchy types stated at the end of §1.1. Steel [18] gives this complete classification of Suslin cardinals and the scale property assuming AD $+ V = L(\mathbb{R})$. Assuming just AD, Martin developed a method for analyzing the next Suslin cardinal. Using this method one can recover almost all, but not completely, all of these results. In particular, this suffices to analyze the Suslin cardinals (the one missing fact is mentioned below). Martin's method and the AD analysis of the Suslin cardinals is presented in [3]. More recently, Woodin (unpublished) using inner-model theoretic techniques has shown how to recover all of these results from just AD. We next state this classification result. We refer the reader to [3] for a proof and the definition of the Martin pointclass $\Lambda(\boldsymbol{\Gamma}, \kappa)$ appearing in the statement.

THEOREM 1.25 (Classification of Scales and Suslin Cardinals). *Suppose κ is a limit of Suslin cardinals, and κ is below the supremum of the Suslin cardinals (so κ is a Suslin cardinal). We have the following cases.*

- $\mathrm{cof}(\kappa) = \omega$ *(type I).*
  Let $\boldsymbol{\Sigma}_0 = \bigcup_\omega S_{<\kappa}$. *Then* $\mathrm{scale}(\boldsymbol{\Sigma}_0)$ *with norms into* κ, *and* $\mathrm{scale}(\boldsymbol{\Pi}_{2n+1})$, $\mathrm{scale}(\boldsymbol{\Sigma}_{2n+2})$ *with norms into* $\delta_{2n+1} \doteq \delta(\boldsymbol{\Pi}_{2n+1})$.
  $\delta_{2n+1} = (\lambda_{2n+1})^+$, *where* $\mathrm{cof}(\lambda_{2n+1}) = \omega$ $(\lambda_1 = \kappa)$. $\delta_1, \lambda_3, \delta_3, \ldots$ *are the next ω Suslin cardinals after κ.* $S(\lambda_{2n+1}) = \boldsymbol{\Sigma}_{2n+1}$, $S(\delta_{2n+1}) = \boldsymbol{\Sigma}_{2n+2}$.
- $\mathrm{cof}(\kappa) > \omega$ *and* $\boldsymbol{\Gamma}_0$ *(the Steel pointclass) is not closed under* $\exists^{\omega^\omega}$ *(types II and III).*
  Let $\boldsymbol{\Sigma}_0 = \exists^{\omega^\omega}\boldsymbol{\Gamma}_0$. *Then* $\mathrm{scale}(\boldsymbol{\Gamma}_0)$, $\mathrm{scale}(\boldsymbol{\Sigma}_0)$ *with norms into* κ *and* $\mathrm{scale}(\boldsymbol{\Pi}_{2n+1})$, $\mathrm{scale}(\boldsymbol{\Sigma}_{2n+2})$ *with norms into* $\delta_{2n+1} = \delta(\boldsymbol{\Pi}_{2n+1})$. $\lambda_1, \delta_1, \lambda_3, \ldots$ *are the next ω Suslin cardinals after κ.* $\delta_{2n+1} = (\lambda_{2n+1})^+$, *where* $\mathrm{cof}(\lambda_{2n+1}) = \omega$ *and* $S(\lambda_{2n+1}) = \boldsymbol{\Sigma}_{2n+1}$, $S(\delta_{2n+1}) = \boldsymbol{\Sigma}_{2n+2}$.
- $\mathrm{cof}(\kappa) > \omega$ *and* $\boldsymbol{\Gamma}_0$ *is closed under* $\exists^{\omega^\omega}$ *(type IV).*
  Then $\mathrm{scale}(\boldsymbol{\Gamma}_0)$ *with norms into* κ. *Let* $\Lambda = \Lambda(\boldsymbol{\Gamma}_0, \kappa)$ *be the Martin pointclass. Let* $\boldsymbol{\Sigma}_0 = \bigcup_\omega \Lambda$. *Let* $\lambda_1 = o(\Lambda)$. *Then* $\mathrm{scale}(\boldsymbol{\Pi}_{2n+1})$, $\mathrm{scale}(\boldsymbol{\Sigma}_{2n+2})$ *with norms into* $\delta_{2n+1} = \delta(\boldsymbol{\Pi}_{2n+1})$. $\lambda_1, \delta_3, \lambda_3$ *are the next ω Suslin cardinals after κ.* $\delta_{2n+1} = (\lambda_{2n+1})^+$, *where* $\mathrm{cof}(\lambda_{2n+1}) = \omega$ *and* $S(\lambda_{2n+1}) = \boldsymbol{\Sigma}_{2n+1}$, $S(\delta_{2n+1}) = \boldsymbol{\Sigma}_{2n+2}$.

The one fact missing from the above statements which the $V = L(\mathbb{R})$ analysis gives is that in the case of a type IV hierarchy we have the scale property at $\boldsymbol{\Sigma}_0$.

§2. The inductive analysis: the overall plan. The previous section reviewed the "global" theory of pointclasses and scales assuming AD. We now wish to descend to a smaller scale and consider the much finer analysis of the cardinal structure up through $\aleph_{\omega_1}$. The results of the previous section describe the scaled pointclasses, but give only a little information about the corresponding cardinals. For example, down at the projective level, the projective ordinals $\boldsymbol{\delta}^1_{2n+1}$ are not computed, nor is information given about the structure of the cardinals (e.g., their cofinalities) in between. As we said before, our goal in this paper is not to describe how descriptions give the complete analysis, but rather to introduce them in a simpler context and see how they determine the cardinal structure. Nevertheless, we make some brief comments here about the nature of the overall inductive analysis. We will continue this discussion in §6 after we have seen the notion of a description.

First note that below $\aleph_{\omega_1}$ all of the limit Suslin cardinals have cofinality ω. That is, we are always in the case of a type I hierarchy as described in Theorem 1.25. Let $\boldsymbol{\delta}^1_\alpha = \delta(\boldsymbol{\Sigma}^1_\alpha)$ for $\alpha < \omega_1$ (recall $\boldsymbol{\Sigma}^1_\alpha$ is the αth Lévy class closed under $\exists^{\omega^\omega}$). For α limit, $\boldsymbol{\Sigma}^1_\alpha$ is the pointclass $\boldsymbol{\Sigma}_0$ described in Theorem 1.25 corresponding to the limit Suslin cardinal $\kappa = \sup_{\beta<\alpha} \delta^1_\beta$. In this case, $\delta(\boldsymbol{\Sigma}_0) = \delta(\boldsymbol{\Sigma}_1) = \kappa^+$. So, as an artifact of our notation we have that $\boldsymbol{\delta}^1_\alpha = \boldsymbol{\delta}^1_{\alpha+1} = \kappa^+$. As described in Theorem 1.25, the projective hierarchy above $\boldsymbol{\Sigma}_0$ behaves similarly to the usual projective hierarchy. In our notation, $\boldsymbol{\delta}^1_{\alpha+n}$ is analogous to $\boldsymbol{\delta}^1_n$. The general analysis proceeds by induction on α. The main task to analyze the measures on the $\boldsymbol{\delta}^1_\alpha$ and use this to prove the strong partition relation on the $\boldsymbol{\delta}^1_\alpha$ for α an odd ordinal (this includes $\boldsymbol{\delta}^1_\alpha = \boldsymbol{\delta}^1_{\alpha+1}$ when α is a limit). Along the way we compute the values of the $\boldsymbol{\delta}^1_\alpha$ and determine the cardinal structure in between the $\boldsymbol{\delta}^1_\alpha$.

Figure 1 shows the projective hierarchy along with the hierarchy above a typical limit Suslin cardinal. It will turn out that $\sup_{\alpha<\omega_1} \boldsymbol{\delta}^1_\alpha = \aleph_{\omega_1}$.

At limit Suslin cardinals κ no new measures arise since $\mathrm{cof}(\kappa) = \omega$. Thus, limit stages below $\aleph_{\omega_1}$ are trivial, and the arguments all take place at the successor steps where they are identical to those of the projective hierarchy. So, to simplify notation let us consider the $\boldsymbol{\delta}^1_n$. Only the odd projective ordinals $\boldsymbol{\delta}^1_{2n+1}$ are relevant in the inductive analysis. Assume inductively we have analyzed all the measures on the $\boldsymbol{\delta}^1_{2m+1}$ for $m < n$, and have shown the strong partition relation on these $\boldsymbol{\delta}^1_{2m+1}$ and the weak partition relation on $\boldsymbol{\delta}^1_{2n+1}$. Assume also we know the values of $\boldsymbol{\delta}^1_m$ for all $m \leq 2n+1$. We then define the notion of a level-$2n+1$ description, which is a finitary object built

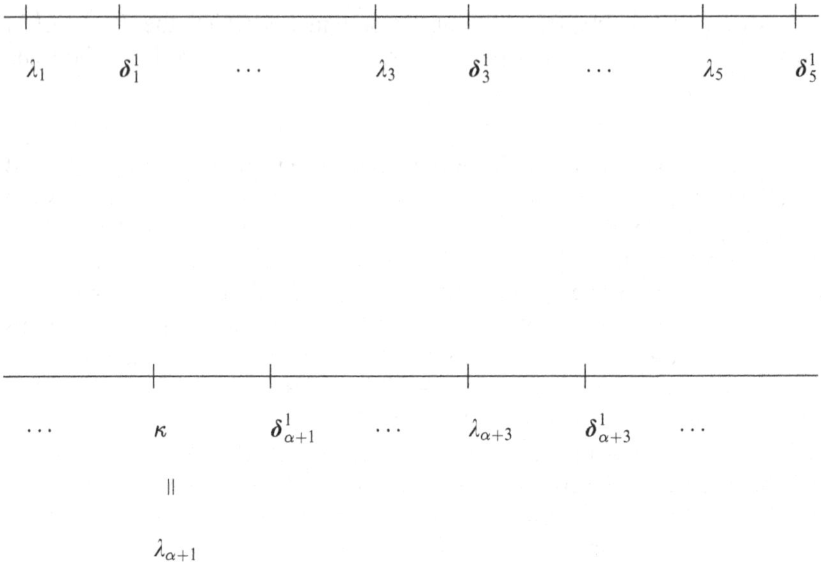

FIGURE 1. The first ω_1 Suslin cardinals.

out of the lower level descriptions (the complete definition will not be given here, it can be found in [4]). Roughly speaking, these objects describe how to generate ordinals in the iterated ultrapowers by certain canonical measures on δ^1_{2n+1} and the smaller cardinals. These level-$2n+1$ descriptions are used to do three things: compute δ^1_{2n+3}, prove the strong partition relation on δ^1_{2n+1}, and prove the weak partition relation on δ^1_{2n+3}. The partition relations are proved by analyzing the measures on δ^1_{2n+1} and λ_{2n+3} respectively. The arguments for all three parts are similar and use the same set of level-$2n+1$ descriptions. This completes this step of the induction.

The argument for converting an analysis of the measures to a proof of a partition property uses two important ingredients. The first is the general method of Martin for proving partition properties in general. Martin used this originally for establishing the strong partition property on ω_1. Martin used the theory of indiscernibles for the $L[x]$, but his argument can be recast into trivial descriptions as well. Martin's method is given in [8] and in a somewhat more general form in [3]. Martin's theorem basically says to prove a partition property on κ it suffices to have a coding of the functions into κ which have sufficiently nice definabilty properties. The second ingredient is an argument of Kunen which shows that one can convert an analysis of the measures on κ into a coding of the subsets of κ. This argument appears in [14], and is also stated explicitly in [3].

In this paper, we will introduce the descriptions through different considerations and simply assume the partition properties where needed.

§3. Partition properties and types. We first recall the Erdős-Rado partition notation.

Definition 3.1. We write $\kappa \to (\kappa)^\lambda$ to mean:: for every partition $\mathcal{P}: \kappa^\lambda \to \{0,1\}$ of the increasing function from λ to κ into two pieces, there is a homogeneous set $H \subseteq \kappa$ of size κ. That is, $\mathcal{P} \restriction (H)^\lambda$ is constant.

We say κ has the *weak* partition property if $\kappa \to (\kappa)^\lambda$ for all $\lambda < \kappa$. We say κ has the *strong* partition property if $\kappa \to (\kappa)^\kappa$.

A more useful working reformulation of the partition property will be given next, and introduces the notion of *type* of a function.

Definition 3.2. A function $f: \lambda \to \mathrm{On}$ is of *uniform cofinality* ω if there is a function $g: \lambda \times \omega \to \mathrm{On}$ which is increasing in the second argument and such that for all $\alpha < \lambda$ we have $f(\alpha) = \sup_n g(\alpha, n)$.

We next specify a canonical type (we give a more general definition of types below). We call this the "correct type" since it is the correct type for partition relations.

Definition 3.3. We say f is of the *correct type* if f is increasing, everywhere discontinuous (i.e., $f(\alpha) > \sup_{\beta<\alpha} f(\beta)$ for limit α), and of uniform cofinality ω.

We next give the c.u.b. reformulation of the partition property.

Definition 3.4. We say $\kappa \overset{\text{c.u.b.}}{\longrightarrow} (\kappa)^\lambda$ if for every partition $\mathcal{P}$ of the function $f: \lambda \to \kappa$ of the correct type, there is a c.u.b. $C \subseteq \kappa$ homogeneous for $\mathcal{P}$.

Fact 3.5. We have the following implications, which show the essential equivalence of the two partition notions:

$$\kappa \overset{\text{c.u.b.}}{\longrightarrow} (\kappa)^\lambda \Rightarrow \kappa \to (\kappa)^\lambda,$$

$$\kappa \to (\kappa)^{\omega\cdot\lambda} \Rightarrow \kappa \overset{\text{c.u.b.}}{\longrightarrow} (\kappa)^\lambda.$$

Proof. For the first fact, assume $\kappa \overset{\text{c.u.b.}}{\longrightarrow} (\kappa)^\lambda$ and let $\mathcal{P}$ be a partition of the functions $g: \lambda \to \kappa$. In particular, $\mathcal{P}$ partitions functions of the correct type. Let C be c.u.b. and homogeneous for this subpartition. Define $h: \kappa \to \kappa$ by $h(\alpha) = \omega^{\text{th}}$ element of C greater than $\sup_{\beta<\alpha} h(\beta)$. Let $H = \mathrm{ran}(h)$. Then H is homogeneous for $\mathcal{P}$, since any increasing $g: \lambda \to H$ is of the correct type.

For the second fact, assume $\kappa \to (\kappa)^{\omega\cdot\lambda}$ and let $\mathcal{P}$ be a partition of the functions $f: \lambda \to \kappa$ of the correct type. Let $\mathcal{P}'$ be the partition of increasing $g: \lambda \times \omega \to \kappa$ (with lexicographic ordering on $\lambda \times \omega$) given by $\mathcal{P}'(g) = \mathcal{P}(f)$

where $f(\alpha) = \sup_n g(\alpha, n)$. Let H be homogeneous for $\mathcal{P}'$ and let C be the set of limit points of H. C is homogeneous for $\mathcal{P}$. ⊣

A more general notion of type is given in the following definition.

Definition 3.6. Let $g : \lambda \to \mathrm{On}$. We say $f : \lambda \to \mathrm{On}$ is of *uniform cofinality* g if there is a function

$$f' : \{(\alpha, \beta) : \alpha < \lambda \wedge \beta < g(\alpha)\} \to \mathrm{On}$$

which is increasing in the second argument and such that $f(\alpha) = \sup_\beta f'(\alpha, \beta)$.

If g is the constant ρ function, then we say f has uniform cofinality ρ. Also, if μ is a measure on λ, we have the natural notion of f being of uniform cofinality g almost everywhere with respect to μ.

If κ has the strong partition property, we have the partition property for functions $f : \kappa \to \kappa$ of type g (increasing, discontinuous, of uniform cofinality g) by an argument similar to that of the above fact.

Given a measure μ on κ with the strong partition property, we have a natural measure ν on the ultrapower $j_\mu(\kappa)$. Namely, $A \subseteq j_\mu(\kappa)$ has ν measure one iff there is a c.u.b. $C \subseteq \kappa$ such that for all $f : \kappa \to C$ of the correct type, $[f]_\mu \in A$. We can likewise use any other type to define a measure.

§4. **Trivial descriptions and the first exemplary problem.** In this section we define a natural family of measures on the ω_n which we call W_1^n, and S_1^n, which play an important role in the first level of the inductive analysis. Their definitions, however, are straightforward assuming the weak and strong partition relations on ω_1 (hence the notation). We consider the problem of computing the ultrapowers $j_{S_1^n}(\omega_m)$. To do this we introduce the very simplest instance of descriptions, so simple in fact that they are trivial objects (just integers). Nevertheless, we show how we can use the trivial descriptions to analyze these ultrapowers and solve the problem. In the next section, we will consider a second problem which will force us to introduce non-trivial descriptions. There the descriptions will be more complicated combinatorial objects (though still hereditarily finite sets), though the rest of the apparatus is essentially the same. The two exemplary problems will in fact be related, which will provide a check on the computations.

In the actual inductive analysis, the starting point is to show the weak partition relation on ω_1, which is shown directly. Using the trivial descriptions one next analyzes the measures on ω_1 and thereby establishes the strong partition relation on ω_1 (in Theorem 2.36 and Corollary 4.16 of [3] these arguments are given). Again, here we will assume the strong partition relations on the $\boldsymbol{\delta}^1_{2n+1}$, and in particular on $\omega_1 = \boldsymbol{\delta}^1_1$.

From the weak partition relation on ω_1 we get immediately that sets of the form C^n form a base for a measure on $(\omega_1)^n$ (use the c.u.b. version of the weak partition property with exponent n). We denote this measure W_1^n. It is the n-fold product of the normal measure W_1^1 on ω_1.

A trivial description will "describe" how to generate an equivalence class of a function with respect to the measure W_1^n.

DEFINITION 4.1. A *trivial description* is an integer $d \in \omega - \{0\}$. The set $\mathcal{D}$ of trivial descriptions is ordered by the usual ordering on ω. The *lowering operator* $\mathcal{L}$ is defined on all $d \in \mathcal{D}$ except the minimal description $d = 1$. We define $\mathcal{L}(d) = d - 1$.

Let $\mathcal{D}_n$ be those trivial descriptions $d \leq n$. We next define the "interpretation" function and some related notation. The interpretation function interprets the description, that is, it uses the description to generate an equivalence class with respect to W_1^n.

DEFINITION 4.2 (Interpretation Function). If $f : n \to \omega_1$ (i.e., $f = (\alpha_1, \dots, \alpha_n)$), and $d \in \mathcal{D}_n$, define $h(f;d) = f(d) = \alpha_d$. For $d \in \mathcal{D}_n$, Let $(W_1^n; d) = [f \mapsto (f;d)]_{W_1^n}$. If $g : \omega_1 \to \omega_1$, let also $(g;f;d) = g(f(d))$, and $(g; W_1^n; d) = [f \mapsto (g;f;d)]_{W_1^n}$.

So, $(W_1^n; d)$ represents an ordinal less than $j_{W_1^n}(\omega_1)$ ($= \omega_{n+1}$ as shown below), and does $(g; W_1^n; d)$ which can be viewed as a "lift" of the ordinal $(W_1^n; d)$.

The main tool we need to work with the trivial descriptions is the *Kunen tree*. Since the argument is short, we give the construction. First some fairly standard notation. For $x \in \omega^\omega$, let $<_x = \{(n,m) : x(\langle n,m\rangle) = 1\}$ be the binary relation coded by x. Let $\mathrm{LO} = \{x : <_x$ is a linear order$\}$, and let $\mathrm{WO} = \{x : <_x$ is a wellorder$\}$. LO is $\mathbf{\Pi}_1^0$ (i.e., closed), and WO is $\mathbf{\Pi}_1^1$.

Let S be the Shoenfield tree on $\omega \times \omega_1$ with $\mathrm{WO} = p[S]$. We can take S to be the tree defined by: $(s, \vec{\alpha}) \in S$ iff s doesn't violate being in LO and

$$\forall i, j \leq |s| \, (s(\langle i, j\rangle) = 1 \to \alpha_i < \alpha_j).$$

With minor modifications we can, if desired, assume S is *homogeneous*, i.e., s determines the order-type of $\vec{\alpha}$. Also, we can require that $\alpha_0 > \alpha_1, \dots, \alpha_{n-1}$.

THEOREM 4.3 (Kunen). *There is a tree T on $\omega \times \omega_1$ with the following property. For any $f : \omega_1 \to \omega_1$ there is an $x \in \omega^\omega$ such that T_x is wellfounded and for all infinite ordinals $\alpha < \omega_1$ we have $f(\alpha) < |T_x \restriction \alpha|$.*

PROOF. If $f : \omega_1 \to \omega_1$, play the game where I plays out x, II plays out y, and II wins the run iff

$$x \in \mathrm{WO} \to (y \in \mathrm{WO} \wedge |y| \geq f(|x|)).$$

II has a winning strategy σ by boundedness. This suggests the following definition.

Let V be the tree on $\omega \times \omega \times \omega_1 \times \omega \times \omega$ given by:

$$(s, t, \vec{\alpha}, u, v) \in V \leftrightarrow \exists \sigma, x, y, z \text{ extending } s, t, u, v\ [\sigma(x) = y \wedge (t, \vec{\alpha}) \in S$$
$$\wedge \forall i\ y(\langle z(i), z(i+1)\rangle) = 1]$$

For f and σ as above, V_σ is wellfounded and for any infinite α, $|V_\sigma \restriction \alpha| \geq f(\alpha)$. Finally, we can identify V with a tree on $\omega \times \omega_1$ by weaving the last four coordinates into a single coordinate (which keeps the previous inequality). $\dashv$

We now analyze functions from $(\omega_1)^n$ to ω_1 using the trivial descriptions and the Kunen tree. We let $\forall^*$ denote "for almost all" with respect to the measure W_1^n (the value of n will be clear from the context). Likewise, we write $\forall^*_\mu$ to denote "for almost all ordinals with respect to the measure μ."

For $C \subseteq \omega_1$, let $N_C(\beta)$ be the least element of C greater than β.

An easy partition argument shows the following. If $f : (\omega_1)^n \to \omega_1$ is such that $f(\vec{\alpha}) < \alpha_i$ for W_1^n almost all $\vec{\alpha}$, then there is a c.u.b. $C \subseteq \omega_1$ such that $\forall^* \vec{\alpha}\ f(\vec{\alpha}) < N_C(\alpha_{i-1})$ [to see this, partition the $n+1$ tuples $(\alpha_0, \dots, \alpha_{i-1}, \beta, \alpha_i, \dots, \alpha_{n-1})$ according to whether $f(\vec{\alpha}) < \beta$. On the homogeneous side the stated property holds, and a c.u.b. $C \subseteq \omega_1$ homogeneous for the partition works.]

Translating the above fact into the language of trivial descriptions we immediately get the following result, which we could call the "main theorem" for trivial descriptions.

THEOREM 4.4. *If* $\alpha < (W_1^n; d)$, *then there is a* $g : \omega_1 \to \omega_1$ *such that* $\alpha < (g; W_1^n; \mathcal{L}(d))$.

Using the Kunen tree, fix $x \in \omega^\omega$ such that $g(\beta) \leq |T_x \restriction \beta|$ for almost all β. We then have:

$$(g; W_1^n; \mathcal{L}(d)) \leq (\beta \mapsto |T_x \restriction \beta|; W_1^n; \mathcal{L}(d))$$
$$< (W_1^n; \mathcal{L}(d))^+$$

The last inequality follows since we can map $(W_1^n; \mathcal{L}(d))$ onto $(\beta \to |T_x \restriction \beta|; W_1^n; \mathcal{L}(d))$ as follows: if $[g]_{W_1^n} < (W_1^n; \mathcal{L}(d))$, map $[g]$ to $[f \mapsto |T_x \restriction (f; \mathcal{L}(d))(g(f))|]$, where $|T_x \restriction \alpha(\beta)|$ denotes the rank of β in the tree $T_x \restriction \alpha$.

So, we get that $(W_1^n; d) \leq (W_1^n; \mathcal{L}(d))^+$. As an immediate corollary we have the following upper bound.

COROLLARY 4.5. $j_{W_1^n}(\omega_1) \leq \omega_{n+1}$.

To get the lower bound we use the following general result of Martin (see Theorem 7.1 of [5] or Theorem 4.17 of [3] for a proof).

THEOREM 4.6 (Martin). *Assume* $\kappa \to (\kappa)^\kappa$. *Then for any measure* μ *on* κ, $j_\mu(\kappa)$ *is a cardinal.*

Also, if it is easy to see that if $m < n$ then $j_{W_1^m}(\omega_1) < j_{W_1^n}(\omega_1)$. In fact, $j_{W_1^m}(\omega_1)$ embeds into $(W_1^{m+1}; d)$, where $d = m + 1$. So we have the exact computation of the ultrapower:

Corollary 4.7. $j_{W_1^n}(\omega_1) = \omega_{n+1}$.

This computation can be interpreted as giving a computation of the projective ordinal $\boldsymbol{\delta}^1_3$. Briefly, this is because the Shoenfield tree for a $\boldsymbol{\Sigma}^1_2$ set is weakly homogeneous with measures of the form W_1^n. The homogeneous tree construction then gives that every $\boldsymbol{\Pi}^1_2$, and hence also every $\boldsymbol{\Sigma}^1_3$ set is λ-Suslin where $\lambda = \sup_n j_{W_1^n}(\omega_1) = \omega_\omega$ from the above computation. So, $\boldsymbol{\delta}^1_3 = \lambda_3^+ \leq \omega_{\omega+1}$. Since $\operatorname{cof}(\omega_n) > \omega$, clearly, we must have $\lambda_3 \geq \omega_\omega$, and thus $\boldsymbol{\delta}^1_3 = \omega_{\omega+1}$.

Types of functions on ω_1. Before turning to the first exemplary problem, we mention the analysis of types of functions $f\colon (\omega_1)^n \to \omega_1$.

Fix n, and a permutation π of $\{1, 2, \ldots, n\}$ beginning with n. Say,

$$\pi = (n, i_2, \ldots, i_n).$$

Let $<_\pi$ be the ordering on $(\omega_1)^n$ given by:

$$(\alpha_1, \ldots, \alpha_n) <_\pi (\beta_1, \ldots, \beta_n) \leftrightarrow (\alpha_n, \alpha_{i_2}, \ldots, \alpha_{i_n}) <_{\text{lex}} (\beta_n, \beta_{i_2}, \ldots, \beta_{i_n}).$$

We say $f\colon (\omega_1)^n \to \omega_1$ is ordered by π if on a measure one set f is order-preserving from $<_n$ to ω_1. Let π_n denote the particular permutation $\pi_n = (n, 1, 2, \ldots, n-1)$. In this case we let $<_n$ abbreviate $<_{\pi_n}$.

Fact 4.8. If on every measure one set f depends on all its arguments, then for some π, f is ordered by π.

In this case, the type of f is determined by π and the possible uniform cofinalities. We have the following possibilities:

- There is a measure one set on which f is continuous (i.e., $f \restriction C^n$ is continuous).
- f has uniform cofinality ω (i.e., f is of the correct type).
- $f(\alpha_1, \ldots, \alpha_n)$ has uniform cofinality α_i.

We omit the easy proof, but instead, as an example, show that the possible uniform cofinalities mentioned in the lase case are all distinct.

Example 4.9. Show that the uniform cofinalities $g(\vec{\alpha}) = \alpha_i$ and $g(\vec{\alpha}) = \alpha_j$ are distinct for $i \neq j$.

Proof. Suppose $f_1\colon \{(\vec{\alpha}, \beta)\colon \beta < \alpha_i\} \to \omega_1$ induces f as in the definition of uniform cofinality, and likewise $f_2\colon \{(\vec{\alpha}, \beta)\colon \beta < \alpha_j\} \to \omega_1$ also induces f.

Consider the partition $\mathcal{P}$ where we partition tuples

$$\alpha_1 < \cdots < \alpha_{i-1} < \beta_1 < \alpha_i < \cdots < \alpha_{j-1} < \beta_2 < \alpha_j < \cdots < \alpha_n$$

according to whether $f_1(\vec{\alpha}, \beta_1) < f_2(\vec{\alpha}, \beta_2)$. Neither side can be homogeneous, a contradiction. ⊣

Lastly, we introduce the measures S_1^n mentioned earlier. In the full analysis, we introduce a canonical family of measures for each regular cardinal below $\aleph_{\omega_1}$.

Recall π_n is the permutation $\pi_n = (n, 1, 2, \dots, n-1)$.

DEFINITION 4.10. S_1^n is the measure on ω_{n+1} induced by functions $f : (\omega_1)^n \to \omega_1$ ordered by π_n and of the correct type, and the measure W_1^n. That is, $A \subseteq \omega_{n+1}$ has S_1^n measure one if there is a c.u.b. $C \subseteq \omega_1$ such that for all $f : (\omega_1)^n \to C$ ordered by π_n and of the correct type we have $[f]_{W_1^n} \in A$.

It is easy to see that S_1^1 is the ω-cofinal normal measure on ω_2.

The family of measures S_1^n plays a distinguished role in that their ultrapowers dominate those for the other measures on ω_ω. More precisely, for any measure μ on ω_ω there is an n and a c.u.b. $C \subseteq \boldsymbol{\delta}_3^1$ such that for all $\alpha \in C$ with $\mathrm{cof}(\alpha) = \omega_2$,

$$j_\mu(\alpha) \leq j_{S_1^n}(\alpha).$$

This fact is proved by embedding arguments [5]. Rather than give the full proof, we give an example.

EXAMPLE 4.11. We show this for $\mu = \nu \times \nu$, where ν is the measure corresponding to $\pi = (3, 2, 1)$.

PROOF. We take $n = 3$. We consider the auxiliary measure $\mathcal{M} = W_1^1 \times W_1^1 \times S_1^2$.

Given $f \colon <_3 \to \omega_1$ of the correct type and given $\eta_1 < \eta_2 < \omega_1$ and h_3 representing $(\eta_1, \eta_2, [h_3]) \in \omega_1 \times \omega_1 \times \omega_3$, we define $g_1, g_2 \colon (\omega_1)^3 \to \omega_1$ by:

$$g_i(\alpha_1, \alpha_2, \alpha_3) = f(\eta_i, h_3(\alpha_1, \alpha_2), \alpha_3).$$

Let $\alpha \in C$, the set of ordinals $< \boldsymbol{\delta}_3^1$ closed under ultrapowers by measures on ω_ω (one can compute directly for the specific measures involved here that ω_ω, and hence $\boldsymbol{\delta}_3^1$ are fixed points. The argument is similar to the computation in the first exemplary problem to follow. The analysis of measures on ω_ω actually shows this true for any measure on ω_ω).

Define $\pi \colon j_\mu(\alpha) \to j_{S_1^3}(\alpha)$ as follows.

Given an equivalence class $[F]_\mu$, for $F \colon \mathrm{dom}(\mu) \to \alpha$, let $\pi([F]_\mu) = [G]_{S_1^3}$, where:

$$G([f]_{W_1^3}) = [(\eta_1, \eta_2, [h_3]) \mapsto F([g_1], [g_2])]_{\mathcal{M}}$$

and $g_1 = g(f, \eta_1, h_3)$ is the g function defined above using f, η_1, and h_3, and likewise for g_2 (using η_2).

The following observations complete the proof.

(1) For any fixed f of the correct type, fixed η_1, η_2, h_3 with $\eta_1 < \eta_2 < h_3(\gamma, \beta)$ for all $\gamma < \beta$ and all $\alpha_1 < \alpha_2 < \alpha_3$ in a c.u.b. set we have that g_i is well-defined and of order π.

Also, $g_1(\vec{\alpha}) < g_2(\vec{\beta})$ iff $\alpha_3 \leq \beta_3$ (this gives the order-type of the product measure μ).

(2) For any fixed f, $[g_i]$ only depends on $\eta_1, \eta_2, [h_3]$.

(3) If $[f_1] = [f_2]$, then $\mathcal{M}$ almost all $\eta_1, \eta_2, [h_3]$ we have that $[g_i^1] = [g_i^2]$, where g_i^1 uses f_1 and likewise for g_i^2.

(4) If $\mu(A) = 1$, then

$$\forall^* f \, \forall^* \eta_1, \eta_2, h_3 \, ([g_1], [g_2]) \in A.$$

(5) $\forall^* f \, \exists (G_1, G_2) \, \forall^* \eta_1, \eta_2, h_3 \, (g_1 \leq G_1 \wedge g_2 \leq G_2)$.

(1)-(4) give that π is well-defined, (5) and $\alpha \in C$ give that $\pi([F]) < j_{S_1^3}(\alpha)$. ⊣

We introduce one more bit of notation.

Definition 4.12. If $f : (\omega_1)^n \to \omega_1$ is order-preserving with respect to $<_n$, then we define the ℓth *invariant* of f by:

$$f(\ell)(\alpha_1, \ldots, \alpha_\ell) = \sup\{f(\alpha_1, \ldots, \alpha_{\ell-1}, \beta_\ell, \ldots, \beta_{n-1}, \alpha_\ell)\},$$

where the supremum ranges over all $\vec{\beta}$ such that the arguments to f are increasing.

A notational convention. We introduce a useful notational convention for dealing with iterated ultrapowers. We use it in a relatively minor way in the first exemplary problem, but in a more serious way for the second problem.

Suppose $\theta \in \mathrm{On}$, $\mu_1, \ldots, \mu_n$ are measures, and $P \subseteq \mathrm{On}$. We write

$$\forall^*_{\mu_1} \alpha_1 \cdots \forall^*_{\mu_n} \alpha_n \, P(\theta(\alpha_1, \ldots, \alpha_n))$$

to abbreviate the following:

If we fix $\alpha_1 \mapsto \theta(\alpha_1)$ representing θ in the ultrapower by μ_1, then for μ_1 almost all α_1 it is the case that:

If we fix $\alpha_2 \mapsto \theta(\alpha_1, \alpha_2)$ representing $\theta(\alpha_1)$ with respect to μ_2, then for μ_2 almost all α_2 it is the case that:

$$\vdots$$

If we fix $\alpha_n \mapsto \theta(\alpha_1, \ldots, \alpha_n)$ representing $\theta(\alpha_1, \ldots, \alpha_{n-1})$ with respect to μ_n, then for μ_n almost all α_n we have $P(\theta(\alpha_1, \ldots, \alpha_n))$.

The first exemplary problem. We now consider the first of the two problems we will use to illustrate the theory of descriptions. The first problem, however, requires only the trivial descriptions.

First exemplary problem. Compute the ultrapower $j_{S_1^m}(\omega_n)$. We show the answer is ω_k where:

$$k = 1 + 2\left[\binom{n-1}{1} + \cdots + \binom{n-1}{m}\right].$$

To solve this problem, we describe explicitly the cardinals structure below $j_{S_1^m}(\omega_n)$. Although the argument only uses the trivial descriptions, we will cast it notationally as a special instance of the non-trivial descriptions that we will use for the second problem. We call this special set of descriptions $\mathcal{S}$. The definition follows.

DEFINITION 4.13. For $m, n \geq 1$, let $\mathcal{S}_{m,n}$ be the set of tuples of the form

$$d = (d_m, d_1, d_2, \dots, d_\ell) \text{ or } d = (d_m, d_1, d_2, \dots, d_\ell)^s,$$

where the d_i are trivial descriptions in $\mathcal{D}_{n-1}$ (i.e., $1 \leq d_i \leq n-1$), $\ell < m$, and $d_1 < d_2 < \cdots < d_\ell < d_m$. Here the s is a formal symbol (which stands for "sup").

We write $d = (d_m, d_1, d_2, \dots, d_\ell)^{(s)}$ to denote that s may or may not appear. It is easy to see that there are $2\left[\binom{n-1}{1} + \cdots + \binom{n-1}{m}\right]$ many $d \in \mathcal{S}_{m,n}$ (the factor of 2 since the symbol s may or may not appear).

We extend our notational conventions for trivial descriptions to the set $\mathcal{S}$. Given $d \in \mathcal{S}_{m,n}$, we associate an ordinal $(d; S_1^m)$ to d defined as follows.

DEFINITION 4.14. Let $d \in \mathcal{S}_{m,n}$. Then $(d; S_1^m)$ is the ordinal represented with respect to the measure S_1^m by the function $[f]_{W_1^m} \mapsto (d; f) < \omega_n$, where for $f : (\omega_1)^m \to \omega_1$ of the correct type, $(d; f) < \omega_n$ is the ordinal represented with respect to W_1^{n-1} by the function $(\alpha_1, \dots, \alpha_{n-1}) \mapsto (d; f)(\vec{\alpha})$. Finally,

$$(d; f)(\vec{\alpha}) = f(\ell + 1)(\alpha_{d_1}, \alpha_{d_2}, \dots, \alpha_{d_\ell}, \alpha_{d_m})$$

if s does not appear in d. If s appears in d, let

$$\begin{aligned}(d; f)(\vec{\alpha}) &= f^s(\ell + 1)(\alpha_{d_1}, \alpha_{d_2}, \dots, \alpha_{d_\ell}, \alpha_{d_m}) \\ &\doteq \sup_{\beta < \alpha_{d_\ell}} f(\ell + 1)(\alpha_{d_1}, \alpha_{d_2}, \dots, \beta, \alpha_{d_m}).\end{aligned}$$

To finish the computation of the first exemplary problem, we prove the following.

THEOREM 4.15. *The $(d; S_1^m)$ for $d \in \mathcal{S}_{m,n}$ are precisely the cardinals below $j_{S_1^m}(\omega_n)$.*

The descriptions in $\mathcal{S}_{m,n}$ are ordered as follows. Set

$$(d_m, d_1, \dots, d_\ell)^{(s)} < (d'_m, d'_1, \dots, d'_{\ell'})^{(s)}$$

iff one of the following holds:

(1) There is a least place of disagreement in the description sequences, say $d_i \neq d'_i$, and $d_i < d'_i$.
(2) $\vec{d}$ is an initial segment of $\vec{d'}$, and s appears in $\vec{d}$.
(3) $\vec{d'}$ is an initial segment of $\vec{d}$, and s does not appear in $\vec{d'}$.

We define a lowering operator on $\mathcal{L}_{m,n}$ as we did for the trivial descriptions. Let $\mathcal{L}(d_m, d_1, \dots, d_\ell)^{(s)}$ be the largest sequence which is less than $(d_m, d_1, \dots, d_\ell)^{(s)}$. We describe the $\mathcal{L}$ operation explicitly:

- If $\ell = m - 1$ then $\mathcal{L}(d_m, d_1, \dots, d_\ell) = (d_m, d_1, \dots, d_\ell)^s$.
- If $\ell = m - 1$ then $\mathcal{L}(d_m, d_1, \dots, d_\ell)^s = (d_m, d_1, \dots, \mathcal{L}(d_\ell))$ if $\mathcal{L}(d_\ell) = d_\ell - 1$ is defined and $> d_{\ell-1}$ (if $\ell > 1$). Otherwise, $\mathcal{L}(d_m, d_1, \dots, d_\ell)^s = (d_m, d_1, \dots, d_{l-1})^s$.
- If $\ell < m-1$, then $\mathcal{L}(d_m, d_1, \dots, d_\ell) = (d_m, d_1, \dots, d_\ell, d_{\ell+1})$ where $d_{\ell+1} = \mathcal{L}(d_m) = d_m - 1$ provided $d_m - 1 > d_\ell$ (if $\ell \geq 1$). Otherwise set $\mathcal{L}(d_m, d_1, \dots, d_\ell) = (d_m, d_1, \dots, d_\ell)^s$.

EXAMPLE 4.16. $m = 3$, $n = 5$ (i.e., we are considering $j_{S_1^3}(\omega_5)$)

Beginning with the maximal description (4) in $\mathcal{S}_{3,5}$ we successively apply the lowering operator to generate the following sequence (we stop when we reach the minimal sequence $(1)^s$ in $\mathcal{S}_{3,5}$).

$$\begin{array}{ccc}
d = (4) & \mathcal{L}(d) = (4,3) & \mathcal{L}^2(d) = (4,3)^s \\
\mathcal{L}^3(d) = (4,2) & \mathcal{L}^4(d) = (4,2,3) & \mathcal{L}^5(d) = (4,2,3)^s \\
\mathcal{L}^6(d) = (4,2)^s & \mathcal{L}^7(d) = (4,1) & \mathcal{L}^8(d) = (4,1,3) \\
\mathcal{L}^9(d) = (4,1,3)^s & \mathcal{L}^{10}(d) = (4,1,2) & \mathcal{L}^{11}(d) = (4,1,2)^s \\
\mathcal{L}^{12}(d) = (4,1)^s & \mathcal{L}^{13}(d) = (4)^s & \mathcal{L}^{14}(d) = (3) \\
\mathcal{L}^{15}(d) = (3,2) & \mathcal{L}^{16}(d) = (3,2)^s & \mathcal{L}^{17}(d) = (3,1) \\
\mathcal{L}^{18}(d) = (3,1,2) & \mathcal{L}^{19}(d) = (3,1,2)^s & \mathcal{L}^{20}(d) = (3,1)^s \\
\mathcal{L}^{21}(d) = (3)^s & \mathcal{L}^{22}(d) = (2) & \mathcal{L}^{23}(d) = (2,1) \\
\mathcal{L}^{24}(d) = (2,1)^s & \mathcal{L}^{25}(d) = (2)^s & \mathcal{L}^{26}(d) = (1) \\
\mathcal{L}^{27}(d) = (1)^s & &
\end{array}$$

So, granting the theorem, we have $j_{S_1^3}(\omega_5) = \omega_{29}$ as the cardinals below $j_{S_1^3}(\omega_5)$ correspond to the descriptions listed (in the same order).

To prove the theorem, we need to show both upper and lower bounds. The upper bound follows from the following lemma.

LEMMA 4.17. *For $d \in \mathcal{S}_{m,n}$ non-minimal, $(d; S_1^m) \leq (\mathcal{L}(d); S_1^m)^+$. If d is $\mathcal{L}$-minimal, then $(d; S_1^m) \leq \omega_1$.*

PROOF. We consider the case $d = (4,1)$ in the above example. So, $\mathcal{L}(d) = (4,1,3)$. Let $\theta < (d; S_1^m)$. Then,

$$\forall^*_{S_1^m}[f]\, \theta([f]) < (d; f).$$

Thus,

$$\forall^*_{S_1^m}[f]\, \forall^*_{W_1^{n-1}}\vec{\alpha}\ \theta(f)(\vec{\alpha}) < (d; f)(\vec{\alpha}) = f(2)(\alpha_1, \alpha_4)$$
$$= \sup_{\beta < \alpha_4} f(\alpha_2, \beta, \alpha_4).$$

So,

$$\forall^*_{S_1^m}[f]\,\forall^*_{W_1^{n-1}}\vec{\alpha}\,\exists\beta < \alpha_4\;\; \theta(f)(\vec{\alpha}) < f(\alpha_1, \beta, \alpha_4).$$

A partition argument shows that for almost all f there is a $g : \omega_1 \to \omega_1$ such that

$$\forall^*_{S_1^m}[f]\,\forall^*_{W_1^{n-1}}\vec{\alpha}\;\; \theta(f)(\vec{\alpha}) < f(\alpha_1, g(\alpha_3), \alpha_4).$$

Thus, we have:

$$\forall^*_{S_1^m}[f]\,\exists g : \omega_1 \to \omega_1\,\forall^*_{W_1^{n-1}}\vec{\alpha}\;\; \theta(f)(\vec{\alpha}) < f(\alpha_1, g(\alpha_3), \alpha_4).$$

We now partition the functions $f : {<_3\to}\ \omega_1$ of the correct type according to whether:

$$\forall^*_{W_1^{n-1}}\vec{\alpha}\;\; \theta(f)(\vec{\alpha}) < f(\alpha_1, \alpha_3 + 1, \alpha_4).$$

On the homogeneous side of the partition, the stated property holds. For if C were c.u.b. and homogeneous for the contrary side, fix $f : {<_3\to}\ C$ of the correct type and such that for some $g : \omega_1 \to \omega_1$, which we now fix, we have:

$$\forall^*_{W_1^{n-1}}\vec{\alpha}\;\; \theta(f)(\vec{\alpha}) < f(\alpha_1, g(\alpha_3), \alpha_4).$$

Let $D \subseteq \omega_1$ be a c.u.b. set closed under g. Let $\ell : \omega_1 \to \omega_1$ be the continuous enumeration of D. Define $f' : {<_3\to}\ \omega_1$ by $f'(\beta_1, \beta_2, \beta_3) = f(\ell(\beta_1), \ell(\beta_2), \ell(\beta_3))$. Clearly f' is also of the correct type, $[f']_{W_1^3} = [f]_{W_1^w}$, and f' has range in C. So, by the homogeneity of C for contrary side we have (note that $\theta(f) = \theta([f]_{W_1^3})$ only depends on the equivalence class of f):

$$\forall^*_{W_1^{n-1}}\vec{\alpha}\;\; \theta(f)(\vec{\alpha}) = \theta(f')(\vec{\alpha}) \geq f'(\alpha_1, \alpha_3 + 1, \alpha_4).$$

However, the definition of f and ℓ give that $f'(\alpha_1, \alpha_3 + 1, \alpha_4) > f(\alpha_1, g(\alpha_3), \alpha_4)$ almost everywhere, a contradiction.

So, let C be c.u.b. and homogeneous for the stated side of the partition. Fix $x \in \omega^\omega$ such that $|T_x \restriction \alpha| > N_C(\alpha)$ almost everywhere (recall $N_C(\beta)$ is the least element of C greater than β). It follows that there a $\theta' < (\mathcal{L}(d); S_1^m)$ such that

$$\forall^*_{S_1^m}[f]\,\forall^*_{W_1^{n-1}}\vec{\alpha}\;\; \theta(f)(\vec{\alpha}) = f(\alpha_1, |T_x \restriction \alpha_3(\theta'(f)(\vec{\alpha}))|, \alpha_4).$$

where $|T_x \restriction \alpha(\beta)|$ denotes the rank of β in $T_x \restriction \alpha$.

The map $\delta' \mapsto \delta$ defined by

$$\forall^*_{S_1^m}[f]\,\forall^*_{W_1^{n-1}}\vec{\alpha}\,\delta(f)(\vec{\alpha}) = f(\alpha_1, |T_x \restriction \alpha_3(\delta'(f)(\vec{\alpha}))|, \alpha_4).$$

defines a map from $(\mathcal{L}(d); S_1^m)$ onto θ, so $\theta < (\mathcal{L}(d); S_1^m)^+$. ⊣

The lower bound follows from the following lemma.

LEMMA 4.18. $(d; S_1^m) \geq \omega_{|d|+1}$, *where* $|d|$ *denotes the rank of* d *in the* $\mathcal{L}$ *ordering.*

PROOF. Let k be the rank of d in the $\mathcal{L}$-order. Thus there are k elements of $\mathcal{S}_{m,n}$ below d, say $d_1 < \cdots < d_k$.

We define an embedding π from $\omega_{k+1} = j_{W_1^k}(\omega_1)$ into (d, S_1^m).

For $g\colon (\omega_1)^k \to \omega_1$, define $\pi([g]_{W_1^k}) = \theta$ where:

$$\forall^*_{S_1^m} f \; \forall^*_{W_1^{n-1}} \vec{\alpha} \; \theta(f)(\vec{\alpha}) = g((d_1; f)(\vec{\alpha}), \dots, (d_k; f)(\vec{\alpha})).$$

That this works follows from the following observations.

(1) For fixed g, $\theta(f)$ depends only on $[f]_{W_1^m}$, since if $[f_1] = [f_2]$ then almost all $\vec{\alpha}$ will be in a c.u.b. C such that $f_1 \restriction C^{m-1} = f_2 \restriction C^{m-1}$.

(2) If $[g_1]_{W_1^k} = [g_2]_{W_1^k}$, and say $g_1 \restriction C^k = g_2 \restriction C^k$, then S_1^m almost all $[f]$ are represented by $f\colon <_3 \to C$ of the correct type. Thus, $\forall^*_{S_1^m} f \; \forall^*_{W_1^{n-1}} \vec{\alpha}$ we have:

$$g_1((d_1; f)(\vec{\alpha}), \dots, (d_k; f)(\vec{\alpha})) = g_2((d_1; f)(\vec{\alpha}), \dots, (d_k; f)(\vec{\alpha})).$$

Thus, θ depends only on $[g]_{W_1^k}$.

(3) For any fixed g, almost all f have range in a c.u.b. set closed under $g(1)$. Thus,

$$\theta(f)(\vec{\alpha}) = g((d_1; f)(\vec{\alpha}), \dots, (d_k; f)(\vec{\alpha})) < (d; f)(\vec{\alpha}).$$

So, $\pi([g]) < (d; S_1^m)$.

(1) and (2) show π is well-defined, and (3) shows that π embeds $j_{W_1^k}(\omega_1)$ into $(d; S_1^m)$. ⊣

§5. **Non-trivial descriptions and the second exemplary problem.** To motivate non-trivial descriptions, we consider the following problem.

Second exemplary problem. Compute $j_{S_1^{m_1}} \circ j_{S_1^{m_2}} \circ \cdots \circ j_{S_1^{m_t}}(\omega_n)$.

In the first exemplary problem we computed a single ultrapower $j_{S_1^m}(\omega_n)$. Now we compute the iterated ultrapower $j_{S_1^{m_1}}(\cdots(j_{S_1^{m_t}}(\omega_n))\cdots)$.

On the one hand, our previous formula computes this value. We simply iterate the previous formula. To be specific, we consider the following particular example.

EXAMPLE 5.1. $j_{S_1^2} \circ j_{S_1^2}(\omega_3) = j_{S_1^2}(\omega_7) = \omega_{43}$.

We wish, however, to analyze the iterated ultrapower directly. This leads to the general (next level) notion of description.

Set-Up: We are given a finite sequence of measures $S_1^{m_1}, \dots, S_1^{m_t}$ and an integer $n-1$ (which corresponds to the measure W_1^{n-1} which we use since $j_{W_1^{n-1}}(\omega_1) = \omega_n$).

We will define for each such sequence of measures and $n-1$ a set of descriptions

$$\mathcal{D}_{n-1}(S_1^{m_1},\dots,S_1^{m_t}).$$

Slightly more generally (which we need for the actual inductive analysis), we allow a finite sequence of measures $K_1,\dots,K_t$ each of the form S_1^m or W_1^m. So, we define the set of descriptions $\mathcal{D}_{n-1}(K_1,\dots,K_t)$.

Such a d will give an ordinal $(d;K_1,\dots,K_t)$ as follows.

$(d;K_1,\dots,K_t)$ is represented w.r.t. K_1 by the function $[h_1]\mapsto(d;h_1;K_2,\dots,K_t)$. Here $h_1\colon <_{m_1}\to\omega_1$ is of the correct type if $K_1=S_1^{m_1}$, and $h_1\colon m_1\to\omega_1$ if $K_1=W_1^{m_1}$.

$(d;h_1;K_2,\dots,K_t)$ is represented w.r.t. K_2 by the function $[h_2]\mapsto(d;h_1,h_2,K_3,\dots,K_t)$.

We continue in this manner and finally, $(d;h_1,\dots,h_t)<\omega_n$ is represented with respect W_1^{n-1} by the function

$$(\alpha_1,\dots,\alpha_{n-1})\mapsto(d;h_1,\dots,h_t)(\vec{\alpha})<\omega_1.$$

It remains to define $\mathcal{D}_{n-1}(\vec{K})$ and the interpretation $(d;h_1,\dots,h_t)(\vec{\alpha})$.

Main Point: We now allow compositions of the h's. The description will now tell us how to compose these functions. Only composition one way will be allowed. For example $h_1(1)\circ h_2(1)$ will be allowed, but not $h_2(1)\circ h_1(1)$. We need this to get a well-defined ordinal value for the interpretation.

We define the set of descriptions $\mathcal{D}_{n-1}(\vec{K})$ and their interpretation $(d;h_1,\dots,h_t)(\vec{\alpha})$ simultaneously through the following inductive definition. Along the way we also define a "numerical value" $k(d)$ which will have value in $\{1,2,\dots,t\}\cup\{\infty\}$.

Definition 5.2 (Definition of $\mathcal{D}=\mathcal{D}_{n-1}(K_1,\dots,K_t)$).
$\mathcal{D}$ is defined through the following cases.

(1) We allow $d=\cdot_i$ for $1\le i\le n-1$. Set $k(d)=\infty$.
  Interpretation: $(d;\vec{h})(\vec{\alpha})=\alpha_i$.
(2) If $K_k=W_1^{m_k}$, we allow $d=(k;i)$ for $1\le i\le m_k$. Set $k(d)=k$.
  Interpretation: $(d;\vec{h})(\vec{\alpha})=h_k(i)$.
(3) If $K_k=S_1^m$ we allow $d=(k;d_m,d_1,\dots,d_\ell)^{(s)}$ where $k(d_1),\dots,k(d_\ell),k(d_m)>k$ and $d_1<\dots<d_\ell<d_m$ (defined below).
  Interpretation:

$$(d;\vec{h})(\vec{\alpha})=h_k^{(s)}(\ell+1)((d_1;\vec{h})(\vec{\alpha}),\dots,(d_\ell;\vec{h})(\vec{\alpha}),(d_m;\vec{h})(\vec{\alpha})).$$

The descriptions from cases (1) and (2) are called *basic descriptions*. The descriptions from (3) are called *non-basic descriptions*. The non-basic descriptions are built up from other descriptions, having a larger value of $k(d)$.

The descriptions are ordered as follows.

DEFINITION 5.3 (Definition of Ordering). We define $d < d'$ iff

$$\forall^* h_1, \ldots, h_t \, \forall^* \vec{\alpha} \, (d; \vec{h})(\vec{\alpha}) < (d'; \vec{h})(\vec{\alpha}).$$

We can describe the $<$ relation directly. It is also generated by a *lowering operator* $\mathcal{L}$ as before. We give a direct definition of $\mathcal{L}$.

We define for $k \in \{1, \ldots, t\} \cup \{\infty\}$ and $d \in \mathcal{D}$ with $k(d) \geq k$ a *partial lowering* $\mathcal{L}_k(d)$. We will take $\mathcal{L}(d) = \mathcal{L}_1(d)$.

DEFINITION 5.4 (Definition of $\mathcal{L}_k$). We define $\mathcal{L}_k$ through the following cases.

(1) $k = \infty$. In this case, $d = \cdot_i$. We define $\mathcal{L}_k(d) = i - 1$ if $i > 1$ and otherwise d is $\mathcal{L}_k$ minimal.

In the remaining cases, $1 \leq k \leq t$.

(2) $K_k = W_1^m$. If $k(d) = k$, then $d = (k, i)$ for some $1 \leq i \leq m$. We set $\mathcal{L}_k(d) = i - 1$ unless $i = 1$ in which case d is $\mathcal{L}_k$-minimal. If $k(d) > k$, then $\mathcal{L}_k(d) = \mathcal{L}_{k+1}(d)$ unless d in $\mathcal{L}_{k+1}$ minimal in which case $\mathcal{L}_k(d) = (k, m)$.

(3) $K_k = S_1^m$ and $k(d) > k$. If d is $\mathcal{L}_{k+1}$-minimal then d is also $\mathcal{L}_k$-minimal. Otherwise set $\mathcal{L}_k(d) = (k; \mathcal{L}_{k+1}(d))$.

(4) $K_k = S_1^m$ and $k(d) = k$. So, $d = (k; d_m, d_1, \ldots, d_\ell)^{(s)}$.

(a) $\ell = m - 1$ and s does not appear. $\mathcal{L}_k(d) = (k; d_m, d_1, \ldots, d_\ell)^s$.

(b) $\ell = m - 1$ and s does appears. $\mathcal{L}_k(d) = (k; d_m, d_1, \ldots, \mathcal{L}_{k+1}(d_\ell))$ if $\mathcal{L}_{k+1}(d_\ell)$ is defined and $> d_{\ell-1}$ (if $\ell > 1$). Otherwise $\mathcal{L}_k(d) = (k; d_m, d_1, \ldots, d_{\ell-1})^s$.

If $m = 1$ (so $\ell = 0$) or if $\ell = 1$ and $\mathcal{L}_{k+1}(d_\ell)$ is not defined, set $\mathcal{L}_k(d) = d_m$.

(c) $\ell < m - 1$ and s does not appear. $\mathcal{L}_k(d) = (k; d_m, d_1, \ldots, d_\ell, \mathcal{L}_{k+1}(d_m))$ if $\mathcal{L}_{k+1}(d_m)$ is defined and $> d_\ell$ (if $\ell \geq 1$). Otherwise, $\mathcal{L}_k(d) = (k; d_m, d_1, \ldots, d_\ell)^s$.

(d) $\ell < m - 1$ and s appears. $\mathcal{L}_k(d) = (k; d_m, d_1, \ldots, \mathcal{L}_{k+1}(d_\ell))^s$ if $\mathcal{L}_{k+1}(d_\ell)$ is defined and $> d_{\ell-1}$ (if $\ell > 1$). Otherwise, $\mathcal{L}_k(d) = (k; d_m, d_1, \ldots, d_{\ell-1})^s$ if $\ell > 1$ and $= d_m$ for $\ell = 1$.

CLAIM 5.5. The $(d; S_1^{m_1}, \ldots, S_k^{m_t})$ for $d \in \mathcal{D}_{n-1}(S_1^{m_1}, \ldots S_t^{m_t})$ correspond to the cardinals below

$$j_{S_1^{m_1}} \circ j_{S_1^{m_2}} \circ \cdots \circ j_{S_1^{m_t}}(\omega_n).$$

Returning to our example $j_{S_1^2} \circ j_{S_1^2}(\omega_3)$, we illustrate the above definition by listing all the descriptions in $\mathcal{D}_2(S_1^2, S_1^2)$ in decreasing order as generated by the $\mathcal{L}$ operation. These, we will see, generates the cardinals below $j_{S_1^2} \circ j_{S_1^2}(\omega_3)$.

EXAMPLE 5.6. Descriptions corresponding to $j_{S_1^2} \circ j_{S_1^2}(\omega_3)$.

$$d = (1; (2; \cdot_2)) \qquad \mathcal{L}(d) = (1; (2; \cdot_2); (2; \cdot_2, \cdot_1))$$
$$\mathcal{L}^2(d) = (1; (2; \cdot_2); (2; \cdot_2, \cdot_1))^s \qquad \mathcal{L}^3(d) = (1; (2; \cdot_2); (2; \cdot_2, \cdot_1)^s)$$
$$\mathcal{L}^4(d) = (1; (2; \cdot_2); (2; \cdot_2, \cdot_1)^s)^s \qquad \mathcal{L}^5(d) = (1; (2; \cdot_2); \cdot_2)$$
$$\mathcal{L}^6(d) = (1; (2; \cdot_2); \cdot_2)^s \qquad \mathcal{L}^7(d) = (1; (2; \cdot_2); (2; \cdot_1))$$
$$\mathcal{L}^8(d) = (1; (2; \cdot_2); (2; \cdot_1))^s \qquad \mathcal{L}^9(d) = (1; (2; \cdot_2); \cdot_1)$$
$$\mathcal{L}^{10}(d) = (1; (2; \cdot_2); \cdot_1)^s \qquad \mathcal{L}^{11}(d) = (2; \cdot_2)$$
$$\mathcal{L}^{12}(d) = (1; (2; \cdot_2, \cdot_1)) \qquad \mathcal{L}^{13}(d) = (1; (2; \cdot_2, \cdot_1); (2, \cdot_2, \cdot_1)^s)$$
$$\mathcal{L}^{14}(d) = (1; (2; \cdot_2, \cdot_1); (2, \cdot_2, \cdot_1)^s)^s \qquad \mathcal{L}^{15}(d) = (1; (2; \cdot_2, \cdot_1); \cdot_2)$$
$$\mathcal{L}^{16}(d) = (1; (2; \cdot_2, \cdot_1); \cdot_2)^s \qquad \mathcal{L}^{17}(d) = (1; (2; \cdot_2, \cdot_1); (2; \cdot_1))$$
$$\mathcal{L}^{18}(d) = (1; (2; \cdot_2, \cdot_1); (2; \cdot_1))^s \qquad \mathcal{L}^{19}(d) = (1; (2; \cdot_2, \cdot_1); \cdot_1)$$
$$\mathcal{L}^{20}(d) = (1; (2; \cdot_2, \cdot_1); \cdot_1)^s \qquad \mathcal{L}^{21}(d) = (2; \cdot_2, \cdot_1)$$
$$\mathcal{L}^{22}(d) = (1; (2; \cdot_2, \cdot_1)^s) \qquad \mathcal{L}^{23}(d) = (1; (2; \cdot_2, \cdot_1)^s, \cdot_2)$$
$$\mathcal{L}^{24}(d) = (1; (2; \cdot_2, \cdot_1)^s, \cdot_2)^s \qquad \mathcal{L}^{25}(d) = (1; (2; \cdot_2, \cdot_1)^s, (2; \cdot_1))$$
$$\mathcal{L}^{26}(d) = (1; (2; \cdot_2, \cdot_1)^s, (2; \cdot_1))^s \qquad \mathcal{L}^{27}(d) = (1; (2; \cdot_2, \cdot_1)^s, \cdot_1)$$
$$\mathcal{L}^{28}(d) = (1; (2; \cdot_2, \cdot_1)^s, \cdot_1)^s \qquad \mathcal{L}^{29}(d) = (2; \cdot_2, \cdot_1)^s$$
$$\mathcal{L}^{30}(d) = (1; \cdot_2) \qquad \mathcal{L}^{31}(d) = (1; \cdot_2, (2; \cdot_1))$$
$$\mathcal{L}^{32}(d) = (1; \cdot_2, (2; \cdot_1))^s \qquad \mathcal{L}^{33}(d) = (1; \cdot_2, \cdot_1)$$
$$\mathcal{L}^{34}(d) = (1; \cdot_2, \cdot_1)^s \qquad \mathcal{L}^{35}(d) = \cdot_2$$
$$\mathcal{L}^{36}(d) = (1; (2; \cdot_1)) \qquad \mathcal{L}^{37}(d) = (1; (2; \cdot_1), \cdot_1)$$
$$\mathcal{L}^{38}(d) = (1; (2; \cdot_1), \cdot_1)^s \qquad \mathcal{L}^{39}(d) = (2; \cdot_1)$$
$$\mathcal{L}^{40}(d) = (1; \cdot_1) \qquad \mathcal{L}^{41}(d) = \cdot_1$$

So, granting that these are the cardinals below the iterated ultrapower, we again see that $j_{S_1^2} \circ j_{S_1^2}(\omega_3) = \omega_{43}$.

REMARK 5.7. The iterated ultrapower is not the same as the ultrapower by the product measure in the AD context. For example, $j_{S_1^2} \circ j_{S_1^2}(\omega_3) = \omega_{43}$ but a computation similar to that above shows that $j_{S_1^2 \times S_1^2}(\omega_3) = \omega_{11}$.

We must again show upper-bound and lower-bound results. The proof of the lower bound result is exactly as before (using now the non-trivial descriptions), so we omit it. The upper-bound follows from the following lemma, in which we use the following notation.

For $g : \omega_1 \to \omega_1$ define $\theta = (g; d; K_1, \ldots, K_t)$ by:

$$\forall^* h_1, \ldots, h_t \, \forall^* \vec{\alpha} \; \theta([f_1], \ldots, [f_t])(\vec{\alpha}) = g((d; f_1, \ldots, f_t)(\vec{\alpha})).$$

LEMMA 5.8. *If* $\theta < (d; K_1, \ldots, K_t)$, *then there is a* $g : \omega_1 \to \omega_1$ *such that* $\theta < (g; \mathcal{L}(d); K_1, \ldots, K_t)$.

A Kunen tree argument, as in the case of trivial descriptions from the previous section, then shows that $(d; \vec{K}) \leq (\mathcal{L}(d); \vec{K})^+$.

We illustrate the proof of the lemma by considering the following case from our example:

$$d = \mathcal{L}^{16}(-) = (1; (2; \cdot_2, \cdot_1), \cdot_2)^s$$
$$\mathcal{L}(d) = \mathcal{L}^{17}(-) = (1; (2; \cdot_2, \cdot_1), (2; \cdot_1)).$$

In using our notational convention, we will write $\theta(h_1, \ldots, h_t)$ to abbreviate $\theta([h_1], \ldots, [h_t])$. We will also write $\forall^*_{S_1^{m_i}} h_i$ to abbreviate $\forall_{S_1^{m_i}} [h_i]$.

Suppose $\theta < (d; S_1^2, S_1^2)$. Then,

$$\forall^*_{S_1^2} h_1\, \forall^*_{S_1^2} h_2\, \forall^*_{W_1^2} \alpha_1, \alpha_2\, \theta(h_1, h_2)(\vec{\alpha}) < (d; h_1, h_2)(\vec{\alpha})$$
$$= \sup_{\beta < \alpha_2} h_1(\beta, h_2(\alpha_1, \alpha_2))$$

So,

$$\forall^*_{S_1^2} h_1\, \forall^*_{S_1^2} h_2\, \forall^*_{W_1^2} \alpha_1, \alpha_2 \exists \beta < \alpha_2\, \theta([h_1], [h_2])(\vec{\alpha}) < h_1(\beta, h_2(\alpha_1, \alpha_2))$$

Hence,

$$\forall^*_{S_1^2} h_1\, \forall^*_{S_1^2} h_2\, \exists g : \omega_1 \to \omega_1\, \forall^*_{W_1^2} \alpha_1, \alpha_2\, \theta(h_1, h_2)(\vec{\alpha}) < h_1(g(\alpha_1), h_2(\alpha_1, \alpha_2))$$

For any h_1 and ordinal $\theta(h_1)$ such that $\forall^*_{S_1^2}[h_2]\, \exists g \cdots$ (previous line), consider the partition:

$\mathcal{P}(h_1)$: We partition pairs (h_2, g) where $h_2\colon\ <_2\!\to \omega_1$ is of the correct type, $g : \omega_1 \to \omega_1$ is of the correct type, and

$$h_2(1)(\gamma) < g(\gamma) < h_2(0, \gamma + 1)$$

for all γ according to whether:

$$\forall^*_{W_1^2} \alpha_1, \alpha_2\, \theta(h_1, h_2)(\vec{\alpha}) < h_1(g(\alpha_1), h_2(\alpha_1, \alpha_2))$$

An easy argument (as in the proof for trivial descriptions in the previous section) shows that on the homogeneous side the stated property must hold. A homogeneous set C for the partition (which depends on h_1) gives the witnesses for the following:

$$\forall^*_{S_1^2} h_1\, \exists \text{c.u.b. } C \subseteq \omega_1\, \forall^*_{S_1^2} h_2\, \forall^*_{W_1^2} \alpha_1, \alpha_2$$
$$\theta(h_1, h_2)(\vec{\alpha}) < h_1(N_C(h_2(1)(\alpha_1)), h_2(\alpha_1, \alpha_2))$$

where again $N_C(\beta)$ denotes the least element of C greater than β. Next consider the partition (here N_{h_1} denotes the next element in the range of h_1):

$\mathcal{P}$: We partition $h_1, g\colon\ <_2\!\to \omega_1$ of the correct type with $h_1(\alpha_1, \alpha_2) < g(\alpha_1, \alpha_2) < N_{h_1}(\alpha_1, \alpha_2)$ according to whether:

$$\forall^*_{S_1^2}[h_2]\, \forall^*_{W_1^2} \alpha_1, \alpha_2\, \theta(h_1)(h_2)(\vec{\alpha}) < g(h_2(1)(\alpha_1), h_2(\alpha_1, \alpha_2))$$

On the homogeneous side the stated property holds, by an argument similar that used for the proof in the previous section for trivial descriptions. Fixing $C \subseteq \omega_1$ homogeneous we get:

$$\forall_{S_1^2} h_1 \, \forall^*_{S_1^2} h_2 \, \forall^*_{W_1^2} \alpha_1, \alpha_2 \; \theta(h_1, h_2)(\vec{\alpha}) < N_C(h_2(1)(\alpha_1), h_2(\alpha_1, \alpha_2))$$
$$= (N_C; \mathcal{L}(d); h_1, h_2)(\vec{\alpha})$$

and we are done.

§6. The inductive analysis. We have used trivial descriptions to solve the first exemplary problem, and non-trivial descriptions to solve the second. We now indicate briefly how these same descriptions are used in the actual inductive analysis to analyze the cardinal structure between $\boldsymbol{\delta}^1_3$ and $\boldsymbol{\delta}^1_5$. We omit most of the proofs in this section, since they are similar to those of the exemplary problems already given.

Analysis between $\boldsymbol{\delta}^1_3$ and $\boldsymbol{\delta}^1_5$. Recall that for a sequence of measures $K_1, \ldots, K_t$, with each $K_k = W_1^{m_k}$ or $S_1^{m_k}$, and each n, we have a set $\mathcal{D}_n(\vec{K})$ of descriptions defined.

For the next level of the analysis, we extend this set of descriptions of the previous section (the level 2 descriptions in the inductive analysis) in a rather trivial way (to get the level 3 descriptions). Namely, we consider objects of the form (d), where d is a description as before, and also $(d)^s$ (this is the symbol s applied to the entire description, and is not to be confused with the occurrences of the symbol s inside d itself). As before, we write $(d)^{(s)}$ to mean s may or may not appear. The extra set of parentheses only serves to distinguish the same d when considered as a level 2 or a level 3 description. In practice, there will be no confusion, and so we often drop the parentheses and just write d or d^s. We call this slightly extended set of objects $\mathcal{D}'_n(\vec{K})$.

Assume inductively the weak partition relation on $\boldsymbol{\delta}^1_3$ (i.e., we have analyzed the measures on λ_3 and have as a byproduct the weak partition relation on $\boldsymbol{\delta}^1_3$). We extend our canonical family of measures one level higher.

Definition 6.1. W_3^m is the measure on $\boldsymbol{\delta}^1_3$ induced by the weak partition relation on $\boldsymbol{\delta}^1_3$, functions $f : \omega_{m+1} \to \boldsymbol{\delta}^1_3$ of the correct type, and the measure S_1^m on ω_{m+1}.

The measures W_3^m on $\boldsymbol{\delta}^1_3$ dominate all of the measures on $\boldsymbol{\delta}^1_3$ in the following sense.

Fact 6.2. For any measure μ on $\boldsymbol{\delta}^1_3$, there is an m such that $j_\mu(\boldsymbol{\delta}^1_3) < j_{W_3^m}(\boldsymbol{\delta}^1_3)$.

Remark 6.3. We note that this sense of domination is different from that considered before for the measures S_1^m. In the inductive analysis, the first type of domination is called the local embedding theorem, and the above sense is called the global embedding theorem (local refers to pointwise behavior below δ^1_{2n+1} while global refers to the behavior at δ^1_{2n+1}).

The homogeneous tree construction shows that $\lambda_5 \le \sup_\mu(\delta^1_3)$, where μ ranges over the measures on δ^1_3, and so $\lambda_5 \le \sup_m j_{W_3^m}(\delta^1_3)$.

We define an ordinal $((d)^{(s)}; W_3^n; K_1, \dots, K_t) < j_{W_3^n}(\delta^1_3)$ for $(d)^{(s)} \in \mathcal{D}'_n(\vec{K})$. We define this in the usual iterated way, where for $f: \omega_{n+1} \to \delta^1_3$, $h_1, \dots, h_t$ (here $h_i: <_i \to \omega_1$) we have:

$$((d); f; h_1, \dots, h_t) = f((d; h_1, \dots, h_t)).$$

If the symbol s appears, we define:

$$((d)^s; f; h_1, \dots, h_t) = \sup\{f(\beta) : \beta < f((d; h_1, \dots, h_t))\}.$$

These ordinals are well-defined provided $(d)^{(s)}$ satisfies the following.

Definition 6.4. $(d) \in \mathcal{D}'_n(K_1, \dots, K_t)$ satisfies *condition D* if $\forall^* \vec{h}\ (d; \vec{h})$ is almost everywhere of the correct type. $(d)^s$ satisfies condition D if $\forall^* \vec{h}\ (d; \vec{h})$ is the supremum of ordinals represented by functions of the correct type.

We extend the $\mathcal{L}$ operation to this slightly expanded set of descriptions as follows. We set $\mathcal{L}'((d)) = (d)^s$, and $\mathcal{L}'((d)^s) = (\mathcal{L}(d))$. Let $\mathcal{L}((d)^{(s)})$ be the least iterate of $\mathcal{L}'$ satisfying condition D (if one doesn't exist, we declare $d^{(s)}$ minimal with respect to $\mathcal{L}$).

The following theorem gives the cardinal structure between δ^1_3 and δ^1_5 in terms of the descriptions.

Theorem 6.5. *The successor cardinals below λ_5 are precisely the ordinals of the form $((d)^{(s)}; W_3^n, K_1, \dots, K_t)$ for some $d \in \mathcal{D}'_n(\vec{K})$ with $(d)^{(s)}$ satisfying condition D.*

The cardinal corresponding to $((d)^{(s)}; W_3^m; \vec{K})$ is given by the rank of this tuple in the ordering generated by the following relation:

$$((d)^{(s)}; W_3^m, K_1, \dots, K_t) < (\mathcal{L}((d)^{(s)}); W_3^m; K_1, \dots, K_t, K_{t+1}).$$

More precisely, the rank gives the number of cardinals above δ^1_3 (so rank 0 corresponds to δ^1_3). Also, ranks are computed in the slightly non-standard manner by $|x| = (\sup\{|y| : y < x\}) + 1$ (so all ranks above 0 are successor ordinals).

Note that aside from applying the lowering operator to the description, we lengthen the sequence of measures. This is the main difference between the computation of the second exemplary problem and that of the cardinal structure below δ^1_5.

The upper bound for the cardinal corresponding to $(d^{(s)}; W_3^m; \vec{K})$ follows from the following theorem.

For $g : \boldsymbol{\delta}^1_3 \to \boldsymbol{\delta}^1_3$, define $(g; d; W^n_3, K_1, \ldots, K_t)$ by:

$$\forall^*_{W^n_3}[f]\, \forall^* h_1, \ldots h_t\ (g; d; f, h_1, \ldots, h_t) = g(f^{(s)}((d; h_1, \ldots, h_t))).$$

The next theorem is our "main theorem" for non-trivial descriptions.

THEOREM 6.6. *If* $\theta < (d; W^n_3, \vec{K})$, *then there is a* $g : \boldsymbol{\delta}^1_3 \to \boldsymbol{\delta}^1_3$ *such that* $\theta < (g; \mathcal{L}(d); W^n_3; \vec{K})$.

The lower bound follows by embedding arguments roughly similar to those of the previous sections for the exemplary problems.

We illustrate the above results with an example.

EXAMPLE 6.7. Consider $((d); W^2_3; S^2_1, S^2_1)$ where (this is from an example considered previously, and here $d_M = (1; (2; \cdot_2))$ denotes the maximal description in $\mathcal{D}_2(S^2_1, S^2_1)$):

$$d = \mathcal{L}^{26}(1; (2; \cdot_2)) = \mathcal{L}^{26}(d_M) = (1; (2; \cdot_2, \cdot_1)^s, (2; \cdot_1))^s$$

We compute the cardinal corresponding to (d) as follows:

$$|((d); W^2_3, S^2_1, S^2_1)| = \sup_{K_3} |(\mathcal{L}^{26}(d_M))^s; W^2_3, S^2_1, S^2_1, K_3)| + 1$$

$$= \sup_{\vec{K}} |((\mathcal{L}^{27}(d_M)); W^2_3, S^2_1, S^2_1, \vec{K})| + \omega + 1$$

$$= \sup_{\vec{K}} |((\mathcal{L}^{29}(d_M)); W^2_3, S^2_1, S^2_1, \vec{K})| + \omega + \omega + 1$$

$$= \sup_{\vec{K}} |((\mathcal{L}^{30}(d_M)); W^2_3, S^2_1, S^2_1, \vec{K})| + \omega + \omega + \omega + 1$$

$$= \sup_{\vec{K}} |((\mathcal{L}^{31}(d_M)); W^2_3, S^2_1, S^2_1, \vec{K})| + \omega^\omega + \omega + \omega + \omega + 1$$

$$= \sup_{\vec{K}} |((\mathcal{L}^{33}(d_M)); W^2_3, S^2_1, S^2_1, \vec{K})| + \omega + \omega^\omega + \omega + \omega + \omega + 1$$

$$= \sup_{\vec{K}} |((\mathcal{L}^{35}(d_M)); W^2_3, S^2_1, S^2_1, \vec{K})| + \omega + \omega + \omega^\omega + \omega + \omega + \omega + 1$$

$$= \sup_{\vec{K}} |((\mathcal{L}^{36}(d_M)); W^2_3, S^2_1, S^2_1, \vec{K})| + \omega^\omega + \omega + \omega + \omega^\omega + \omega + \omega + \omega + 1$$

$$= \sup_{\vec{K}} |((\mathcal{L}^{37}(d_M)); W^2_3, S^2_1, S^2_1, \vec{K})| + \omega + \omega^\omega + \omega + \omega + \omega^\omega + \omega + \omega + \omega + 1$$

$$= \sup_{\vec{K}} |((\mathcal{L}^{39}(d_M)); W^2_3, S^2_1, S^2_1, \vec{K})| + \omega \cdot 2 + \omega^\omega + \omega + \omega + \omega^\omega + \omega + \omega + \omega + 1$$

$$= \sup_{\vec{K}} |((\mathcal{L}^{40}(d_M)); W^2_3, S^2_1, S^2_1, \vec{K})| + \omega \cdot 3 + \omega^\omega + \omega + \omega + \omega^\omega + \omega + \omega + \omega + 1$$

$$= \sup_{\vec{K}} |((\mathcal{L}^{41}(d_M)); W^2_3, S^2_1, S^2_1, \vec{K})| + \omega \cdot 4 + \omega^\omega + \omega + \omega + \omega^\omega + \omega + \omega + \omega + 1$$

$$= \omega^\omega \cdot 2 + \omega \cdot 3 + 1$$

That is, $((d); W_3^2, S_1^2, S_1^2) = \aleph_{\omega^\omega\cdot 2+\omega\cdot 3+1}$.

More on the Cardinal structure. The three normal measure on $\boldsymbol{\delta}_3^1$ correspond to the three regular cardinals below $\boldsymbol{\delta}_3^1$, namely ω, ω_1, ω_2. These regular cardinals are of the form $j_\mu(\boldsymbol{\delta}_3^1)$ for μ one of these normal measures.

A description computation as above computes these to be:

$$j_{\mu_\omega}(\boldsymbol{\delta}_3^1) = \boldsymbol{\delta}_4^1 = \aleph_{\omega+2},$$
$$j_{\mu_{\omega_1}}(\boldsymbol{\delta}_3^1) = \aleph_{\omega\cdot 2+1},$$
$$j_{\mu_{\omega_2}}(\boldsymbol{\delta}_3^1) = \aleph_{\omega^\omega+1}.$$

Also, $j_{W_3^n}(\boldsymbol{\delta}_3^1) = \aleph_{\omega^{\omega^n}+1}$, and so $\lambda_5 = \aleph_{\omega^{\omega^\omega}}$, and $\boldsymbol{\delta}_5^1 = \aleph_{\omega^{\omega^\omega}} + 1$.

As the nest example shows, we can easily read off the cofinality from the description.

EXAMPLE 6.8. We compute the cofinality of $((d); W_3^2; S_1^2, S_1^2) = \aleph_{\omega^\omega\cdot 2+\omega\cdot 3+1}$ (from the previous example) where $d = \mathcal{L}^{26}(1; (2; \cdot_2)) = (1; (2; \cdot_2, \cdot_1)^s, (2; \cdot_1))^s$.

From the main theorem, ordinals of the form $(g; (d)^s; W_3^2, S_1^2, S_1^2)$ for $g\colon \boldsymbol{\delta}_3^1 \to \boldsymbol{\delta}_3^1$ are cofinal in $((d); W_3^2, S_1^2, S_1^2)$. Now, in evaluating $(g; (d)^s; f, h_1, h_2)$ we are evaluating g at $f^s((d; h_1, h_2)) = \sup_{\beta<(d;h_1,h_2)} f(\beta)$. This has the same cofinality as $(d; h_1, h_2)$. From the form of d it is easy to see that $(d; h_1, h_2)$ has uniform cofinality $(\alpha_1, \alpha_2) \mapsto \alpha_1$, and so $(d; h_1, h_2)$ has cofinality ω_1.

Thus, $(g; (d)^s; W_3^n, S_1^2, S_1^2)$ depends only on $[g]_{\mu_{\omega_1}}$. This gives a cofinal embedding from $j_{\mu_{\omega_1}}(\boldsymbol{\delta}_3^1)$ into $((d); W_3^n, S_1^2, S_1^2)$. It follows that $\mathrm{cof}(\aleph_{\omega^\omega\cdot 2+\omega\cdot 3+1}) = \aleph_{\omega\cdot 2+1}$.

Using the "linear" theory (described in the next section), we can efficiently compute the cofinalities of all the cardinals. For example, below $\boldsymbol{\delta}_5^1$ we have:

THEOREM 6.9. *Suppose* $\boldsymbol{\delta}_3^1 < \aleph_{\alpha+1} < \boldsymbol{\delta}_5^1$. *Let* $\alpha = \omega^{\beta_1} + \cdots + \omega^{\beta_n}$, *where* $\omega^\omega > \beta_1 \geq \cdots \geq \beta_n$ *be the normal form for* α. *Then*:

- *If* $\beta_n = 0$, *then* $\mathrm{cof}(\kappa) = \boldsymbol{\delta}_4^1 = \aleph_{\omega+2}$.
- *If* $\beta_n > 0$, *and is a successor ordinal, then* $\mathrm{cof}(\kappa) = \aleph_{\omega\cdot 2+1}$.
- *If* $\beta_n > 0$ *and is a limit ordinal, then* $\mathrm{cof}(\kappa) = \aleph_{\omega^\omega+1}$.

Moving past $\boldsymbol{\delta}_5^1$, in general we have $\boldsymbol{\delta}_{2n+1}^1 = \aleph_{\omega(2n-1)+1}$ where $\omega(0) = 1$ and $\omega(n+1) = \omega^{\omega(n)}$.

There are $2^{n+1}-1$ many regular cardinals below $\boldsymbol{\delta}_{2n+1}^1$. The regular cardinals between $\boldsymbol{\delta}_{2n-1}^1$ and $\boldsymbol{\delta}_{2n+1}^1$ correspond to the ultrapowers of $\boldsymbol{\delta}_{2n+1}^1$ by the normal measures on $\boldsymbol{\delta}_{2n+1}^1$, which correspond to the regular cardinals below $\boldsymbol{\delta}_{2n+1}^1$.

There is a canonical family of measures associated to each regular cardinal.

Below the projective hierarchy, the families are denoted:

$$W_1^m, S_1^{1,m}, W_3^m, S_3^{1,m}, S_3^{2,m}, S_3^{3,m},$$
$$W_5^m, S_5^{1,m}, \dots, S_5^{7,m}, W_7^m, \dots$$

W_{2n+1}^m is defined using the weak partition relation on $\boldsymbol{\delta}_{2n+1}^1$, functions

$$f\colon \operatorname{dom}(S_{2n-1}^{\ell,m}) \to \boldsymbol{\delta}_{2n+1}^1$$

of the correct type and the measure $S_{2n-1}^{\ell,m}$. Here $\ell = 2^n - 1$.

$S_{2n+1}^{1,m}$ is defined as S_1^m was defined using instead $\boldsymbol{\delta}_{2n+1}^1$.

$S_{2n+1}^{\ell,m}$ for $\ell > 1$ is defined using the strong partition relation on $\boldsymbol{\delta}_{2n+1}^1$, functions $F\colon \boldsymbol{\delta}_{2n+1}^1 \to \boldsymbol{\delta}_{2n+1}^1$ of the correct type and the measure μ on $\boldsymbol{\delta}_{2n+1}^1$. Here μ is the measure induced by the weak partition relation on $\boldsymbol{\delta}_{2n+1}^1$, functions $f\colon \operatorname{dom}(\nu) \to \boldsymbol{\delta}_{2n+1}^1$ and the measure ν, where ν is the mth measure in the $\ell - 1^{\text{st}}$ family.

General descriptions. The general level descriptions are defined inductively, and have *indices* associated to them.

Trivial descriptions are level-1 descriptions. They have empty index. These analyze the measure on $\boldsymbol{\delta}_1^1$ and compute $\boldsymbol{\delta}_3^1$.

Level-2 descriptions are those of the form $d = (k; d_m, d_1, \dots, d_\ell)^{(s)}$ we defined previously. If $d \in \mathcal{D}_n(\vec{K})$, we associate the index W_1^n to d i.e., $(d = d^{W_1^n})$. These analyze the measure on λ_3.

The level-3 descriptions are those of the form (d) or $(d)^s$ for d a level-2 description. These analyze the measures on $\boldsymbol{\delta}_3^1$ and compute $\boldsymbol{\delta}_5^1$. They also have index W_1^n.

As we noted before, there is not much difference between the level-2 and level-3 descriptions.

Level-4 and 5 descriptions analyze the measures on λ_5 and $\boldsymbol{\delta}_5^1$ respectively. They are defined relative to a sequence of measures $\vec{K}$ where $K_i \in W_1^m, \dots, S_3^{3,m}$ for level-4, and $(W_5^m, \vec{K})$ for level-5.

They have indices which are in turn sequences of measures of the form

$$(W_3^m, S_1^{1,m_1}, \dots, S_1^{1,m_t})$$

(we can use $W_1^{m_i}$ in place of a $S_1^{m_i}$ in this sequence).

A basic description at level-4 for such an index is a level-3 description in

$$\mathcal{D}(W_3^m, S_1^{1,m_1}, \dots, S_1^{1,m_t}).$$

We omit further details, and refer the reader to [4] for the complete definitions and analysis.

§7. **Linear analysis.** We now present an alternate way to describe the cardinal structure without using descriptions or iterated ultrapowers. This results in a more elementary presentation of the cardinal structure, however the description analysis is still needed to show it works. The proofs for the cardinals below $\boldsymbol{\delta}^1_5$ can be found in [6]. Also, [7] gives more details for the more general case below the projective ordinals.

Also, we need this analysis to show that all descriptions actually represent cardinals (though we do not need it to get the lower bounds for the projective ordinals).

Definition 7.1. Let $\mathfrak{A}_\alpha$ be the free associative left-distributive algebra on α generators $\{\mathsf{V}_\beta\}_{\beta<\alpha}$ using the operations $\oplus$, $\otimes$. Let $\mathfrak{A} = \bigcup_\alpha \mathfrak{A}_\alpha$.

We assign an ordinal height $o(v)$ to every term in $\mathfrak{A}$ inductively as follows.

Definition 7.2.

$$\begin{aligned} o(\mathsf{v}_0) &= 0, \\ o(\mathsf{v}_\alpha) &= \mathrm{ht}(\mathfrak{A}_\alpha) = \sup\{o(t)+1 \colon t \in \mathfrak{A}_\alpha\}, \\ o(s \oplus t) &= o(s) + o(t), \\ o(s \otimes t) &= o(s) \cdot o(t). \end{aligned}$$

So, $o(\mathsf{v}_0) = 0$, $o(\mathsf{v}_1) = 1$, $o(\mathsf{v}_2) = \omega$, $o(\mathsf{v}_3) = \omega^\omega$, $o(\mathsf{v}_4) = \omega^{\omega^2}$, $o(\mathsf{v}_n) = \omega^{\omega^{n-2}}$, $o(\mathsf{v}_\omega) = \omega^{\omega^\omega}$. For $\alpha \geq \omega$, $o(\mathsf{v}_\alpha) = \omega^{\omega^\alpha}$.

We assign to each generator v_α an order type $\mathrm{ot}(\mathsf{v}_\alpha)$ and a measure $\mu(\mathsf{v}_\alpha)$ on this order-type.

This will generate an assignment of an order-type and a collection of measures (a "germ") to each general term as in the following example.

Example 7.3. Consider the term

$$t = (((\mathsf{v}_3 \oplus \mathsf{v}_2 \oplus \mathsf{v}_1) \otimes \mathsf{v}_4) \otimes \mathsf{v}_2) \oplus ((\mathsf{v}_3 \oplus \mathsf{v}_2 \oplus \mathsf{v}_1) \otimes \mathsf{v}_1).$$

We can represent this as a tree as follows:

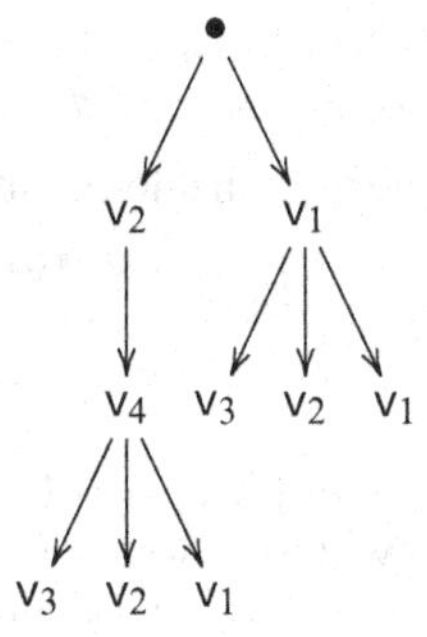

Suppose we know:

$$\begin{array}{ll} \mathrm{ot}(\mathsf{v}_1) = 1, & \mu(\mathsf{v}_1) = \text{principal measure} \\ \mathrm{ot}(\mathsf{v}_2) = \omega_1 & \mu(\mathsf{v}_2) = W_1^1 \\ \mathrm{ot}(\mathsf{v}_3) = \omega_2 & \mu(\mathsf{v}_2) = S_1^1 \\ \mathrm{ot}(\mathsf{v}_4) = \omega_3 & \mu(\mathsf{v}_2) = S_1^2 \end{array}$$

Then $\mathrm{ot}(t) = (\omega_2 + \omega_1 + 1) \cdot \omega_3 \cdot \omega_1 + (\omega_2 + \omega_1 + 1) \cdot 1 = \omega_3 \cdot \omega_1 + \omega_2 + \omega_1$. We identify $\mathrm{ot}(t)$ with the ordering on tuples $\langle i_0, \alpha_0, \ldots, i_k, \alpha_k \rangle$ where $(i_0, \ldots, i_k)$ corresponds to a terminal node in the tree, and for $(i_0, \ldots, i_\ell)$ a node in the tree, $\alpha_\ell < \mathrm{ot}(v)$. Here $v = v^{\vec{i}}$ is the variable corresponding to this node. We let $\mu(t)$ be the collection of measures $\mu(v^{\vec{i}})$.

We set $\mathrm{ot}(\mathsf{v}_{\omega(2n-1)}) = \boldsymbol{\delta}^1_{2n+1}$, $\mu(\mathsf{v}_{\omega(2n-1)}) = \omega$-cofinal normal measure on $\boldsymbol{\delta}^1_{2n+1}$. For example, $\mathrm{ot}(\mathsf{v}_\omega) = \boldsymbol{\delta}^1_3$, $\mathrm{ot}(\mathsf{v}_{\omega^{\omega^\omega}}) = \boldsymbol{\delta}^1_5$.

To $\mathsf{v}_{2+\omega(2n-1)+\alpha}$ (where $\alpha < \omega(2n+1)$) we associate the measure defined by the strong partition relation on $\boldsymbol{\delta}^1_{2n+1}$, functions $F : \boldsymbol{\delta}^1_3 \to \boldsymbol{\delta}^1_3$ of the correct type, and the following measure ν on $\boldsymbol{\delta}^1_{2n+1}$. Let t be the term in the algebra with $o(t) = \alpha$. ν is the measure induced by the weak partition relation on $\boldsymbol{\delta}^1_{2n+1}$, functions $f : \mathrm{ot}(t) \to \boldsymbol{\delta}^1_{2n+1}$ of the correct type, and the measures (germ) $\mu(t)$.

Remark 7.4. The measure ν is a measure on tuples $(\ldots \delta^{\vec{i}} \ldots)$ where $\vec{i} = (i_0, \ldots, i_k)$ is a terminal node in the tree corresponding to t. We identify these tuples of ordinals below $\boldsymbol{\delta}^1_{2n+1}$ with ordinals less than $\boldsymbol{\delta}^1_{2n+1}$ by ordering by the largest ordinal of the tuple first, then the next largest, and so forth; i.e., we use reverse lexicographic order. It turns out, however, that the ordering we use for this identification doesn't matter in the sense that we get the same ultrapower in Theorem 7.6 below (the only restriction is that we must order by the largest $\delta^{\vec{i}}$ first).

Example 7.5. $\mu(\mathsf{v}_{\omega+1}) = \omega$-cofinal normal measure on $\boldsymbol{\delta}^1_4$, $\mu(\mathsf{v}_{\omega+2})$ is measure on $\aleph_{\omega+3}$. $\mu(\mathsf{v}_{\omega\cdot 2})$ is the ω-cofinal normal measure on $\aleph_{\omega\cdot 2+1}$.

This assignment describes the cardinal structure as follows.

Theorem 7.6. *Let $t \in \mathfrak{A}$ with $\mathrm{ot}(t) < \boldsymbol{\delta}^1_{2n+1}$. Then $j_\nu(\boldsymbol{\delta}^1_{2n+1}) = \aleph_{\omega(2n-1)+o(t)+1}$ where $\nu = \nu(t)$ is the measure defined above.*

Example 7.7. Consider the term considered above:

$$t = (((\mathsf{v}_3 \oplus \mathsf{v}_2 \oplus \mathsf{v}_1) \otimes \mathsf{v}_4) \otimes \mathsf{v}_2) \oplus ((\mathsf{v}_3 \oplus \mathsf{v}_2 \oplus \mathsf{v}_1) \otimes \mathsf{v}_1).$$

Then $j_{\nu(t)}(\boldsymbol{\delta}^1_3) = \aleph_{\omega^{\omega^2} \cdot \omega + \omega^\omega + \omega + 2}$.

§8. A collapsing result. As an application of the cardinal structure theory under AD, we present the following theorem on the collapse of cardinals between $L(\mathbb{R})$ and V. This result only uses the structural theory below $\aleph_{\omega_1}$, that is, below the supremum of the first ω_1 many Suslin cardinals. As we mentioned before, this theory is essentially identical to that below the projective ordinals, which we have outlined in the previous sections. In fact, we only need this theory in a general way, namely to show that for $\alpha < \omega_1$ that $\boldsymbol{\delta}^1_\alpha < \aleph_{\omega_1}$. The other determinacy arguments used are of a more general nature. We will only sketch the proof here, as a detailed proof will be given elsewhere.

In addition to the determinacy theory, the proof also uses two other essential ingredients, Shelah's pcf theory (a ZFC theory), and Woodin's covering theorem (Theorem 8.2). The idea of the proof is compare the cardinal structure in $L(\mathbb{R})$ below $\aleph_{\omega_2}$ given by the AD theory with the structure in V given by the pcf theory. Woodin's theorem provides the connection between the two models. The next theorem is our collapsing result.

THEOREM 8.1. *Assume the non-stationary ideal on ω_1 is ω_2-saturated and there are $\omega + 1$ Woodin cardinals in V. Then there is a $\kappa < (\aleph_{\omega_2})^{L(\mathbb{R})}$ such that κ is regular in $L(\mathbb{R})$ but κ is not a cardinal in V.*

We next state Woodin's covering theorem (see Corollary 3.48 of [21] for a proof).

THEOREM 8.2 (Woodin). *Assume the same hypotheses as the previous theorem. If $A \subseteq \lambda < \Theta^{L(\mathbb{R})}$ and $|A| = \omega_1$, then there is a $B \in L(\mathbb{R})$, $|B| = \omega_1$ with $A \subseteq B$.*

A special case of this (which we also use) is the fact that every c.u.b. $C \subseteq \omega_1$ contains a c.u.b. $C_1 \subseteq C$ with $C_1 \in L(\mathbb{R})$ (see Theorems 3.16 and 3.17 of [21]).

From Shelah's pcf theory we need the following (see Lemma 6.3 of [2]).

THEOREM 8.3 (ZFC). *Let A be a set of regular cardinals with $|A| < \inf(A)$ and with $\operatorname{cof}(\sup(A)) > \omega$. Then there is a c.u.b. $C \subseteq \sup(A)$ such that $\max \operatorname{pcf}(C^+) = (\sup(A))^+$, where $C^+ = \{\kappa^+ : \kappa \in C\}$.*

We also need a certain partition property which we now formulate. Let $\vec{\kappa} = \{\kappa_\alpha\}_{\alpha<\rho}$ be an increasing, discontinuous sequence of regular cardinals.

Let $\vec{\theta} = \{\theta_\alpha\}_{\alpha<\rho}$ be a ρ-sequence of ordinals with $\theta_\alpha \leq \kappa_\alpha$. We consider block functions f from $\sum_{\alpha<\rho} \theta_\alpha$ to $\sup_{\alpha<\rho} \kappa_\alpha$. By this we mean a function which maps the ordinals in the copy of θ_α to $\kappa_\alpha - \sup_{\beta<\alpha} \kappa_\beta$. Here we are regarding $\sum_{\alpha<\rho} \theta_\alpha$ as an ordering with domain the disjoint union of copies of the ordinals θ_α.

We say $\vec{\kappa} \to (\vec{\kappa})^{\vec{\theta}}$ if for any partition $\mathcal{P}$ of the block functions f into two pieces, there is a blockwise c.u.b. homogeneous set $\vec{H} = \{H_\alpha\}_{\alpha<\rho}$ for $\mathcal{P}$ (for functions blockwise of the correct type).

Let $\mathcal{R} \subseteq \aleph_{\omega_1}$ be the set of cardinals of the form $\boldsymbol{\delta}^1_{\alpha+1}$ for limit α.

For $\kappa \in \mathcal{R}$, there is pointclass $\boldsymbol{\Sigma}_0$ closed under $\exists^{\omega^\omega}$, $\wedge$, $\vee_\omega$ and $\operatorname{scale}(\boldsymbol{\Sigma}_0)$ such that if $\boldsymbol{\Pi}_1 = \forall^{\omega^\omega} \boldsymbol{\Sigma}_0$, then: $\boldsymbol{\Pi}_1$ is closed under $\forall^{\omega^\omega}$, countable unions and intersections, and we have $\operatorname{scale}(\boldsymbol{\Pi}_1)$, and $\boldsymbol{\delta}^1_{\alpha+1} = \delta(\boldsymbol{\Pi}_1)$.

Theorem 8.4 (AD + DC). *Let* $\vec{\kappa} = \{\kappa_\alpha\}_{\alpha<\omega_1} \subseteq \mathcal{R}$. *Then for all* $\theta < \omega_1$ *we have* $\vec{\kappa} \to (\vec{\kappa})^\theta$.

From Corollary 5.12 of [1] and the inductive analysis of the $\boldsymbol{\delta}^1_\alpha$ for $\alpha < \omega_1$ we actually get a much stronger result, namely the polarized strong partition property for the sequence. We do not need this stronger result here. The proof of Theorem 8.4 is similar to that of Theorem 5.10 of [1], but easier since we are only using codes for countable ordinals and only using a fixed exponent $\theta < \omega_1$ (also, we can uniformly find universal sets for the $\boldsymbol{\Pi}_1$ classes which simplifies the argument of [1]).

In fact, Theorem 8.4 holds for any ω_1 sequence of pointclasses resembling $\boldsymbol{\Pi}^1_1$.

As a consequence of the partition property we have the following.

Theorem 8.5 (AD + DC). *Let* μ *be any measure on* ω_1. *Let* $\vec{\kappa} = \{\kappa_\alpha\}_{\alpha<\omega_1} \subseteq \mathcal{R}$. *Then* $\prod \kappa_\alpha/\mu$ *is regular.*

Proof. Let $\delta = \prod \kappa_\alpha/\mu$. Suppose $\pi \colon \lambda \to \delta$ is cofinal, where $\lambda < \delta$. Consider the partition of pairs of block functions given by $\mathcal{P}([f]_\mu, [g]_\mu) = 1$ iff there is an element of $\mathrm{ran}(\pi)$ between $[f]_\mu$ and $[g]_\mu$. The homogeneous side must be the 1 side. Let S be block homogeneous for $\mathcal{P}$. For $[f]_\mu < \delta$, define $[f']$ by $f'(\alpha) = \omega\cdot(f(\alpha)+1)^{\text{st}}$ element of S. Then $A = \{[f']_\mu : [f]_\mu < \delta\}$ has order-type δ and between any two elements of A is an element of $\mathrm{ran}(\pi)$. ⊣

Proof of Theorem 8.1 (sketch). From the pcf theory, Theorem 8.3, let $C \subseteq \aleph_{\omega_1}$ be a c.u.b. subset of the limit Suslin cardinals such that max $\mathrm{pcf}(C^+) = \aleph_{\omega_1+1}$. Without loss of generality (by the remark after Theorem 8.2) we may assume $C \in L(\mathbb{R})$.

Let $\mu = W^1_1$ be the normal measure on ω_1 (in $L(\mathbb{R})$), and let $\mathcal{U}$ (in V) be an ultrafilter on ω_1 extending μ.

Let $f \colon \omega_1 \to C^+$ be increasing, $f \in L(\mathbb{R})$, and $f(\alpha) > (\boldsymbol{\delta}^1_\alpha)^+$ almost everywhere. Thus, by Theorem 8.5, $\lambda \doteq [f]_\mu$ is regular in $L(\mathbb{R})$. Assume λ is also regular in V, towards a contradiction. We also assume all elements of $C^+ \subseteq \mathcal{R}$ are regular in V.

Also, $\lambda > \rho \doteq \aleph_{\omega_1+1}$. In V, $\mathrm{cof}(\prod f(\alpha)/\mathcal{U}) = \rho$, from the definition of C. In V we define a cofinal map $\pi \colon \rho \to \lambda$ as follows. In V, fix a scale $\{f_\alpha\}_{\alpha<\rho}$ [Recall a scale means the sequence of functions is $\mathcal{U}$ increasing and $\mathcal{U}$ dominating, that is, if $g \in \prod f(\alpha)$, then for some α we have $g <_{\mathcal{U}} f_\alpha$. Actually, we only need $\mathcal{U}$ unboundedness here.]

For $\alpha < \rho$, let $\pi(\alpha) = ([g]_\mu)^{L(\mathbb{R})}$, where $g \in L(\mathbb{R})$ represents the μ least equivalence class such that $g \geq f_\alpha$ almost everywhere w.r.t. $\mathcal{U}$. Such a function exists by Woodin's theorem, and the definition is well-defined.

To show π is cofinal, let $\beta < \lambda$. Let $[g_0]_\mu = \beta$. Pick α such that $f_\alpha >_{\mathcal{U}} g_0$. By definition $\pi(\alpha) = [g]_\mu$ where $g \in L(\mathbb{R})$ and $g >_{\mathcal{U}} f_\alpha$. So, $g >_{\mathcal{U}} f_\alpha >_{\mathcal{U}} g_0$. Since $g, g_0 \in L(\mathbb{R})$, we have $g >_\mu g_0$. Hence, $\pi(\alpha) = [g]_\mu > [g_0]_\mu = \beta$. ⊣

REFERENCES

[1] Arthur W. Apter, James M. Henle, and Stephen C. Jackson, *The calculus of partition sequences, changing cofinalities, and a question of Woodin*, ***Transactions of the American Mathematical Society***, vol. 352 (2000), no. 3, pp. 969–1003.

[2] Maxim R. Burke and Menachem Magidor, *Shelah's* pcf *theory and its applications*, ***Annals of Pure and Applied Logic***, vol. 50 (1990), no. 3, pp. 207–254.

[3] Steve Jackson, *Structural consequences of AD*, ***Handbook of Set Theory*** (Foreman and Kanamori, editors), to appear.

[4] ———, AD *and the projective ordinals*, ***Cabal Seminar 81–85***, Lecture Notes in Mathematics, vol. 1333, Springer, Berlin, 1988, pp. 117–220.

[5] ———, *A computation of* δ^1_5, ***Memoirs of the American Mathematical Society***, vol. 140 (1999), no. 670, pp. 1–94.

[6] Steve Jackson and F. Khafizov, *Descriptions and cardinals below* $\boldsymbol{\delta}^1_5$, preprint.

[7] Steve Jackson and Benedikt Löwe, *Canonical measure assignments*, to appear.

[8] Alexander S. Kechris, AD *and projective ordinals*, ***Cabal Seminar 76–77***, Lecture Notes in Mathematics, vol. 689, Springer, Berlin, 1978, pp. 91–132.

[9] ———, *Souslin cardinals, κ-Souslin sets and the scale property in the hyperprojective hierarchy*, ***Cabal Seminar 77–79***, Lecture Notes in Mathematics, vol. 839, Springer, Berlin, 1981, pp. 127–146.

[10] ———, ***Classical Descriptive Set Theory***, Graduate Texts in Mathematics, vol. 156, Springer-Verlag, New York, 1995.

[11] Alexander S. Kechris, Robert M. Solovay, and John R. Steel, *The axiom of determinacy and the prewellordering property*, ***Cabal Seminar 77-79***, Lecture Notes in Mathematics, vol. 839, Springer, Berlin, 1981, pp. 101–125.

[12] Donald A. Martin and John R. Steel, *A proof of projective determinacy*, ***Journal of the American Mathematical Society***, vol. 2 (1989), no. 1, pp. 71–125.

[13] Yiannis N. Moschovakis, ***Descriptive Set Theory***, Studies in Logic and the Foundations of Mathematics, vol. 100, North-Holland, Amsterdam, 1980.

[14] Robert M. Solovay, *A* Δ^1_3 *coding of the subsets of* ω_ω, ***Cabal Seminar 76–77***, Lecture Notes in Mathematics, vol. 689, Springer, Berlin, 1978, pp. 133–150.

[15] ———, *The independence of DC from AD*, ***Cabal Seminar 76–77***, Lecture Notes in Mathematics, vol. 689, Springer, Berlin, 1978, pp. 171–183.

[16] John R. Steel, *Closure properties of pointclasses*, ***Cabal Seminar 77–79***, Lecture Notes in Mathematics, vol. 839, Springer, Berlin, 1981, pp. 147–163.

[17] ———, *Determinateness and the separation property*, ***The Journal of Symbolic Logic***, vol. 46 (1981), no. 1, pp. 41–44.

[18] ———, *Scales in* $L(\boldsymbol{R})$, ***Cabal Seminar 79–81***, Lecture Notes in Mathematics, vol. 1019, Springer, Berlin, 1983, pp. 107–156.

[19] Robert Van Wesep, *Wadge degrees and descriptive set theory*, ***Cabal Seminar 76–77***, Lecture Notes in Mathematics, vol. 689, Springer, Berlin, 1978, pp. 151–170.

[20] W. Hugh Woodin, *Supercompact cardinals, sets of reals, and weakly homogeneous trees*, ***Proceedings of the National Academy of Sciences of the United States of America***, vol. 85 (1988), no. 18, pp. 6587–6591.

[21] ———, ***The Axiom of Determinacy, Forcing Axioms, and the Nonstationary Ideal***, de Gruyter Series in Logic and its Applications, vol. 1, Walter de Gruyter, Berlin, 1999.

DEPARTMENT OF MATHEMATICS
UNIVERSITY OF NORTH TEXAS
DENTON, P.O. BOX 311430, TX 76203-1430, USA
E-mail: jackson@unt.edu

THREE LECTURES ON AUTOMATIC STRUCTURES

BAKHADYR KHOUSSAINOV AND MIA MINNES

Preface. This paper grew out of three tutorial lectures on automatic structures given at the Logic Colloquium 2007 in Wrocław (Poland). The paper will follow the outline of the tutorial lectures, supplementing some material along the way. We discuss variants of automatic structures related to several models of computation: word automata, tree automata, Büchi automata, and Rabin automata. Word automata process finite strings, tree automata process finite labeled trees, Büchi automata process infinite strings, and Rabin automata process infinite binary labeled trees. Finite automata are the most commonly known in the general computer science community. An automaton of this type reads finite input strings from left to right, making state transitions along the way. Depending on its last state after processing a given string, the automaton either accepts or rejects the input string. Automatic structures are mathematical objects which can be represented by (word, tree, Büchi, or Rabin) automata. The study of properties of automatic structures is a relatively new and very active area of research.

We begin with some motivation and history for studying automatic structures. We introduce definitions of automatic structures, present examples, and discuss decidability and definability theorems. Next, we concentrate on finding natural isomorphism invariants for classes of automatic structures. These classes include well-founded partial orders, Boolean algebras, linear orders, trees, and finitely generated groups. Finally, we address the issue of complexity for automatic structures. In order to measure complexity of automatic structures we involve topology (via the Borel hierarchy), model theory (Scott ranks), computability theory (Σ^1_1-completeness), and computational complexity (the class P).

This paper consists of three sections based on the tutorial lectures. The first lecture provides motivating questions and historical context, formal definitions of the different types of automata involved, examples of automatic structures, and decidability and definability results about automatic structures. The second lecture begins by introducing tree and Rabin automatic structures. We then outline techniques for proving nonautomaticity of sets

Logic Colloquium '07
Edited by Françoise Delon, Ulrich Kohlenbach, Penelope Maddy, and Frank Stephan
Lecture Notes in Logic, 35

and structures and study the algorithmic and structural properties of automatic trees, Boolean algebras, and finitely generated groups. The final lecture presents a framework for reducing certain questions about computable structures to questions about automatic structures. These reductions have been used to show that, in some cases, the sharp bound on complexity of automatic structures is as high as the bounds for computable structures. We conclude by looking at Borel structures from descriptive set theory and connecting them to Büchi and Rabin automatic structures.

§1. Basics.

1.1. Motivating questions. The study of structures has played a central role in the development of logic. In the course of this study, several themes have been pursued. We will see how questions related to each of these themes is addressed in the particular case of automatic structures.

The isomorphism problem. One of the major tasks in the study of structures is concerned with the classification of isomorphism types. The isomorphism problem asks: "given two structures, are they isomorphic?" Ideally, we would like to define invariants which are simpler than their associated structures and yet describe the structures up to isomorphism. An example of such an invariant is the dimension of vector spaces. For algebraically closed fields, such an invariant is the characteristic of a field along with its transendence degree. Despite this, there are well-studied classes of structures (such as abelian groups, Boolean algebras, linearly ordered sets, algebras, and graphs) for which it is impossible to give simple isomorphism invariants. The absence of these invariants means that the isomorphism problem can't be reduced to checking the equality of a simpler object. Thus, in general, the isomorphism problem is highly undecidable.

The elementary equivalence problem. Since there is no general solution to the isomorphism problem, we may approximate it in various ways. One natural approximation comes from logic in the form of elementary equivalence. While elementary equivalence may be defined with respect to any logic, we refer to elementary equivalence with respect to either the first-order logic or the monadic second-order (MSO) logic. There are several good examples of positive results in this setting: elementarily equivalent (in the first-order logic) Boolean algebras can be characterized via Ershov-Tarski invariants [16, 57]; elementarily equivalent abelian groups (in the first-order logic) can be characterized via Shmelev invariants [55]. Moreover, there is a body of work devoted to understanding when elementary equivalence (or any of its weaker versions) imply isomorphism. An example here is Pop's conjecture: if two finitely generated fields are elementarily equivalent then they are isomorphic. See [53] for a solution of this problem.

The model checking problem. Instead of looking at the full theory of a structure (as in the elementary equivalence problem), the model checking problem asks whether a particular sentence is satisfied by the structure. This question is of particular interest in the computer science community. The problem asks, for a given sentence φ and a given structure, whether the structure satisfies φ. For example, the sentence φ can be the axiomatization of groups or can state that a graph is connected (in which case a stronger logic is used than the first-order logic).

Deciding the theories of structures. This is a traditional topic in logic that seeks algorithms to decide the full first-order theory (or MSO theory) of a given structure. Clearly, the problem can be thought of as a uniform version of the model checking problem. Often, to prove that a structure has a decidable first-order theory, one translates formulas into an equivalent form that has a small number of alternations of quantifiers (or no quantifiers at all), and then shows that the resulting simpler formulas are easy to decide. This approach is intimately related to the next theme in our list.

Characterization of definable relations. Here, we would like to understand properties of definable relations in a given structure. A classical example is real closed fields, where a relation is definable in the first-order logic if and only if it is a Boolean combination of polynomial equations and inequations [56]. Usually, but not always, a good understanding of definable relations yields a quantifier elimination procedure and shows that the theory of the structure is decidable.

It is apparent that all of the above problems are interrelated. We will discuss these problems and their refinements with respect to classes of automatic structures.

1.2. Background. The theory of automatic structures can be motivated from the standpoint of computable structures. The theory of computable structures has a rich history going back to van der Waerden who informally introduced the concept of an explicitly given field in the 1930s. This concept was formalized by Frölich and Shepherdson [20], and later extended by Rabin [47] in the 1950s. In the 1960s, Mal'cev initiated a systematic study of the theory (see, for example [40]). Later, computability theoretic techniques were introduced in order to study the effective content of algebraic and model theoretic results and constructions. See [17, 18] for the current state and historical background of the area.

A computable structure is one whose domain and basic relations are all computable by Turing machines. As a point of comparison with finite automata, we note that the Turing machine represents unbounded resources and unbounded read-passes over the data. The themes outlined in Subsection 1.1 have been recast with computability considerations in mind and have given rise to the development of the theory of computable structures. For example, one may ask whether a given structure is computable. This is clearly a question about the isomorphism type of the structures since it asks whether the isomorphism type of the structure contains a computable structure. So, the study of the computable isomorphism types of structures is a refinement of the isomorphism problem for structures. Another major theme is devoted to understanding the complexity of natural problems like the isomorphism problem and the model checking problem. In this case, complexity is often measured in terms of Turing degrees. We again refer the reader to [17, 18] and the survey paper [23].

In the 1980s, as part of their feasible mathematics program, Nerode and Remmel [43] suggested the study of polynomial-time structures. A structure is said to be polynomial-time if its domain and relations can be recognized by Turing machines that run in polynomial time. An important yet simple result (Cenzer and Remmel, [12]) is that every computable purely relational structure is computably isomorphic to a polynomial-time structure. While this result is positive, it implies that solving questions about the class of polynomial-time structures is as hard as solving them for the class of computable structures. For instance, the problem of classifying the isomorphism types of polynomial-time structures is as hard as that of classifying the isomorphism types of computable structures.

Since polynomial-time structures and computable structures yielded similar complexity results, greater restrictions on models of computations were imposed. In 1995, Khoussainov and Nerode suggested bringing in models of computations that have less computational power than polynomial-time Turing machines. The hope was that if these weaker machines were used to represent the domain and basic relations, then perhaps isomorphism invariants could be more easily understood. Specifically, they suggested the use of finite state machines (automata) as the basic computation model. Indeed, the project has been successful as we discuss below.

The idea of using automata to study structures goes back to the work of Büchi. Büchi [10, 11] used automata to prove the decidability of a theory called $S1S$ (monadic second-order theory of the natural numbers with one successor). Rabin [48] then used automata to prove that the monadic second-order theory of two successor functions, $S2S$, is also decidable. In the realm of logic, these results have been used to prove decidability of first-order or MSO theories. Büchi knew that automata and Presburger arithmetic (the first-order theory of the natural numbers with addition) are closely connected. He used automata to give a simple (non quantifier elimination) proof of the decidability of Presburger arithmetic. Capturing this notion, Hodgson [27] defined automaton decidable theories in 1982. While he coined the definition of automatic structures, throughout the 1980s his work remained largely unnoticed. In 1995, Khoussainov and Nerode [31] rediscovered the concept of automatic structure and initiated the systematic study of the area.

Thurston observed that many finitely generated groups associated with 3-manifolds are finitely presented groups with the additional property that finite automata recognize equality of words and the graphs of the operations of left multiplication by a generator; these are the Thurston automatic groups. These groups yield rapid algorithms [15] for computing topological and algebraic properties of interest in geometric group theory (such as the word problem). In this development, a group is regarded as a unary algebra including one unary operation for each generator, the operation being left multiplication of a word by that generator. Among these groups are Coxeter groups, braid groups, Euclidean groups, and others. The literature is extensive, and we do not discuss it here. We emphasize that Thurston automatic groups differ from automatic groups in our sense; in particular, the vocabulary of the associated structures is starkly different. Thurston automatic groups are represented as structures whose relations are unary operations (corresponding to left multiplication by each generator). On the other hand, an automatic group in our sense deals with the group operation itself (a binary relation) and hence must satisfy the constraint that the graph of this operation be recognizable by a finite automaton. The Thurston requirement for automaticity applies only to finitely generated groups but includes a wider class of finitely generated groups than what we call automatic groups. Unlike our definition, Thurston's includes groups whose binary operation of multiplication is not recognizable by a finite automaton.

In the computer science community, an interest in automatic structures comes from problems related to model checking. Model checking is motivated by the quest to prove correctness of computer programs. This subject allows infinite state automata as well as finite state automata. Consult [2, 1, 9] for current topics of interest. Examples of infinite state automata include concurrency protocols involving arbitrary number of processes, programs manipulating some infinite sets of data (such as the integers or reals), push-down automata, counter automata, timed automata, Petri-nets, and rewriting systems. Given such an automaton and a specification (formula) in a formal system, the model checking problem asks us to compute all the states of the system that satisfy the specification. Since the state space is infinite, the process of checking the specification may not terminate. For instance, the standard fixed point computations very often do not terminate in finite time. Specialized methods are needed to cover even the problems encountered in practice. Abstraction methods try to represent the behavior of the system in finite form. Model checking then reduces to checking a finite representation of states that satisfy the specification. Automatic structures arise naturally in infinite state model checking since both the state space and the transitions of infinite state systems are usually recognizable by finite automata. In 2000, Blumensath and Grädel [8] studied definability problems for automatic structures and the computational complexity of model checking for automatic structures.

There has been a series of PhD theses in the area of automatic structures (e.g. Rubin [51], Blumensath [7], Bárány [3], Minnes [41], and Liu [39]).

A recently published paper of Khoussainov and Nerode [32] discusses open questions in the study of automatic structures. There are also survey papers on some of the areas in the subject by Nies [44] and Rubin [52]. This avenue of research remains fruitful and active.

1.3. Automata recognizable languages and relations. The central models of computation in the development of the theory of automatic structures are all finite state machines. These include finite automata, Büchi automata, tree automata, and Rabin automata. The main distinguishing feature of the different kinds of automata is the kind of input they read.

DEFINITION 1. A *Büchi automaton* $\mathcal{M}$ is a tuple (S, ι, Δ, F), where S is a finite set of states, ι is the initial state, $\Delta \subseteq S \times \Sigma \times S$ (with Σ a finite alphabet) is the transition table, and $F \subseteq S$ is the set of accepting states.

A Büchi automaton can be presented as a finite directed graph with labelled edges. The vertices of the graph are the states of the automaton (with designated initial state and accepting states). An edge labeled with σ connects a state q to a state q' if and only if the triple (q, σ, q') is in Δ. The inputs of a Büchi automaton are *infinite strings* over the alphabet Σ. Let $\alpha = \sigma_0\sigma_1\dots$ be such an infinite string. The string labels an infinite path through the graph in a natural way. Such a path is called a run of the automaton on α. Formally, a run is an infinite sequence $q_0, q_1, \dots,$ of states such that q_0 is the initial state and $(q_i, \sigma_i, q_{i+1}) \in \Delta$ for all $i \in \omega$. The run is accepting if some accepting state appears in the run infinitely often. Note that the automaton may have more than one run on a single input α. We say that the Büchi automaton $\mathcal{M}$ accepts a string α if $\mathcal{M}$ has an accepting run on α. Thus, acceptance is an existential condition.

DEFINITION 2. The set of all infinite words accepted by a Büchi automaton $\mathcal{M}$ is called the language of $\mathcal{M}$. A collection of infinite words is called a *Büchi recognizable language* if it is the language of some Büchi automaton.

EXAMPLE 1. The following two languages over $\Sigma = \{0, 1\}$ are Büchi recognizable:

- $\{\alpha \mid \alpha$ has finitely many 1s$\}$,
- $\{\alpha \mid \alpha$ has infinitely many 1s and infinitely many 0s$\}$.

There are efficient algorithms to decide questions about Büchi recognizable languages. A good reference on basic algorithms for Büchi automata is Thomas' survey paper [59]. A central building block in these algorithms is the linear-time decidability of the emptiness problem for Büchi automata. The emptiness problem asks for an algorithm that, given a Büchi automaton, decides if the automaton accepts at least one string. An equally important result in the development of the theory of Büchi automata says that Büchi recognizable languages are closed under union, intersection, and complementation.

THEOREM 1 (Büchi [10]). 1. *There is an algorithm that, given a Büchi automaton, decides (in linear time in the size of the automaton) if there is some string accepted by the automaton.*
2. *The collection of all Büchi recognizable languages is closed under the operations of union, intersection, and complementation.*

PROOF. The first part of the theorem is easy: the Büchi automaton accepts an infinite string if and only if there is a path from the initial state to an accepting state which then loops back to the accepting state. Thus, the emptiness algorithm executes a breadth-first search for such a "lasso" . Closure under union and intersection follows from a standard product construction. It is considerably more difficult to prove closure under complementation. See [59] for a discussion of the issues related to the complementation problem for Büchi automata. ⊣

Theorem 1 is true effectively: given two automata we can construct an automaton that recognizes all the strings accepted by either of the automata (the union automaton), we can also construct an automaton that recognizes all the strings accepted by both automata (the intersection automaton). Likewise, there is an algorithm that given an automaton builds a new automaton (called the complement automaton) that accepts all the strings rejected by the original automaton. We emphasize that complementation constructions have deep significance in modern automata theory and its applications.

Since we will be interested in using automata to represent structures, we need to define what it means for a relation to be recognized by an automaton. Until now, we have discussed Büchi automata recognizable *sets* of infinite strings. It is easy to generalize this notion to relations on Σ^ω. Basically, to process a tuple of infinite strings, we read each string in parallel. More formally, we define the *convolution* of infinite strings $\alpha_1, \dots, \alpha_k \in \Sigma^\omega$ as the infinite string $c(\alpha_1, \dots, \alpha_k) \in (\Sigma^k)^\omega$ whose value at position i is the tuple $\langle \alpha_1(i), \dots, \alpha_k(i) \rangle$. The convolution of a k-ary relation R, denoted by $c(R)$, is the set of infinite strings over Σ^k which are the convolutions of tuples of R. We say that the relation R is Büchi recognizable if and only if the set $c(R)$ is Büchi recognizable.

We will now describe some operations which preserve recognizability (proving that recognizability is preserved is similar to Theorem 1). Let A be a language of infinite strings over Σ and R be a relation of arity k on Σ^ω. The *cylindrification* operation on a relation R (with respect to A) produces the new relation

$$cyl(R) = \{\langle a_1, \dots, a_k, a \rangle : \langle a_1, \dots, a_k \rangle \in R \text{ and } a \in A\}.$$

The *projection*, or $\exists$, operation (with respect to A) is defined by

$$\exists x_i R = \{\langle a_1, \dots, a_{i-1}, a_{i+1}, \dots, a_k \rangle : \\ \exists a \in A(\langle a_1, \dots, a_{i-1}, a, a_{i+1}, \dots, a_k \rangle \in R)\}.$$

The *universal projection*, or $\forall$, operation (with respect to A) is defined as

$$\forall x_i R = \{\langle a_1, \dots, a_{i-1}, a_{i+1}, \dots, a_k\rangle : \forall a \in A(\langle a_1, \dots, a_{i-1}, a, a_{i+1}, \dots, a_k\rangle \in R)\}.$$

In all these operations ($cyl, \exists, \forall$), if R and A are both Büchi recognizable then the resulting relations are also Büchi recognizable. The *instantiation* operation is defined by

$$I(R, c) = \{\langle x_1, \dots, x_{k-1}\rangle : \langle x_1, \dots, x_{k-1}, c\rangle \in R\}.$$

If R is Büchi automatic then $I(R, c)$ is Büchi automatic if and only if c is an ultimately periodic infinite string. (An ultimately periodic word is one of the form $uv^\omega = uvvv\cdots$ where u and v are finite strings.) The *rearrangement* operations permute the coordinates of a relation. If $\pi : k \to k$ is a permutation on the set of k elements and R is a k-ary Büchi automatic relation, then

$$\pi R = \{\langle x_{\pi(1)}, \dots, x_{\pi(k)}\rangle : \langle x_1, \dots, x_k\rangle \in R\}$$

is Büchi automatic. The *linkage* operation identifies the last coordinates of some relation with the first coordinates of another. Given the relations R (of arity m_1) and S (of arity m_2) and index $i < m_1$, the linkage of R and S on i is the relation of arity $m_2 + i - 1$ defined by

$$L(R^{m_1}, S^{m_2}; i) = \{\langle a_1, \dots, a_{m_2+i-1}\rangle : \langle a_1, \dots, a_{m_1}\rangle \in R \,\&\, \langle a_i, \dots, a_{m_2+i-1}\rangle \in S\}.$$

For example, $L(R^3, S^4; 2) = \{\langle a_1, a_2, a_3, a_4, a_5\rangle : \langle a_1, a_2, a_3\rangle \in R \,\&\, \langle a_2, a_3, a_4, a_5\rangle \in S\}$. If R and S are Büchi recognizable relations then so is $L(R, S; i)$.

The closure of Büchi recognizable relations under Boolean operations (Theorem 1 on page 138) connects Büchi recognizability and propositional logic. The $\exists$ and $\forall$ operations bring us to the realm of first-order logic. We now connect automata with the monadic second-order (MSO) logic of the successor structure $(\omega; S)$. The MSO logic is built on top of the first-order logic as follows. There is one nonlogical membership symbol $\in$, and there are set variables $X, Y, \dots$ that range over subsets of ω. Formulas are defined inductively in a standard way via the Boolean connectives and quantifiers over set variables as well as over individual variables.

EXAMPLE 2. In MSO of $(\omega; S)$, the following relations and properties are expressible: the subset relation $X \subseteq Y$, the natural order $\leq$, finiteness of sets, and whether a set is a singleton. For example,

$$Single(X) := \exists x\big(x \in X \,\&\, \forall y(y \in X \iff x = y)\big).$$

The definability of singletons allows us to transform any MSO formula into an equivalent MSO formula *all* of whose variables are set variables. We can also interpret addition on natural numbers as follows. Associate with every finite set X the binary string $\sigma(X)$ that has 1 at position i if and only if $i \in X$.

We use the rules of binary addition to express the statement that X, Y, Z are finite sets and $\sigma(X) +_2 \sigma(Y) = \sigma(Z)$ in the MSO logic of $(\omega; S)$.

There is a natural correspondence between k-ary relations on $\mathcal{P}(\omega)$ (the power set of the natural numbers) and k-ary relations on $\{0, 1\}^\omega$. For example, consider the case of $k = 2$. Any binary relation R on $\mathcal{P}(\omega)$ is a collection of pairs (X, Y) of subsets. Identify X and Y with their characteristic functions α_X and α_Y, each of which is an infinite string over $\{0, 1\}$. Recall that the convolution of (α_X, α_Y) is an infinite string over $\{0, 1\}^2$ such that $c(\alpha_X, \alpha_Y)(i) = (\alpha_X(i), \alpha_Y(i))$. The convolution of R is the language $c(R) = \{c(\alpha_X, \alpha_y) : (X, Y) \in R\}$. With this correspondence between relations on $\mathcal{P}(\omega)$ and relations on $\{0, 1\}^\omega$, Büchi proved a general characterization theorem that links the MSO definable relations of the successor structure $(\omega; S)$ and Büchi automata.

THEOREM 2 (Büchi [10]). *A relation $R \subseteq \mathcal{P}(\omega)^k$ is definable in MSO logic if and only if R is Büchi recognizable. Given an MSO definable relation R, there is an algorithm which builds a Büchi automaton recognizing R. In particular, the MSO theory of $(\omega; S)$, denoted as $S1S$, is decidable.*

We illustrate the theorem on a simple example: if R is a MSO definable relation, we check if $\exists X \forall Y\ R(X, Y)$ is true in $(\omega; S)$. Theorem 2 says that R is definable in MSO logic if and only if $c(R)$ is Büchi recognizable. Using Theorem 2, we build an automaton recognizing R. This is done inductively based on the complexity of the formula defining R. We then construct an automaton for the set of infinite strings $\{X : \forall Y\ R(X, Y)\}$. We use Theorem 1 from page 138 again to check if the set $\{X : \forall Y\ R(X, Y)\}$ is empty. If it is empty, the sentence is false; otherwise, it is true. Büchi's theorem is a quintessential example of how understanding the definable relations in a structure helps us decide the theory of the structure (see Subsection 1.1).

We briefly turn our attention to *word automata*. The underlying graph definition of a word automaton is identical to that of a Büchi automaton. As mentioned earlier, the difference between these two models of computation lies in that word automata process *finite* strings rather than infinite strings. Let $\mathcal{M} = (S, \iota, \Delta, F)$ be a word automaton over the finite alphabet Σ, and let u be a finite string over Σ. Then u labels a path in the underlying directed graph of $\mathcal{M}$, starting from the initial state. This path is called a run of $\mathcal{M}$ on input u. If the last state in the run is accepting then the run is an accepting run. The automaton $\mathcal{M}$ accepts a string if there is some accepting run of $\mathcal{M}$ on the string. The collection of all finite strings accepted by a word automaton is called a *regular* (or, equivalently, a FA recognizable) language. As in the setting of Büchi automata, we have the following theorem:

THEOREM 3 (Kleene [36]).

1. *There is an algorithm that, given a word automaton, decides (in linear time in the size of the automaton) if some string is accepted by the automaton.*

2. *The collection of all regular languages is closed under the operations of union, intersection, and complementation.*

EXAMPLE 3. The following sets of strings are regular languages.

- $\{w101 : w \in \{0,1\}^*$ has no subword of the form $101\}$
- $\{w : w \in \{0,1\}^*$ is a binary representation of some positive integer with the least significant bit first$\}$.

EXAMPLE 4. Regular languages can be naturally embedded into Büchi recognizable sets. That is, W is regular if and only if $W\diamond^\omega$ is Büchi recognizable (where $\diamond$ is a new symbol not in the alphabet Σ).

Previously, we defined Büchi recognizable relations. We would like an analogous notion for word automata. To define regular relations, we need to define the *convolution* of finite strings. Let $x_1, \dots, x_k \in \Sigma^*$. If $x_1, \dots, x_k$ are all of the same length, the i-th element of the convolution $c(x_1, \dots, x_k)$ is the tuple $\langle x_1(i), \dots, x_k(i)\rangle$ (as in the infinite string case). Otherwise, assume without loss of generality that x_n is the longest string among $x_1, \dots, x_k$. For each $m \neq n$, append a new symbol $\diamond$ to the end of x_m as many times as necessary to make the padded version of x_m have the same length as x_n. Then $c(x_1, \dots, x_k)$ is the convolution of the padded strings. If R is a k-ary relation on Σ^*, R is called *regular* if its convolution $c(R)$ is a regular language.

As before, we can develop a calculus of regular relations. Given a regular relation R and regular set A, the cylindrification $c(R)$, projection $\exists x_i R$, and universal projection $\forall x_i R$ are all recognizable by finite automaton. Likewise, given a regular relation R and any finite string c, the instantiation $I(R, c)$ is a regular relation. Similarly, the rearrangement and linkage operations preserve regularity of relations.

1.4. Büchi and word automatic structures. The main focus of this tutorial is the study of structures defined by automata. We now give a formal definition of this concept and provide several examples of such structures. Recall that a structure $\mathcal{A}$ is a tuple $(A; R_1, \dots, R_m)$ where A is a nonempty set called the domain and $R_1, \dots, R_m$ are basic (or atomic) relations on A.

DEFINITION 3. A structure is *word automatic* if its domain and basic relations are regular. A structure is *Büchi automatic* if its domain and basic relations are Büchi recognizable.

Often, we refer to word automatic structures and Büchi automatic structures simply as automatic structures. The type of automaticity will be clear from the context. We present some examples of word automatic structures. We begin with structures whose domains are $\{1\}^*$ (automatic structures over the one letter alphabet $\{1\}$ are called *unary automatic structures*; they have been studied in [7, 51, 29]).

EXAMPLE 5. The structure $(\{1\}^*; \leq, S)$, where $1^m \leq 1^n \iff m \leq n$ and $S(1^n) = 1^{n+1}$, is word automatic.

EXAMPLE 6. The structure $(\{1\}^*;\ \bmod_1,\ \bmod_2, \ldots,\ \bmod_n)$, where n is a fixed positive integer, is word automatic. The word automata recognizing the modular relations contain cycles of appropriate lengths.

Next, we move to structures with a binary alphabet $\{0, 1\}$. It is not too hard to see that any automatic structure over a finite alphabet is isomorphic to an automatic structure over a binary alphabet [51]. Clearly, any word automatic structure has a countable domain.

EXAMPLE 7. The structure $(\{0, 1\}^*; \vee, \wedge, \neg)$ is word automatic because bitwise operations on binary strings can be recognized by finite automata.

EXAMPLE 8. Presburger arithmetic, the structure $(\{0, 1\}^* \cdot 1; +_2, \leq)$, where $+_2$ is binary addition if the binary strings are interpreted as the least significant bit first base-2 expansion of natural numbers. The usual algorithm for adding binary numbers involves a single carry bit, and therefore a small word automaton can recognize the relation $+_2$.

EXAMPLE 9. Instead of Presburger arithmetic, we may consider the structure $(\{0, 1\}^* \cdot 1; +_2, |_2)$. This is arithmetic with weak divisibility: $w |_2 v$ if w represents a power of 2 which divides the number represented by v. Since we encode natural numbers by their binary representation, weak divisibility is a regular relation.

EXAMPLE 10. If we take binary strings at face value rather than as representations of natural numbers, we arrive at a different automatic structure:

$$(\{0, 1\}^*; \preceq, S_0, S_1, EqL),$$

where $\preceq$ is the prefix relation, S_0 and S_1 denote the functions which append a 0 or 1 to the binary string (respectively), and EqL is the equal length relation. It is easy to show that this structure is automatic. We will see later that this structure has a central role in the study of automatic structures.

EXAMPLE 11. A useful example of a word automatic structure is the configuration space of a Turing machine. The configuration space is a graph whose nodes are the configurations of the machine (the state, the contents of the tape, and the position of the read/write head). An edge exists between two nodes if there is a one-step transition of the machine which moves it between the configurations represented by these nodes.

EXAMPLE 12. Any word automatic structure is Büchi automatic, but the converse is not true. In the next subsection, we will see examples of Büchi automatic structures which have uncountable domains. These structures cannot be word automatic.

We mention a theorem which has some of the flavour of the downward Löwenheim-Skolem theorem. The classical theorem states that any infinite structure over a countable language has a countable elementary substructure. Recall that an elementary substructure is one which has the same first-order theory as the original structure in the language expanded by naming all the elements of the substructure. For Büchi automatic structures, the analogue of a countable substructure is the substructure consisting of ultimately periodic words. The following theorem shows that this substructure is elementarily equivalent to the original structure (however, it is computable and not necessarily automatic).

THEOREM 4 (Hjorth, Khoussainov, Montalbán, and Nies [26]). *Let $\mathcal{A}$ be a Büchi automatic structure and consider the substructure $\mathcal{A}'$ whose domain is*

$$A' = \{\alpha \in A : \alpha \text{ is ultimately periodic}\}.$$

Then $\mathcal{A}'$ is a computable elementary substructure of $\mathcal{A}$. Moreover, there is an algorithm which from a first-order formula $\varphi(\bar{a}, \bar{x})$ with parameters $\bar{a}$ from A' produces a Büchi automaton which accepts exactly those tuples $\bar{x}$ of ultimately periodic words which make the formula true in the structure.

Theorem 4 has also been independently proved by Kaiser, Rubin, and Bárány in [4].

1.5. Presentations and operations on automatic structures. The isomorphism type of a structure is the class of all structures isomorphic to it. We single out those isomorphism types that contain automatic structures.

DEFINITION 4. A structure $\mathcal{A}$ is called (word or Büchi) *automata presentable* if it is isomorphic to some (word or Büchi) automatic structure $\mathcal{B}$. In this case, $\mathcal{B}$ is called an *automatic presentation* of $\mathcal{A}$.

We sometimes abuse terminology and call automata presentable structures automatic. Let $\mathcal{B} = (D^{\mathcal{B}}; R_1^{\mathcal{B}}, \dots, R_s^{\mathcal{B}})$ be an automatic presentation of $\mathcal{A}$. Since $\mathcal{B}$ is automatic, the sets $D^{\mathcal{B}}, R_1^{\mathcal{B}}, \dots, R_s^{\mathcal{B}}$ are all recognized by automata, say by $\mathcal{M}, \mathcal{M}_1, \dots, \mathcal{M}_s$. Often the automatic presentation of $\mathcal{A}$ is identified with the finite sequence $\mathcal{M}, \mathcal{M}_1, \dots, \mathcal{M}_s$ of automata. From this standpoint, automata presentable structures have finite presentations.

Examples of automata presentable structures arise as specific mathematical objects of independent interest, or as the result of closing under automata presentability preserving relations. For example, if a class of structures is defined inductively and we know that the base structures have automata presentations and that each of the closure operations on the class preserve automaticity, then we may conclude that each member of the class is automata presentable. To aid in this strategy, we present some automaticity preserving operations on structures. The automaton constructions are often slight modifications of those used to show that recognizable sets form a Boolean algebra. We will also make use of the following relations on strings over a finite alphabet Σ. The lexicographic ordering $x \leq_{lex} y$ for $x, y \in \Sigma^*$ holds if and only if x is a prefix of y or $x = w0u$ and $y = w1v$ for some $w, u, v \in \Sigma^*$. The length-lexicographic ordering is given by $x \leq_{llex} y$ if and only if $|x| < |y|$ or $|x| = |y|$ and $|x| \leq_{lex} |y|$. Both $\leq_{lex}$ and $\leq_{llex}$ are automatic linear orders for any finite alphabet Σ.

PROPOSITION 1. *If $\mathcal{A}$ and $\mathcal{B}$ are (word or Büchi) automatic structures then so is their Cartesian product $\mathcal{A} \times \mathcal{B}$. If $\mathcal{A}$ and $\mathcal{B}$ are (word or Büchi) automatic structures then so is their disjoint union $\mathcal{A} + \mathcal{B}$.*

PROPOSITION 2. *If $\mathcal{A}$ is word automatic and E is a regular equivalence relation, then the quotient $\mathcal{A}/E$ is word automatic.*

PROOF. To represent the structure $\mathcal{A}/E$, consider the set

$$Rep = \{x \in A \mid \forall y (y <_{llex} x \; \& \; y \in A \to (x, y) \notin E)\},$$

where $<_{llex}$ is the length-lexicographic linear order. Since Rep is first-order definable in $(\mathcal{A}, E)$, it is regular and constitutes the domain of the quotient structure. Restricting the basic relations of $\mathcal{A}$ to the set Rep yields regular relations. Hence, the quotient structure is automatic. ⊣

A straightforward analysis of countable Büchi recognizable languages shows that a countable structure has a word automatic presentation if and only if it has Büchi automatic presentation. It is natural to ask whether Proposition 2 can be extended to countable Büchi automatic structures. This has recently been answered positively in [4] by a delicate analysis of Büchi recognizable equivalence relations with countably many equivalence classes.

THEOREM 5 (Bárány, Kaiser, and Rubin [4]). *For a Büchi automatic structure $\mathcal{A}$ and a Büchi recognizable equivalence relation E with countably many equivalence classes, the quotient structure $\mathcal{A}/E$ is Büchi automatic.*

A long-standing open question had asked whether Büchi automatic structures behave as nicely as word automatic structures with respect to Büchi recognizable equivalence relations. In other words, whether the countability assumption can be removed from the theorem above. A counterexample in [26] recently settled this question. We will outline a proof of the following theorem in the last lecture.

THEOREM 6 (Hjorth, Khoussainov, Montalbán, and Nies [26]). *The class of Büchi automatic structures is not closed under the quotient operation with respect to Büchi recognizable equivalence relations.*

We remark that in some recent papers, the closure of the Büchi automatic structures under quotients with respect to Büchi recognizable equivalence relations and isomorphism is studied. By Theorem 6, this closure is a larger class than the class of Büchi automatic structures. For example, Kuske and Lohrey show that structures in this large class have decidable theories with respect to first-order logic extended by generalized counting quantifiers [38]. In this survey, we focus on the truly Büchi automatic structures. We now give some natural examples of automata presentable structures. Note that many of the automatic structures that we mentioned earlier arise as the presentations of the following automata presentable structures.

EXAMPLE 13. The natural numbers under addition and order have a word automata presentation. The automatic presentation here is the word structure $(\{0,1\}^* \cdot 1; +_2, \leq)$.

EXAMPLE 14. The real numbers under addition have a Büchi automata presentation. The presentation is $(\{0,1\}^* \cdot \{\star\} \cdot \{0,1\}^\omega; +_2)$.

EXAMPLE 15. All finitely generated abelian groups are word automata presentable. Recall that a group $\mathcal{G}$ is finitely generated if there is a finite set S such that $\mathcal{G}$ is the smallest group containing S; a group $\mathcal{G}$ is called abelian if the group operation is commutative $a \cdot b = b \cdot a$. Since every such group is isomorphic to a finite direct sum of $(\mathcal{Z}; +)$ and $(\mathcal{Z}_n; +)$ [50], and since each of these has an automatic presentation, any finitely generated abelian group has an automatic presentation.

Recall that a *Boolean algebra* is a structure $(D; \vee, \wedge, \neg, 0, 1)$ which satisfies axioms relating the join $(\vee)$, meet $(\wedge)$, and complement $(\neg)$ operations and the constants 0 and 1. An archetypal Boolean algebra is $\mathcal{B}_\omega$, the collection of all finite and co-finite subsets of ω. In this case, the Boolean operations are the Boolean set operations: $\vee = \cup, \wedge = \cap, \neg = {}^c$.

EXAMPLE 16. $\mathcal{B}_\omega$ is word automata presentable. There is an automatic presentation of $\mathcal{B}_\omega$ with domain $\{0,1\}^* \cup \{2,3\}^*$ where words in $\{0,1\}^*$ represent finite sets and words in $\{2,3\}^*$ represent cofinite sets.

EXAMPLE 17. The Boolean algebra of all subsets of ω is Büchi automata presentable. A presentation using infinite strings treats each infinite string as the characteristic function of a subset of ω. The union, intersection, and complementation operations act bitwise on the infinite strings and are recognizable by Büchi automata.

EXAMPLE 18. The linear order of the rational numbers $(\mathcal{Q}; \leq)$ has a word automata presentation: $(\{0,1\}^* \cdot 1; \leq_{lex})$, where $u \leq_{lex} v$ is the lexicographic ordering.

1.6. Decidability results for automatic structures. Theorem 2 on page 140 uses automata to prove the decidability of the MSO theory of the successor structure $(\omega; S)$. We now explore other decidability consequences of algorithms for automata. The foundational theorem of Khoussainov and Nerode [31] uses the closure of regular relations (respectively, Büchi recognizable relations) under Boolean and projection operations to prove the decidability of the first-order theory of any automatic structure.

THEOREM 7 (Khoussainov, Nerode [31]; Blumensath, Grädel [8]). *There is an algorithm that, given a (word or Büchi) automatic structure* $\mathcal{A}$ *and a first-order formula* $\varphi(x_1, \ldots, x_n)$, *produces an automaton recognizing those tuples* $\langle a_1, \ldots, a_n \rangle$ *that make the formula true in* $\mathcal{A}$.

PROOF. We go by induction on the complexity of the formula, using the fact that automata recognizable relations are closed under the Boolean operations and the projection operations as explained in Subsection 1.3. ⊣

The Khoussainov and Nerode decidability theorem can be applied to individual formulas to yield Corollary 1, or uniformly to yield Corollary 2.

COROLLARY 1. *Let* $\mathcal{A}$ *be a word automatic structure and* $\varphi(\bar{x})$ *be a first-order formula in the language of* $\mathcal{A}$. *There is a linear-time algorithm that, given* $\bar{a} \in A$, *checks if* $\varphi(\bar{a})$ *holds in* $\mathcal{A}$.

PROOF. Let $\mathcal{M}_\varphi$ be an automaton for φ. Given $\bar{a}$, the algorithm runs through the state space of $\mathcal{M}_\varphi$ and checks if $\mathcal{M}_\varphi$ accepts the tuple. This can be done in linear time in the size of the tuple. ⊣

COROLLARY 2 (Hodgson [27]). *The first-order theory of any automatic structure is decidable.*

The connection between automatic structures and first-order formulas goes in the reverse direction as well. That is, first-order definability can produce new automatic structures. We say that a structure $\mathcal{B} = (B; R_1, \ldots R_n)$ is first-order definable in a structure $\mathcal{A}$ if there are first-order formulas φ_B and $\varphi_1, \ldots, \varphi_n$ (with parameters from $\mathcal{A}$) which define $B, R_1, \ldots, R_n$ (respectively) in the structure $\mathcal{A}$. Khoussainov and Nerode's theorem immediately gives the following result about first-order definable structures.

COROLLARY 3. *If* $\mathcal{A}$ *is (word or Büchi) automatic and* $\mathcal{B}$ *is first-order definable in* $\mathcal{A}$, *then* $\mathcal{B}$ *is also (word or Büchi) automatic.*

In fact, automatic structures can yield algorithms for properties expressed in logics stronger than first-order. We denote by $\exists^\infty$ the "there are countably infinitely many" quantifer, and by $\exists^{n,m}$ the "there are m many mod n" quantifiers. Then $(FO + \exists^\infty + \exists^{n,m})$ is the first-order logic extended with these quantifiers. The following theorem from [35] generalizes the Khoussainov and Nerode theorem to this extended logic. We note that Blumensath observed the $\exists^\infty$ case first, in [7].

THEOREM 8 (Khoussainov, Rubin, and Stephan [34]). *There is an algorithm that, given a word automatic structure* $\mathcal{A}$ *and a* $(FO + \exists^\infty + \exists^{n,m})$ *formula* $\varphi(x_1, \ldots, x_n)$ *with parameters from* $\mathcal{A}$, *produces an automaton recognizing those tuples* $\langle a_1, \ldots, a_n \rangle$ *that make the formula true in* $\mathcal{A}$.

COROLLARY 4. *The* $(FO + \exists^\infty + \exists^{n,m})$*-theory of any word automatic structure is decidable.*

The next corollary demonstrates how the extended logic can be used in a straightforward way.

COROLLARY 5. *If L is a word automatic partially ordered set, the set of all pairs $\langle x, y \rangle$ such that the interval $[x, y]$ has an even number of elements is regular.*

Another interesting application is an automata theoretic version of König's lemma. Recall that the classical version of König's Lemma says that every infinite finitely branching tree contains an infinite path. In Lecture 2, we discuss König's lemma in greater detail.

COROLLARY 6. *Let $\mathcal{T} = (T; \leq)$ be a word automatic infinite finitely branching tree. There exists an infinite regular path in $\mathcal{T}$.*

PROOF. We assume that the order $\leq$ is such that the root of the tree is the least element. We use the auxiliary automatic relations $\leq_{llex}$ (the length lexicographic order) to give a $(FO + \exists^\infty)$ definition of an infinite path of $\mathcal{T}$. Note that the immediate successor relation S is FO definable from the partial order.

$$P = \{x : \exists^\infty y(x \leq y) \,\&\, \forall(y \leq x)\forall(z \neq z' \in S(y))[z \leq x \Longrightarrow z <_{llex} z']\}$$

In words, P is the left-most infinite path in the tree. The first clause of the definition restricts our attention to those nodes of the tree which have infinitely many descendants. Since $\mathcal{T}$ is finitely branching, these nodes are exactly those which lie on infinite paths. It is easy to see that the above definition guarantees that P is closed downward, linearly ordered, and infinite. Moreover, it is regular by Theorem 7 on page 146. ⊣

Recently, Kuske and Lohrey generalized the decidability theorem for $(FO + \exists^\infty + \exists^{n,m})$ to Büchi automatic structures [38].

1.7. Definability in automatic structures. Throughout computer science and logic, there are classifications of problems or sets into hierarchies. Some examples include time complexity, relative computability, and proof-theoretic strength. In each case, there is a notion of reducibility between members in the hierarchy. One often searches for a complete, or typical, member at each level of the hierarchy. More precisely, an element of the hierarchy is called complete for a particular level if it is in that level, and if all other elements of the level are reducible to it. We may view automatic structures as a complexity class. In that context, we would like to find complete structures. For this question to be well-defined, we must specify a notion of reducibility. In light of the results of the previous section, it seems natural to consider logical definability of structures as the notion of reducibility. For Büchi automatic structures, Theorem 2 (Büchi's theorem) on page 140 immediately gives a complete structure with respect to MSO definability.

COROLLARY 7. *A structure is Büchi automatic if and only if it is definable in the MSO logic of the successor structure $(\omega; S)$.*

For word automatic structures, it turns out that first-order definability suffices. Blumensath and Grädel identify the following complete structure [8]. (Each of the basic relations in the following structure is defined in Subsection 1.3.)

THEOREM 9 (Blumensath and Grädel [8]). *A structure is word automatic if and only if it is first-order definable in the word structure* $(\{0,1\}^*; \preceq, S_0, S_1, EqL)$.

PROOF. One direction is clear because automatic structures are closed under first-order definability. For the converse, suppose that $\mathcal{A}$ is an automatic structure. We will show that it is definable in $(\{0,1\}^*; \preceq, S_0, S_1, EqL)$.

It suffices to show that every regular relation on $\{0,1\}^*$ is definable in the word structure. By changing the alphabet, we assume that L is a set recognized by the word automaton $\mathcal{M} = (S, \iota, \Delta, F)$. Assume $S = \{1, \dots, n\}$ with $1 = \iota$ (the initial state). We define a formula $\varphi_{\mathcal{M}}$ in the language of $(\{0,1\}^*; \preceq, S_0, S_1, EqL)$ which will hold of the string u if and only if u is accepted by $\mathcal{M}$. We use the following auxiliary definable relations in our definition of $\varphi_{\mathcal{M}}$.

- Length order, $|p| \leq |x|$, is defined by $\exists y(y \preceq x \,\&\, EqL(y, p)))$;
- The digit test relation (asserting that the digit of x at position $|p|$ is 0) is defined by $\exists y \exists z(z \preceq y \preceq x \,\&\, EqL(y, p) \,\&\, S_0(z, y)))$;
- The distinct digits relation states that the digits of x_1 and x_2 at position $|p|$ are distinct.

The following paragraph describing a run of $\mathcal{M}$ on input u can now be translated into a single first-order formula $\varphi_{\mathcal{M}}$ with free variable u and parameters corresponding to Δ. "There are strings $x_1, \dots, x_n$ each of length $|u| + 1$. For each p, if $|p| \leq |u| + 1$, there is exactly one x_j with digit 1 at position $|p|$ (the strings $x_1, \dots, x_n$ describe which states we're in at a given position in the run). If x_i has digit 1 at position $|p|$, and σ is the digit of u at position $|p|$, and $\Delta(i, \sigma) = j$ then x_j has digit 1 at positions $|p| + 1$. The digit in the first position of x_1 is 1. There is some x_j such that $j \in F$ and for which the digit in the last position is 1." Therefore, L is first-order definable in $(\{0,1\}^*; \preceq, S_0, S_1, EqL)$. Since the domain and each of the relations of an automatic structure are regular, they are also first-order definable. ⊣

If we use weak monadic second-order, WMSO, logic (where set quantification is only over finite sets) instead of first-order definability as our notion of reducibility, we arrive at a more natural complete structure for word automatic structures. Blumensath and Grädel [8] show that, in this case, the successor structure is complete. This result has nice symmetry with the Büchi case, where a structure is Büchi automatic if and only if it is MSO definable in $(\omega; S)$.

COROLLARY 8. *A structure is word automatic if and only if it is WMSO definable in the successor structure* $(\omega; S)$.

PROOF. By Theorem 9, it suffices to give a WMSO definition of the structure $(\{0,1\}^*; \preceq, S_0, S_1, EqL)$ in $(\omega; S)$. To do so, interpret each $v \in \{0,1\}^*$ by the set $Rep(v) = \{i : v(i) = 1\} \cup \{|v|+1\}$. Then $Rep(v)$ is a finite set and for each nonempty finite set X there is a string v such that $Rep(v) = X$. Moreover, under this representation, each of $S_0, S_1, \preceq, EqL$ is a definable predicate. ⊣

We refer the reader to [8, 51, 52] for issues related to definability in automatic structures.

§2. Characterization results.

2.1. Automata on trees and tree automatic structures. The two flavours of automata we have presented so far operate on linear inputs: finite and infinite strings. We now take a slight detour and consider labelled trees as inputs for automata. Our archetypal tree is the two successor structure,

$$\mathcal{T} = (\{0,1\}^*; S_0, S_1).$$

The root of this binary tree is the empty string, denoted as λ. Paths in $\mathcal{T}$ are defined by infinite strings in $\{0,1\}^\omega$. Let Σ be a finite alphabet. A Σ-labelled tree $(\mathcal{T}, v)$ associates a mapping $v : \mathcal{T} \to \Sigma$ to the binary tree. The set of all Σ-labelled trees is denoted by $Tree(\Sigma)$.

A *Rabin automaton* $\mathcal{M}$ is specified by $\mathcal{M} = (S, \iota, \Delta, \mathcal{F})$, where S and ι are the finite set of states and the initial state, the transition relation is $\Delta \subset S \times \Sigma \times (S \times S)$, and the accepting condition is given by $\mathcal{F} \subset \mathcal{P}(S)$. An input to a Rabin automaton is a labelled tree $(\mathcal{T}, v)$. A run of $\mathcal{M}$ on $(\mathcal{T}, v)$ is a mapping $r : \mathcal{T} \to S$ which respects the transition relation in that

$$r(\lambda) = \iota, \qquad \forall x \in \mathcal{T}\big[\big(r(x), v(x), r(S_0(x)), r(S_1(x))\big) \in \Delta\big].$$

The run r is accepting if for every path η in $\mathcal{T}$, the set

$$Inf(\eta) = \{s \in S : s \text{ appears on } \eta \text{ infinitely many times}\}$$

is an element of $\mathcal{F}$. The language of a Rabin automaton $\mathcal{M}$, denoted as $L(\mathcal{M})$, is the set of all Σ-labelled trees $(\mathcal{T}, v)$ accepted by $\mathcal{M}$.

EXAMPLE 19. Here are a few examples of Rabin automata recognizable sets of $\{0,1\}$-labelled trees.

- $\{(\mathcal{T}, v) : v(x) = 1 \text{ for only finitely many } x \in \mathcal{T}\}$
- $\{(\mathcal{T}, v) : \text{each path has infinitely many nodes labelled } 1\}$
- $\{(\mathcal{T}, v) : \forall x \in \mathcal{T}(v(x) = 1 \Longrightarrow \text{the subtree rooted at } x \text{ is labelled by 0s only})\}$
- $\{(\mathcal{T}, v) : \exists x \in \mathcal{T}(v(x) = 1)\}$.

As you may recall from Subsection 1.3, fundamental facts about Büchi automata gave us algorithms for checking emptiness and for constructing new automata from old ones. Rabin's breakthrough theorems in [48] yield analogous results for Rabin automata.

THEOREM 10 (Rabin [48]). 1. *There is an algorithm that, given a Rabin automaton $\mathcal{M}$, decides if $L(\mathcal{M})$ is empty.*

2. *The class of all Rabin recognizable tree languages is effectively closed under the operations of union, intersection, and complementation.*

In the setting of sequential automata, Theorem 2 on page 2 connected automata and logic. In particular, the logic used was MSO, where we allow quantification both over elements of the domain and over subsets of the domain. Properties of sets such as being a path or being open or clopen (in the natural topology on the paths through the tree) are all definable in MSO of the binary tree. Rabin's theorem [48] connects MSO definability and automaton recognizability. Note that, as in the Büchi case, convolutions can be used to define Rabin recognizability of relations on trees.

THEOREM 11 (Rabin [48]). *A relation $R \subseteq (Tree(\Sigma))^k$ is definable in the MSO logic of the two successor structure if and only if R is recognizable by a Rabin automaton. In particular, the MSO theory of $\mathcal{T}$, denoted as $S2S$, is decidable.*

This theorem has led to numerous applications in logic and theoretical computer science. Many of these applications involve proving the decidability of a particular theory by reducing it to the MSO theory of the binary tree.

Rabin automata have natural counterparts which work on finite binary trees. A finite binary tree X is a finite, prefix-closed, and rooted subset of $\mathcal{T}$. Let $L(X)$ denote the set of leaves (terminal nodes) of a finite binary tree X. A finite Σ-tree is a pair (X, v) where X is a finite binary tree and $v : X \setminus L(X) \to \Sigma$ is a mapping which labels nonleaf nodes in X with elements of the alphabet. A (top-down) *tree automaton* is $\mathcal{M} = (S, \iota, \Delta, F)$ where S, ι, F are as in the word automatic case, and $\Delta \subset S \times \Sigma \times (S \times S)$ is the transition relation. A run of $\mathcal{M}$ on a finite Σ-tree (X, v) is a map $r : X \to S$ which satisfies $r(\lambda) = \iota$ and $(r(x), v(x), r(S_0(x)), r(S_1(x))) \in \Delta$. The run is accepting if each leaf node $x \in L(x)$ is associated to an accepting state $r(x) \in F$. The language of a tree automaton is the set of finite Σ-trees it accepts. As before, tree automata have pleasant algorithmic properties (see the discussions in [14, 58, 48]).

THEOREM 12 (Doner [14]; Thatcher and Wright [58]; Rabin [48]).

1. *There is an algorithm that, given a tree automaton $\mathcal{M}$, decides if $L(\mathcal{M})$ is empty.*
2. *The class of all tree automata recognizable languages is effectively closed under the operations of union, intersection, and complementation.*

Automata which work on trees can be used to define tree automatic structures. In this context, domain elements of structures are represented as trees rather than strings.

DEFINITION 5. A structure is *tree automatic* if its domain and basic relations are recognizable by tree automata. A structure is *Rabin automatic* if its domain and basic relations are recognizable by Rabin automata.

Every tree automatic structure is Rabin automatic (we can pad finite trees into infinite ones). Since strings embed into trees, it is easy to see that every word automatic structure is tree automatic and that every Büchi automatic structure is Rabin automatic. However, this inclusion is strict. For example, the natural numbers under multiplication $(\omega; \times)$ is a tree automatic structure but (as we will see in the next subsection) it is not word automatic. Similarly, the countable atomless Boolean algebra is tree automatic but not word automatic [33]. In Lecture 3, we will discuss a recent result which separates Büchi and Rabin structures. Although there is a strict separation between the classes of sequential-input and branching-input automatic structures, their behaviour in terms of definability is very similar.

THEOREM 13 (Rabin [48]). *A structure is Rabin automatic if and only if it is MSO definable in the binary tree* $\mathcal{T}$.

THEOREM 14 (Rabin [48]). *A structure is tree automatic if and only if it is WMSO definable in the binary tree* $\mathcal{T}$.

Similarly, there is a theorem in the spirit of Löwenheim-Skolem for Rabin automatic structures akin to Theorem 4 (page 143) for Büchi automatic structures. Recall that a Σ-labelled tree is called *regular* if it has only finitely many isomorphic subtrees.

THEOREM 15 (Hjorth, Khoussainov, Montalbán and Nies [26]). *Let $\mathcal{A}$ be a Rabin automatic structure and consider the substructure $\mathcal{A}'$ whose domain is*

$$\mathcal{A}' = \{\alpha \in A : \alpha \text{ is a regular tree}\}.$$

Then $\mathcal{A}'$ is a computable elementary substructure of $\mathcal{A}$. Moreover, there is an algorithm that from a first-order formula $\varphi(\bar{a}, \bar{x})$ with parameters $\bar{a}$ from A' produces a Rabin automaton which accepts exactly those regular tree tuples $\bar{x}$ which make the formula true in the structure.

2.2. Proving nonautomaticity. Thus far, our toolbox contains several ways to prove that a given structure is automatic (explicitly exhibiting the automata, using extended first-order definitions, and using interpretations in the complete automatic structures). However, the only proof we've seen so far of nonautomaticity is restricted to word automata and uses cardinality considerations (cf. Subsections 1.3, 1.5). We could also use general properties of automatic structures to prove that a given structure is not automatic; for example, if a structure has undecidable first-order theory then it is not automatic by Corollary 2 on page 146. We will now see a more careful approach to proving nonautomaticity.

In finite automata theory, the Pumping Lemma is a basic tool for showing nonregularity. Recall that the lemma says that if a set L is regular then there is some n so that for every $w \in L$ with $|w| > n$, there are strings x, u, v with $|u| > 0$ such that $w = xuv$ and for all m, $xu^m v \in L$. The constant n is the number of states of the automaton recognizing L. The following Constant Growth Lemma uses the Pumping Lemma to arrive at an analogue for automatic structures [31].

LEMMA 1 (Khoussainov and Nerode [31]). *If $f : D^n \to D$ is a function whose graph is a regular relation, there is a constant C (which is the number of states of the automaton recognizing the graph of f) such that for all $x_1, \dots, x_n \in D$*

$$|f(x_1, \dots, x_n)| \leq \max\{|x_1|, \dots, |x_n|\} + C.$$

PROOF. Suppose for a contradiction that $|f(x_1, \dots, x_n)| - \max\{|x_1|, \dots, |x_n|\} > C$. Therefore, the convolution of the tuple $(x_1, \dots, x_n, f(x_1, \dots, x_n))$ contains more than C $\diamond$'s appended to each x_i. In particular, some state in the automaton is visited more than once after all the x_i's have been read. As in the Pumping Lemma, we can use this to obtain infinitely many tuples of the form $(x_1, \dots, x_n, y)$ with $y \neq f(x_1, \dots x_n)$ accepted by the automaton, or it must be the case that the automaton accepts strings which do not represent convolutions of tuples. Both of these cases contradict our assumption that the language of the automaton is the graph of the function f. ⊣

The Constant Growth Lemma can be applied in the settings of automatic monoids and automatic structures in general to give conditions on automaticity [33]. Recall that a monoid is a structure $(M; \cdot)$ whose binary operation $\cdot$ is associative.

LEMMA 2 (Khoussainov, Nies, Rubin, and Stephan [33]). *If $(M; \cdot)$ is an automatic monoid, there is a constant C (the number of states in the automaton recognizing $\cdot$) such that for every n and every $s_1, \dots, s_n \in M$*

$$|s_1 \cdot s_2 \cdot \dots \cdot s_n| \leq \max\{|s_1|, |s_2|, \dots, |s_n|\} + C \cdot \lceil \log(n) \rceil.$$

PROOF. Let C be the number of the states in the automaton recognizing the graph of the monoid multiplication. We proceed by induction on n. In the base case, the inequality is trivial: $|s_1| \leq |s_1|$. For $n > 1$, write $n = u + v$ such that $u = \lfloor \frac{n}{2} \rfloor$. Note that $u < n$ and $v < n$. Let $x_1 = s_1 \cdot \dots \cdot s_u$ and $x_2 = s_{u+1} \cdot \dots \cdot s_n$. By the induction hypothesis, $|x_1| \leq \max\{|s_1|, \dots, |s_u|\} + C \cdot \lceil \log(u) \rceil$ and $|x_2| \leq \max\{|s_{u+1}|, \dots, |s_n|\} + C \cdot \lceil \log(v) \rceil$. Applying Lemma 1,

$$\begin{aligned} |s_1 \cdot \dots \cdot s_n| = |x_1 \cdot x_2| &\leq \max\{|x_1|, |x_2|\} + C \\ &\leq \max\{|s_1|, \dots, |s_n|\} + C \max\{\lceil \log(u) \rceil, \lceil \log(v) \rceil\} + C \\ &\leq \max\{|s_1|, \dots, |s_n|\} + C \lceil \log(n) \rceil. \end{aligned}$$ ⊣

Let $\mathcal{A} = (A; F_0, F_1, \dots, F_n)$ be an automatic structure. Let $X = \{x_1, x_2, \dots\}$ be a subset of A. The generations of X are the elements of A which can be obtained from X by repeated applications of the functions of $\mathcal{A}$. More precisely,

$$G_1(X) = \{x_1\} \text{ and } G_{n+1}(X) = G_n(X) \cup \{F_i(\bar{a}) : \bar{a} \in G_n(X)\} \cup \{x_{n+1}\}.$$

The Constant Growth Lemma dictates the rate at which generations of X grow and yields the following theorem.

THEOREM 16 (Khoussainov and Nerode [31]; Blumensath [7]). *Suppose $X \subset A$ and there is a constant C_1 so that in the length lexicographic listing of X ($x_1 <_{llex} x_2 <_{llex} \cdots$) we have $|x_n| \leq C_1 \cdot n$ for all $n \geq 1$. Then there is a constant C such that $|y| \leq C \cdot n$ for all $y \in G_n(X)$. In particular,*

$$G_n(X) \subseteq \Sigma^{\leq C \cdot n}$$

if $|\Sigma| > 1$, and $|G_n(X)| \leq C \cdot n$ if $|\Sigma| = 1$.

Just as the Pumping Lemma allows immediate identification of certain nonregular sets, the above theorem lets us determine that certain structures are not automatic.

COROLLARY 9. *The free semigroup $(\{0, 1\}^*; \cdot)$ is not word automatic. Similarly, the free group $F(n)$ with $n > 1$ generators is not word automatic.*

PROOF. We give the proof for the free semigroup with two generators. Consider $X = \{0, 1\}$. By induction, we see that for each n, $\{0, 1\}^{<2^n} \subseteq G_{n+1}(\{0, 1\})$. Therefore, $|G_{n+1}(X)| \geq 2^{2^n - 1} - 1$ and hence can't be bounded by $2^{C \cdot n}$ for any constant C. ⊣

Similarly, one can prove the following.

COROLLARY 10. *For any bijection $f : \omega \times \omega \to \omega$, the structure $(\omega; f)$ is not word automatic.*

The proofs of the next two corollaries require a little more work but employ growth arguments as above.

COROLLARY 11. *The natural numbers under multiplication $(\omega; \times)$ is not word automatic.*

COROLLARY 12. *The structure $(\omega; \leq, \{n! : n \in \omega\})$, where the added unary predicate picks out the factorials, is not word automatic.*

We now switch our focus to subclasses of (word) automatic structures. For some of these classes, structure theorems have been proved which lead to good decision methods for questions like the isomorphism problem. Such structure theorems must classify both the members of a class and the nonmembers. Hence, techniques for proving nonautomaticity become very useful. In other cases, as we will see in the next lecture, complexity results give evidence that no nice structure theorems exist. The classes we consider below are partial orders, linear orders, trees, Boolean algebras, and finitely generated groups.

2.3. Word automatic partial orders. The structure $(A; \leq)$ is a *partially ordered set* if the binary relation $\leq$ is reflexive $\forall x(x \leq x)$, anti-symmetric $\forall x \forall y(x \leq y \;\&\; y \leq x \rightarrow x = y)$, and transitive $\forall x \forall y \forall z(x \leq y \;\&\; y \leq z \rightarrow x \leq z)$. A partially ordered set $(A; \leq)$ is word automatic if and only if A and $\leq$ are both recognized by word automata. For the rest of this section, we deal only with word automatic partial orders and for brevity, we simply call them automatic.

EXAMPLE 20. We have already seen some examples of automatic partial orders: the full binary tree under prefix order $(\{0,1\}^*; \preceq)$; the finite and co-finite subsets of ω under subset inclusion; the linear order of the rational numbers.

Recall that a linear order is one where $\leq$ is also total: for any x, y in the domain, either $x \leq y$ or $y \leq x$.

EXAMPLE 21. Small ordinals (such as ω^n for n finite) [31] are automatic partial orders. In fact, Delhommé showed that automatic ordinals are exactly all those ordinals below ω^ω [13]. We will see the proof of this fact later in this subsection.

EXAMPLE 22. The following example was one of the first examples of a nontrivial automatic linear order [51]. Given an automatic linear order L and a polynomial $f(x)$ with positive integer coefficients, consider the linear order

$$\Sigma_{x \in \omega}(L + f(x))$$

where we have a copy of L, followed by a finite linear order of length $f(0)$, followed by another copy of L, followed by a finite linear order of length $f(1)$, etc. This linear order is automatic. In fact, the linear order obtained by the same procedure where the function f is an exponential $f(x) = a^{b \cdot x + c}$ with $a, b, c \in \omega$ is also automatic.

The last example involves the addition of linear orders. The sum of orders L_1 and L_2 is the linear order in which we lay down the order L_1 and then we place all of L_2. Since $L_1 + L_2$ is first-order definable from the disjoint union of L_1 and L_2, Proposition 1 on page 144 and Corollary 3 on page 146 imply that the sum operation preserves automaticity. Another basic operation on linear orders also preserves automaticity: the product linear order $L_1 \cdot L_2$ is one where a copy of L_1 is associated with each element of L_2; each of the copies of L_1 is ordered as in L_1, while the order of the copies is determined by L_2. The order $L_1 \cdot L_2$ is first-order definable from the Cartesian product union of L_1 and L_2, and hence is automatic if L_1 and L_2 are automatic.

We use several approaches to study the class of automatic partial orders. First, we restrict to well-founded partial orders and consider their ordinal heights. We will see that automatic partial orders are exactly those partial orders with relatively low ordinal heights. This observation parallels Delhommé's previously mentioned result that automatic ordinals are exactly those below ω^ω [13]. Next, we study automatic linear orders. We present results about the Cantor-Bendixson ranks of automatic linear orders, and see the implications of these results for decidability questions. In particular, we see that the isomorphism problem for automatic ordinals is decidable. Finally, we consider partial orders as trees and consider the branching complexity of automatic partial order trees. We present several automatic versions of König's famous lemma about infinite trees.

We now introduce well-founded partial orders and ordinal heights. A binary relation R is called *well-founded* if there is no infinite chain of elements $x_0, x_1, x_2, \ldots$ such that $(x_{i+1}, x_i) \in R$ for all i. For example, $(\mathcal{Z}^+; S)$ is a well-founded relation but $(\mathcal{Z}^-; S)$ is not well-founded (we use $\mathcal{Z}^+$ and $\mathcal{Z}^-$ to denote the positive and negative natural numbers). Given a well-founded structure $\mathcal{A} = (A; R)$ with domain A and binary relation R, a *ranking function* for $\mathcal{A}$ is an ordinal-valued function f on A such that $f(y) < f(x)$ whenever $(y, x) \in R$. We define $ord(f)$ as the least ordinal larger than or equal to all values of f. It is not hard to see that $\mathcal{A} = (A; R)$ is well-founded if and only if $\mathcal{A}$ has a ranking function.

Given a well-founded structure $\mathcal{A}$, its *ordinal height* (denoted $r(\mathcal{A})$) is the least ordinal α which is $ord(g)$ for some ranking function g for $\mathcal{A}$. An equivalent definition of the ordinal height uses an assignment of rank to each element in the domain of $\mathcal{A}$. If x is an R-minimal element of A, set $r_{\mathcal{A}}(x) = 0$. For any other element in A, put $r_{\mathcal{A}}(z) = \sup\{r(y) + 1 : (y, z) \in R\}$. Then, we define $r(\mathcal{A}) = \sup\{r_{\mathcal{A}}(x) : x \in A\}$. The following property of ordinal heights is useful when we work with substructures of well-founded relations.

LEMMA 3. *Given a well-founded structure $\mathcal{A} = (A; R)$, if $r(\mathcal{A}) = \alpha$ and $\beta < \alpha$ then there is $x \in A$ such that $r_{\mathcal{A}}(x) = \beta$.*

The ordinal height can be used to measure the depth of a structure. In our exploration of automatic structures, we study the ordinal heights attained by automatic well-founded relations. As a point of departure, recall that any automatic structure is also a computable structure (see Subsection 1.2). We therefore begin by considering the ordinal heights of computable structures. An ordinal is called computable if it is the order-type of some computable well-ordering of the natural numbers.

LEMMA 4. *Each computable ordinal is the ordinal height of some computable well-founded relation. Conversely, the ordinal height of each computable well-founded relation is a computable ordinal.*

Since any automatic structure is a computable structure, Lemma 4 gives us an upper bound on the ordinal heights of automatic well-founded relations. We now ask whether this upper bound is sharp. We will consider this question both in the setting of all automatic well-founded relations (in Lecture 3), and in the setting of automatic well-founded partial orders (now). The following theorem characterizes automatic well-founded partial orders in terms of their ordinal heights.

THEOREM 17 (Khoussainov and Minnes [30]). *An ordinal α is the ordinal height of an automatic well-founded partial order if and only if $\alpha < \omega^\omega$.*

One direction of the proof of the characterization theorem is easy: each ordinal below ω^ω is automatic (Example 21 on page 154) and is an automatic well-founded total order. Moreover, the ordinal height of an ordinal is itself.

For the converse, we will use a property of the natural sum of ordinals. The *natural sum* of α and β, denoted $\alpha +' \beta$, is defined recursively as $\alpha +' 0 = \alpha$, $0 +' \beta = \beta$, and $\alpha +' \beta$ is the least ordinal strictly greater than $\gamma +' \beta$ for all $\gamma < \alpha$ and strictly greater than $\alpha +' \gamma$ for all $\gamma < \beta$ (see, for example [19]; note that the natural sum of ordinals is commutative whereas the usual sum is not). An equivalent definition of the natural sum uses the *Cantor normal form* of ordinals. Recall that any ordinal can be written in this normal form as

$$\alpha = \omega^{\beta_1} n_1 + \omega^{\beta_2} n_2 + \cdots + \omega^{\beta_k} n_k,$$

where $\beta_1 > \beta_2 > \cdots > \beta_k$ and $k, n_1, \ldots, n_k \in \mathcal{N}$. We define

$$(\omega^{\beta_1} a_1 + \cdots + \omega^{\beta_k} a_k) +' (\omega^{\beta_1} b_1 + \cdots + \omega^{\beta_k} b_k) = \\ \omega^{\beta_1}(a_1 + b_1) + \cdots + \omega^{\beta_k}(a_k + b_k).$$

The following lemma gives subadditivity of ordinal heights of substructures with respect to the natural sum of ordinals.

LEMMA 5. *Suppose $\mathcal{A} = (A; \leq)$ is a well-founded partial order and A_1, A_2 form a partition of A ($A_1 \sqcup A_2 = A$, a disjoint union). Let $\mathcal{A}_1 = (A_1; \leq_1)$, $\mathcal{A}_2 = (A_1; \leq_2)$ be obtained by restricting $\leq$ to A_1, A_2. Then $r(\mathcal{A}) \leq r(\mathcal{A}_1) +' r(\mathcal{A}_2)$.*

PROOF. For each $x \in A$, consider the sets $A_{1,x} = \{z \in A_1 : z < x\}$ and $A_{2,x} = \{z \in A_2 : z < x\}$. The structures $\mathcal{A}_{1,x}, \mathcal{A}_{2,x}$ are substructures of $\mathcal{A}_1, \mathcal{A}_2$ respectively. Define a ranking function of $\mathcal{A}$ by $f(x) = r(\mathcal{A}_{1,x}) +' r(\mathcal{A}_{2,x})$. The range of f is contained in $r(\mathcal{A}_1) +' r(\mathcal{A}_2)$. Therefore, $r(\mathcal{A}) \leq r(\mathcal{A}_1) +' r(\mathcal{A}_2)$. ⊣

We now outline the proof of the nontrivial direction of the characterization theorem of automatic well-founded partial orders. Note that this proof follows Delhommé's proof that ordinals larger than ω^ω are not automatic [13]. We assume for a contradiction that there is an automatic well-founded partial order $\mathcal{A} = (A; \leq)$ such that $r(\mathcal{A}) = \alpha \geq \omega^\omega$. Let $\mathcal{M}_A = (S_A, \iota_A, \Delta_A, F_A)$, $\mathcal{M}_\leq = (S_\leq, \iota_\leq, \Delta_\leq, F_\leq)$ be the word automata which recognize A and $\leq$ (respectively). For each $u \in A$, the set of predecessors of u, denoted by $u\downarrow$, can be partitioned into finitely many disjoint pieces as

$$u\downarrow = \{x \in A : |x| < |u| \,\&\, x < u\} \sqcup \bigsqcup_{v \in \Sigma^* : |v| = |u|} X_v^u$$

where $X^u_v = \{vw \in A : vw < u\}$ (extensions of v which are predecessors of u). Since $r(\mathcal{A}) \geq \omega^\omega$, Lemma 3 guarantees that for each n, there is an element $u_n \in A$ with $r_{\mathcal{A}}(u) = \omega^n$. Moreover, Lemma 5 implies that if a structure has ordinal height ω^n, any finite partition of the structure contains a set of ordinal height ω^n. In particular, for each u_n there is v_n such that $|u_n| = |v_n|$ and $r(X^{u_n}_{v_n}) = r(u_n) \downarrow= \omega^n$. We now use the automata $\mathcal{M}_A, \mathcal{M}_\leq$ to define an equivalence relation of finite index on pairs (u, v): $(u, v) \sim (u', v')$ if and only if $\Delta_A(\iota_A, v) = \Delta_A(\iota_A, v')$ and $\Delta_\leq(\iota_\leq, \binom{v}{u}) = \Delta_\leq(\iota_\leq, \binom{v'}{u'})$. Suppose that $(u, v) \sim (u', v')$. Then the map $f : X^u_v \to X^{u'}_{v'}$ defined as $f(vw) = v'w$ is an order-isomorphism. Hence, $r(X^u_v) = r(X^{u'}_{v'})$. Also, there are at most $|S_A| \times |S_\leq|$ $\sim$equivalence classes. Therefore, the sequence $\{(u_n, v_n)\}$ contains some m, n $(m \neq n)$ such that $(u_m, v_m) \sim (u_n, v_n)$. But, $\omega^m = r(u_m) = r(X^{u_m}_{v_m}) = r(X^{u_n}_{v_n}) = r(u_n) = \omega^n$. This is a contradiction with $m \neq n$. Thus, there is no automatic well-founded partial order whose ordinal height is greater than or equal to ω^ω.

The above characterization theorem applies to automatic *well-founded* partial orders. We now examine a different class of automatic partial orders: linear orders. We seek a similar characterization theorem for automatic linear orders based on an alternate measure of complexity. Let $(L; \leq)$ be a linear order. Then $x, y \in L$ are called $\equiv_F$-equivalent if there are only finitely many elements between them. We can use $\equiv_F$ equivalence to measure how far a linear order is from "nice" linear orders like $(\omega; \leq)$ or $(\mathcal{Q}; \leq)$. To do so, given a linear order, we take its quotient with respect to the $\equiv_F$ equivalence classes as many times as needed to reach a fixed-point. The first ordinal at which the fixed-point is reached is called the *Cantor-Bendixson rank* of the linear order, denoted $CB(L; \leq)$. Observe that $CB(\mathcal{Q}; \leq) = 0$ and $CB(\omega; \leq) = 1$. Moreover, the fixed-point reached after iteratively taking quotients of any linear order will either be isomorphic to the rational numbers (perhaps with endpoints) or the linear order with a single element. A useful lemma tells us that automaticity is preserved as we take quotients by $\equiv_F$.

LEMMA 6. *If $\mathcal{L} = (L; \leq)$ is an automatic linear order then so is the quotient linear order $\mathcal{L}/\equiv_F$.*

PROOF. By Proposition 2 from page 144, it suffices to show that $\equiv_F$ is definable in the extended logic $(FO + \exists^\infty)$ of $(L; \leq)$. The definition is

$$x \equiv_F y \iff \neg\exists^\infty z\Big[\big(x \leq z \,\&\, z \leq y\big) \vee \big(y \leq z \,\&\, z \leq x\big)\Big] \qquad \dashv$$

Theorem 17 showed that well-founded automatic partial orders have relatively low ordinal heights. In this vein, it is reasonable to expect a low bound on the Cantor-Bendixson rank of automatic linear orders as well. The following characterization theorem does just that. This characterization theorem and its implications for linear orders and ordinals are discussed in [34, 35].

THEOREM 18 (Khoussainov, Rubin, and Stephan [34]). *An ordinal α is the Cantor-Bendixson rank of an automatic linear order if and only if it is finite.*

The proof of Theorem 18 has many common features with the proof of Theorem 17. One direction is easy: for $n < \omega$, ω^n is automatic and $CB(\omega^n) = n$. The hard direction relies on understanding the Cantor-Bendixson ranks of suborders of a given linear order. In particular, we make use of suborders determined by intervals of a linear order: sets of the form $\{z : x \leq z \leq y\}$ for some x and y.

LEMMA 7. *For any linear order $\mathcal{L}$ and ordinal α, if $CB(\mathcal{L}) = \alpha$ and $\beta \leq \alpha$ then there is an interval $[x, y]$ of $\mathcal{L}$ with $CB([x, y]) = \beta$.*

To prove that any linear order with infinite Cantor-Bendixson rank is not automatic, we go by contradiction. We suppose that such a linear order exists, and use the associated automata to define an equivalence relation on intervals which has only finitely many equivalence classes. Moreover, intervals in the same equivalence class have the same Cantor-Bendixson rank. However, since we assume that the linear order has infinite Cantor-Bendixson rank, Lemma 7 allows us to pick out intervals with every finite Cantor-Bendixson rank. Therefore, two such intervals must be in the same equivalence class. But this contradicts our choice of intervals with distinct Cantor-Bendixson ranks.

Theorem 18 has been productive for decidability results. In particular, it yields algorithms for computing the Cantor-Bendixson rank of a given automatic linear order, and for studying scattered linear orders and ordinals. A linear order is called *dense* if for each x and y with $x \leq y$, there is some z such that $x \leq z \leq y$. The linear orders with zero elements or one element are trivially dense. For countable linear orders, there are exactly four isomorphism types of nontrivial dense linear orders (the rational numbers restricted to $(0, 1)$, the rational numbers restricted to $[0, 1)$, the rational numbers restricted to $(0, 1]$, the rational numbers restricted to $[0, 1]$). Note that being dense is a first-order definable property, so Corollary 2 on page 146 tells us we can decide if a given automatic structure is a dense linear order. A linear order is called *scattered* if it contains no nontrivial dense suborder. A linear order is an ordinal if it is well-founded.

COROLLARY 13. *There is an algorithm which, given an automatic linear order $\mathcal{L}$, computes the Cantor-Bendixson rank of $\mathcal{L}$.*

PROOF. Check if $\mathcal{L}$ is dense. If it is, output $CB(\mathcal{L}) = 0$. Otherwise, Lemma 6 from page 157 tells us that the quotient $\mathcal{L}/\equiv_F$ is automatic. We iterate checking if the quotient is dense and, if it is not, constructing the next quotient and incrementing the counter. Each of these steps is effective because denseness is a first-order question. Moreover, this procedure eventually stops by Theorem 18 and we can count how many iterations are required. Once we reach a dense quotient structure, we output this number. ⊣

The following two corollaries about automatic scattered linear orders use trivial modifications of the above algorithm.

Corollary 14. *It is decidable if a given automatic linear order is scattered.*

Corollary 15. *Given an automatic linear order $\mathcal{L}$ that is not scattered, there is an algorithm which computes an automatic dense suborder of L.*

We now apply Theorem 18 to the subclass of linear orders which are well-founded, the ordinals. We will see that we can effectively check if a given automatic linear order is an ordinal; and given two automatic ordinals, we can check if they are isomorphic. The isomorphism question is one of the central motivating questions in the study of automatic structures (see Subsection 1.1). The class of automatic ordinals was one of the first contexts in which a positive answer to this question was found.

Corollary 16. *If $\mathcal{L}$ is an automatic linear order, there is an algorithm which checks if $\mathcal{L}$ is an ordinal.*

Proof. To check if a given automatic linear order $\mathcal{L}$ is an ordinal, we need to check if it has an infinite descending sequence. Note that infinite descending sequences can occur either within an $\equiv_F$ equivalence class or across such classes. We begin by checking whether $\mathcal{L}$ is not dense and $\forall(x \in L)\exists^\infty y(x \equiv_F y \,\&\, y < x)$. If this condition holds, we form the quotient $\mathcal{L}/\equiv_F$ and check the condition again for the quotient linear order. We iterate until the condition fails, which must occur after finitely many iterations because $CB(\mathcal{L})$ is finite. If the resulting linear order has exactly one element, output that $\mathcal{L}$ is an ordinal, and otherwise output that it is not. If $\mathcal{L}$ is an ordinal then the $\equiv_F$ equivalence classes are all finite or isomorphic to ω and all quotient linear orders of $\mathcal{L}$ are also ordinals. Therefore, the algorithm will stop exactly when the quotient is dense, in which case it will have exactly one element. If $\mathcal{L}$ is not an ordinal, then either there will be a stage of the algorithm at which there is an infinite descending chain within a single $\equiv_F$ equivalence class or the final dense linear order will contain an infinite descending chain. In either case, the algorithm will recognize that $\mathcal{L}$ is not an ordinal. ⊣

Corollary 17. *The Cantor normal form of a given automatic ordinal is computable.*

Proof. Given an automatic ordinal $\mathcal{L}$, we use first-order definitions of maximal elements and the set of limit ordinals in $\mathcal{L}$ to iteratively determine the coefficients in the Cantor normal form. The set of limit ordinals play a role because if $\alpha = \omega^m a_m + \cdots + \omega a_1$, then the Cantor normal form of the set of limit ordinals strictly below α is $\omega^{m-1}a_m + \cdots + \omega^1 a_2 + a_1$. ⊣

Since two ordinals are isomorphic if and only if they have the same Cantor normal form, the following corollary is immediate.

Corollary 18. *The isomorphism problem for automatic ordinals is decidable.*

We do not know whether the isomorphism problem for automatic linear orders is decidable.

2.4. Word automatic trees. We now focus our interest on partial orders which form trees. We begin with the following definition of trees.

DEFINITION 6. A *(partial order) tree* is $\mathcal{A} = (A; \leq)$ where $\leq$ is a partial order such that A has a $\leq$-least element and the set $x \downarrow$ (the $\leq$-predecessors of x) is linearly ordered and finite for all $x \in A$.

For any regular language L, if L is prefix-closed then the structure $(L; \preceq)$ where $\preceq$ is the prefix relation is a (word) automatic tree. The two most famous such examples are the full binary tree $(\{0,1\}^*; \preceq)$ and the countably branching tree $\omega^{<\omega} \cong (\{0,1\}^* \cdot 1; \preceq)$. In the following, for a given tree $\mathcal{T} = (T; \leq)$ we denote by $\mathcal{T}_x$ the subtree of $\mathcal{T}$ which is rooted at node x; that is, $\mathcal{T}_x$ is the tree $(\{y \in T : x \leq y\}; \leq)$.

There is a natural connection between trees and topological spaces: given a tree $\mathcal{T} = (T; \leq)$, define an associated topological space whose elements are infinite paths of $\mathcal{T}$ and whose basic open sets are collections of infinite paths defined by $\{P : x \in P\}$ for each $x \in T$. Then the topological Cantor-Bendixson rank transfers to trees in a straightforward way. As in the case of linear orders, we can compute this rank via an iterative search for a fixed-point. The derivative is defined to be $d(\mathcal{T})$, the subtree of $\mathcal{T}$ containing all nodes $x \in T$ such that x belongs to two distinct infinite paths of $\mathcal{T}$. The derivative can be carried along the ordinals by letting $d^{\alpha+1}(\mathcal{T}) = d(d^\alpha(\mathcal{T}))$ and, for limit ordinals γ, $d^\gamma(\mathcal{T}) = \cap_{\beta<\gamma} d^\beta(\mathcal{T})$. The first ordinal α for which $d^\alpha(\mathcal{T}) = d^{\alpha+1}(\mathcal{T})$ is called the Cantor-Bendixson rank of $\mathcal{T}$ and is denoted by $CB(\mathcal{T})$. The proof of the following lemma gives some intuition about Cantor-Bendixson ranks and is left to the reader.

LEMMA 8. *If a tree has only countably many paths, its Cantor-Bendixson rank is* 0 *or a successor ordinal.*

As in our discussion of linear orders, it is natural to look for low bounds on the complexity of automatic trees. In particular, the Cantor-Bendixson ranks of trees and linear orders are intimately connected. A linear order can be associated with each automatic finitely branching tree. The *Kleene-Brouwer ordering* of the tree $\mathcal{T} = (T; \leq)$ is given by $x \leq_{KB_\mathcal{T}} y$ if and only if $x = y$ or $y \leq x$ or there are $u, v, w \in T$ such that v, w are immediate successors of u and $v \leq_{llex} w$ and $v \leq x$ and $w \leq y$. Note that the definition of the Kleene-Brouwer order uses the length-lexicographic ordering inherited by T as a subset of Σ^* for some finite alphabet Σ. It is not hard to see that the Kleene-Brouwer order is a linear order. For example, the Kleene-Brouwer order associated to the tree which has exactly two infinite paths $(0^* \cup 1^*; \preceq)$ is isomorphic to $\mathcal{Z}^- + \mathcal{Z}^-$. If $\mathcal{T}$ is an automatic tree then the relation $\leq_{KB_\mathcal{T}}$ is first-order definable in it. Hence, the results in Subsection 2.3 imply that the Kleene-Brouwer ordering of an automatic finitely branching tree is automatic.

LEMMA 9 (Khoussainov, Rubin, and Stephan [34]). *If $\mathcal{T}$ is an automatic finitely branching tree with countably many paths then $CB(\mathcal{T}) = CB(KB_{\mathcal{T}})$.*

PROOF. The outline of the proof is as follows; for details, see [34]. We claim that if $\mathcal{T}$ is an automatic finitely branching tree with countably many paths then $KB_{\mathcal{T}}$ is scattered. Given any infinite path of $\mathcal{T}$, the linear order $KB_{\mathcal{T}}$ can be expressed as a countable sum of linear orders which are either trivial or $KB_{\mathcal{S}}$ for $\mathcal{S}$ a subtree of $\mathcal{T}$. By induction, we will show that $CB(\mathcal{T}) = CB(KB_{\mathcal{T}})$. If $CB(\mathcal{T}) = 0$ then it is empty so $KB_{\mathcal{T}}$ is also empty and $CB(KB_{\mathcal{T}}) = 0$. For the inductive step, we suppose that $CB(\mathcal{T}) = \beta + 1$ and notice that the subtree of $\mathcal{T}$ whose domain is $X = \{x \in T : CB(\mathcal{T}_x) = \beta + 1\}$ has a finite and nonzero number of infinite paths. Each of these paths give rise to a sum expression of $KB_{\mathcal{T}_x}$, where $CB(KB_{\mathcal{T}_x}) = \beta + 1$. Then $KB_{\mathcal{T}}$ is the sum of these finitely many linear orders $KB_{\mathcal{T}_x}$ and so $KB_{\mathcal{T}} = \beta + 1$ as well. ⊣

The lemma above allows us to transfer the bound on Cantor-Bendixson ranks of automatic linear orders to a bound on the Cantor-Bendixson ranks of automatic finitely branching trees with countably many paths.

THEOREM 19 (Khoussainov, Rubin, and Stephan [34]). *If $\mathcal{T}$ is an automatic finitely branching tree with countably many paths then $CB(\mathcal{T})$ is finite.*

PROOF. Suppose $\mathcal{T}$ is as above. Then $KB_{\mathcal{T}}$ is also automatic. By Theorem 18, $CB(KB_{\mathcal{T}})$ is finite. Lemma 9 then gives that $CB(\mathcal{T})$ is finite. ⊣

We will improve Theorem 19 by weakening its hypotheses. Our goal is to show that the Cantor-Bendixson rank of each automatic partial order tree is finite.

THEOREM 20 (Khoussainov, Rubin, and Stephan [34]). *If $\mathcal{T}$ is an automatic finitely branching tree then $CB(\mathcal{T})$ is finite.*

PROOF. We consider two kinds of nodes on $\mathcal{T}$: those which lie on at most countably many infinite paths, and those which lie on uncountably many infinite paths. For each x which lies on at most countably many infinite paths consider $\mathcal{T}_x$, the subtree of $\mathcal{T}$ rooted at x. Note that $KB_{\mathcal{T}_x}$ is an interval of $KB_{\mathcal{T}}$ for each x. By Lemma 9 and Theorem 18, $CB(\mathcal{T}_x) = CB(KB_{\mathcal{T}_x}) \leq CB(KB_{\mathcal{T}}) = n$ for some n. Therefore, $d^n(\mathcal{T})$ contains only nodes which lie on uncountably many infinite paths. In particular, this implies that $d^{n+1}(\mathcal{T}) = d^n(\mathcal{T})$ so $CB(\mathcal{T}) = n$. ⊣

Finally, we remove the requirement that the tree is finitely branching.

THEOREM 21 (Khoussainov, Rubin, and Stephan [34]). *If $\mathcal{T}$ is an automatic tree then $CB(\mathcal{T})$ is finite.*

PROOF. Given an automatic tree $\mathcal{T} = (T; \leq)$, we consider a finitely branching tree which is first-order definable in $\mathcal{T}$ and whose Cantor-Bendixson rank is no less than that of $\mathcal{T}$. The tree $\mathcal{T}'$ has domain T and partial order $\leq'$ defined by

$$x \leq' y \iff x \leq y \vee \exists v \exists w \left[x \in S(v) \,\&\, w \in S(v) \,\&\, x \leq_{llex} w \,\&\, w \leq y\right),$$

where S is the immediate successor function with respect to $\leq$. To picture $\mathcal{T}'$, note that each node has at most two immediate successors: its $\leq_{llex}$-least successor from $\mathcal{T}$, and its $\leq_{llex}$-least sibling in $\mathcal{T}$. We can define a continuous and injective map from the infinite paths of $\mathcal{T}$ to the infinite paths of $\mathcal{T}'$. Analysing the derivatives of $\mathcal{T}$ and $\mathcal{T}'$ via this map leads to the conclusion that $CB(\mathcal{T}) \leq CB(\mathcal{T}')$. Applying Theorem 20 to $\mathcal{T}'$, $CB(\mathcal{T})$ is finite. ⊣

COROLLARY 19. *An ordinal α is the Cantor-Bendixson rank of an automatic (partial order) tree if and only if it is finite.*

One of the most fundamental tools available for the study of finitely branching trees is König's lemma: every finitely branching infinite tree has an infinite path. It is natural to wonder whether such a result holds in the context of automatic structures. There are several ways one can translate results from mathematics to include feasibility constraints. For example, we can ask whether every automatic finitely branching infinite tree has a regular infinite path. A stronger result would be to find a regular relation which picks out the infinite regular paths through automatic trees. We will now develop several automatic versions of König's lemma in this spirit. As a starting point, recall Corollary 6 from page 147: every infinite automatic finitely branching tree has a regular infinite path. A surprising feature of the landscape of automatic structures is that a version of König's lemma holds even when we remove the finitely branching assumption.

THEOREM 22 (Khoussainov, Rubin, and Stephan [34]). *If an infinite automatic tree has an infinite path, it has a regular infinite path.*

PROOF. The first-order definition of an infinite path in Corollary 6 relied on a definition of the set of nodes which lie on any infinite path. However, in the context of trees which may have nodes with infinite degree, such a definition is harder to find. To show that this set of nodes is regular, we define an auxiliary Büchi recognizable set which tracks nodes on infinite paths. Our desired regular set is then achieved via applications of the projection operation and decoding of the Büchi automaton. For more details, see [34]. ⊣

A full automatic version of König's lemma would produce a regular relation which codes all infinite paths through a given automatic tree. The following theorem from [34] does just this.

THEOREM 23 (Khoussainov, Rubin, and Stephan [34]). *Given an automatic tree $\mathcal{T}$, there is a regular relation $R(x, y, z)$ satisfying the following properties:*

- $\exists y \exists z\ (R(x, y, z) \iff$ *$\mathcal{T}_x$ has at most countably many infinite paths*);

– *for each* $y \in \Sigma^*$ *and for each* x *such that* $\mathcal{T}_x$ *has at most countably many infinite paths, the set* $R_{x,y} = \{z : R(x, y, z)\}$ *is either empty or is an infinite path through* $\mathcal{T}_x$*; and*

– *for each* x *such that* $\mathcal{T}_x$ *has at most countably many infinite paths, if* α *is an infinite path through* $\mathcal{T}_x$ *then there is a* $y \in \Sigma^*$ *such that* $R_{x,y} = \alpha$.

In this section and in the preceding one, we gave tight bounds on the ranks and heights of automatic well-founded partial orders, trees, ordinals, and partial order trees. Such bounds indicate that the level of complexity of the structures in these classes is relatively low. However, it does not give us concrete information on the members of these classes. In the next section, we present classification theorems for various collections of automatic structures.

2.5. Word automatic Boolean algebras. We now use the results about nonautomaticity from Subsection 2.2 to give a classification of word automatic Boolean algebras. An *atom* of a Boolean algebra is a minimal non-0 element. As we saw in Subsection 1.5, the Boolean algebra of finite and co-finite subsets of ω, $\mathcal{B}_\omega$, has a word automata presentation. For each $n \geq 1$, we define $\mathcal{B}_\omega^n$ to be the n-fold Cartesian product of $\mathcal{B}_\omega$. Since finite Cartesian products preserve automaticity, $\mathcal{B}_\omega^n$ is automatic. The main theorem of this section says that each automatic Boolean algebra must be isomorphic to $\mathcal{B}_\omega^n$ for some n.

THEOREM 24 (Khoussainov, Nies, Rubin, and Stephan [33]). *A Boolean algebra is word automatic if and only if it is isomorphic to* $\mathcal{B}_\omega^n$ *for some* $n \geq 1$.

The discussion above shows that any Boolean algebra isomorphic to $\mathcal{B}_\omega^n$ is word automatic. We now prove the converse. Let $\mathcal{B}$ be an automatic Boolean algebra. Assume for a contradiction that $\mathcal{B}$ is not isomorphic to any $\mathcal{B}_\omega^n$. We will construct a set of elements of B which will contradict the Growth Lemma for Monoids (Lemma 2 on page 152). To do so, we recall some terminology for elements of Boolean algebras. We say that $a, b \in B$ are F-equivalent, $a \equiv_F b$, if the element $(a \wedge \bar{b}) \vee (\bar{a} \wedge b)$ is a union of finitely many atoms of $\mathcal{B}$. Note that $\mathcal{B}/\equiv_F$ is itself a Boolean algebra. Moreover, $\mathcal{B}/\equiv_F$ is automatic because the equivalence relation $\equiv_F$ is $(FO + \exists^\infty)$ definable in $\mathcal{B}$ (Proposition 2 on page 144). Moreover, since $\mathcal{B}$ is not isomorphic to $\mathcal{B}_\omega^n$ for any n, $\mathcal{B}/\equiv_F$ is infinite. We call $x \in B$ *infinite* if there are infinitely many elements $y \in B$ such that $y \wedge x = y$ (i.e. $y \leq x$ in the partial order induced by the Boolean algebra). We say that x *splits* y in $\mathcal{B}$ if $x \wedge y \neq 0$ and $\bar{x} \wedge y \neq 0$. We call $x \in B$ *large* if the $\equiv_F$ equivalence class of x is not a finite union of atoms of the quotient algebra. The following lemma collects properties of infinite and large elements of a Boolean algebra.

LEMMA 10. *If* y *is large then there is* x *that splits* y *such that* $y \wedge x$ *is large and* $y \wedge \bar{x}$ *is infinite. If* y *is infinite then there is* x *that splits* y *such that either* $x \wedge y$ *is infinite or* $\bar{x} \wedge y$ *is infinite.*

We use Lemma 10 to inductively define a sequence of trees $\{\mathcal{T}_n\}$ whose nodes are binary strings. To each node σ of the tree $\mathcal{T}_n$ we associate an element b_σ of the Boolean algebra. We denote the set of elements of $\mathcal{B}$ associated to the leaves of the tree $\mathcal{T}_n$ by X_n. At each step of the induction, we verify that the following properties hold:

- There is a leaf $\sigma \in \mathcal{T}_n$ such that b_σ is large in $\mathcal{B}$.
- There are n leaves $\sigma_1, \dots, \sigma_n \in \mathcal{T}_n$ such that b_{σ_i} is infinite.
- The tree $\mathcal{T}_n$ has as least $\frac{n(n+1)}{2}$ many leaves.
- If σ and τ are distinct leaves of $\mathcal{T}_n$, $b_\sigma \wedge b_\tau = 0$.

The base tree is $T_0 = \{\lambda\}$ (where λ is the empty binary string) and $b_\lambda = 1$. Given $\mathcal{T}_n$, we define $\mathcal{T}_{n+1}$ as the extension of $\mathcal{T}_n$ obtained by doing the following for each leaf $\sigma \in \mathcal{T}_n$. If b_σ is large, let $a \in B$ be the length lexicographically first element that splits b_σ such that both $b_\sigma \wedge a$ and $b_\sigma \wedge \bar{a}$ are infinite and one of them is large. Such an a exists by Lemma 10. Add $\sigma 0$ and $\sigma 1$ to $\mathcal{T}_{n+1}$ (as leaves) and let $b_{\sigma 0} = b_\sigma \wedge a$ and $b_{\sigma 1} = b_\sigma \wedge \bar{a}$. If b_σ is one of the n leaves from $\mathcal{T}_n$ for which b_σ is infinite but is not large, let $a \in B$ be the length lexicographically first element of B that splits b_σ such that one of $b_\sigma \wedge a$ and $b_\sigma \wedge \bar{a}$ is infinite. Again, this a exists by Lemma 10. Add $\sigma 0$ and $\sigma 1$ to $\mathcal{T}_{n+1}$ and let $b_{\sigma 0} = b_\sigma \wedge a$ and $b_{\sigma 1} = b_\sigma \wedge \bar{a}$. For any leaf $\sigma \in \mathcal{T}_n$ that does not fall into either of these cases, let $\sigma \in \mathcal{T}_{n+1}$ also be a leaf.

The first induction hypothesis holds for $\mathcal{T}_{n+1}$ because any leaf in $\mathcal{T}_n$ associated to a large element in $\mathcal{B}$ is extended by at least one node associated to a large element in $\mathcal{B}$. Likewise, the second hypothesis is preserved because each of the n nodes associated to infinite elements are extended by a node associated to an infinite element, plus the node associated to a large element is extended by a node associated to an infinite element. The third hypothesis holds because at least $n + 1$ leaves of $\mathcal{T}_n$ are replaced by two leaves each, so if $\mathcal{T}_n$ has at least $\frac{n(n+1)}{2}$ leaves then $\mathcal{T}_{n+1}$ has at least $\frac{(n+1)(n+2)}{2}$ many leaves. Finally, the last hypothesis is preserved because if $\tau \in \mathcal{T}_n$ is a leaf, $\tau \wedge (b_\sigma \wedge a) = (\tau \wedge b_\sigma) \wedge a = 0 \wedge a = 0$.

Each clause in the inductive definition can be formalized in $(FO + \exists^\infty)$. Hence, for all n, $\mathcal{T}_n$ is word automatic. In particular, the functions $f_0, f_1 : X_n \to X_n$ which map $f_0(b_\sigma) = b_{\sigma 0}$ and $f_1(b_\sigma) = b_{\sigma 1}$ have regular graphs. Therefore, the Constant Growth Lemma (Lemma 1 on page 152) implies that there are constants C_0, C_1 such that

$$|b_{\sigma 0}| \leq |b_\sigma| + C_0 \text{ and } |b_{\sigma 1}| \leq |b_\sigma| + C_1.$$

In particular, there is a constant C_2 such that for all $x \in X_n$, $|x| \leq C_2 \cdot n$. In other words, $X_n \subseteq \Sigma^{C_2 \cdot n}$. Lemma 2 then gives that the Boolean algebra generated by X_n is also a subset of $\Sigma^{O(n)}$ and therefore has at most $2^{O(n)}$ elements. On the other hand, since there are at least $\frac{n(n+1)}{2}$ many leaves in $\mathcal{T}_n$

and distinct leaves yield disjoint elements of X_n, the Boolean algebra generated by X_n has size at least $2^{\frac{n(n+1)}{2}}$. This is a contradiction, and proves Theorem 24.

The classification of Boolean algebras in Theorem 24 gives rise to an algorithm solving the isomorphism problem for automatic Boolean algebras.

COROLLARY 20. *It is decidable whether two automatic Boolean algebras are isomorphic.*

PROOF. By Theorem 24, two automatic Boolean algebras are isomorphic if and only if they are isomorphic to B_ω^n for the same n. Given an automata presentation of a Boolean algebra, it satisfies the first-order expressible property "there are n disjoint elements each with infinitely many atoms below, and for each $m > n$ there aren't m disjoint elements each with infinitely many atoms below them" if and only if it is isomorphic to B_ω^n. To decide if two automatic Boolean algebras $\mathcal{B}_1, \mathcal{B}_2$ are isomorphic, search for n_1 and n_2 such that $\mathcal{B}_1 \cong \mathcal{B}_\omega^{n_1}$ and $\mathcal{B}_2 \cong \mathcal{B}_\omega^{n_2}$ and then reply "yes" if and only if $n_1 = n_2$. ⊣

We have seen explicit descriptions for which ordinals and which Boolean algebras have automata presentations. We have also seen theorems which give conditions on invariants (ordinal height, Cantor-Bendixson rank) of structures with automatic presentations. In the following subsection, we will continue to accumulate results of both kinds.

2.6. Word automatic finitely generated groups. We now turn to finitely generated groups which have automata presentations. As in the previous subsections, we consider only word automata presentations. Recall that a group is a structure $(G; \cdot, ^{-1}, e)$ where the group operation is associative and the inverse behaves as expected with respect to the identity. In Subsection 1.5, we saw that finitely generated abelian (commutative) groups have automata presentations. In Subsection 2.2, we saw that the free group $F(n)$ with $n > 1$ is not automatic. These two examples represent extreme behaviours with respect to automata. We now work towards finding the exact boundary between automaticity and nonautomaticity for groups.

Recall that a *subgroup* is a subset of a group which is itself a group; the *index* of a subgroup is the number of left cosets of the subgroup. A good introduction to basic group theory is [50].

DEFINITION 7. A group is *virtually abelian* if it has an abelian subgroup of finite index. Similarly, for any property X of groups, we say that a group is *virtually* X if it has a subgroup of finite index which has property X.

DEFINITION 8. A group is *torsion-free* if the only element of finite order (the least number of times it must be multiplied by itself to yield the identity) is the identity. A subgroup N of G is *normal* if for all $a \in G$, $aN = Na$.

The following basic fact from group theory will play a key part in our analysis of automata presentable finitely generated groups. It says that the abelian subgroup of a virtually abelian finitely generated group can be assumed to have a special form.

FACT 25. Every virtually abelian finitely generated group has a torsion-free normal abelian subgroup of finite index.

We are now ready to prove that each member of a large class of finitely generated groups has an automata presentation. This lemma significantly extends Example 15 which dealt with abelian finitely generated groups.

LEMMA 11. *Any virtually abelian finitely generated group is automatic.*

PROOF. Let G be a virtually abelian finitely generated group. By Fact 25, let $N = \langle x_1, \dots, x_k \rangle$ be an abelian torsion-free normal subgroup of finite index of G. Thus, there are $a_1, \dots, a_n \in G$ such that $G = \sqcup_{i=1}^{n} a_i N$. Since N is normal, for any $j = 1, \dots, n$ there are $h_1, \dots, h_k \in N$ such that

$$x_1 a_j = a_j h_1 = a_j x_1^{m_{1,1}(j)} \cdots x_k^{m_{k,1}(j)}, \dots, x_k a_j = a_j h_k = a_j x_1^{m_{1,k}(j)} \cdots x_k^{m_{k,k}(j)}.$$

Moreover, there are functions $f : \{1, \dots, n\}^2 \to \{1, \dots, n\}$, $h : \{1, \dots, n\}^2 \to N$, and $m_\ell : \{1, \dots, n\}^2 \to \mathcal{N}$ $(1 \leq \ell \leq k)$ such that

$$a_i a_j = a_{f(i,j)} h(i, j) = a_{f(i,j)} x_1^{m_1(i,j)} \cdots x_k^{m_k(i,j)}.$$

For each $g \in G$ there is $h \in N$ and $i \in 1, \dots, n$ so that $g = a_i h = a_i x_1^{m_1} \cdots x_k^{m_k}$. Therefore, we can express the product of two elements of the group as

$$\left(a_i x_1^{p_1} \cdots x_k^{p_k}\right) \cdot \left(a_j x_1^{q_1} \cdots x_k^{q_k}\right) = a_{f(i,j)} x_1^{q_1 + m_1(i,j) + \Sigma_{\ell=1}^{k} p_\ell m_{1,\ell}(j)} \cdots \cdot x_k^{q_k + m_k(i,j) + \Sigma_{\ell=1}^{k} p_\ell m_{k,\ell}(j)}$$

Expressing group multiplication in this way is well-suited to automata operations and leads to an automata presentation of G. ⊣

In fact, [46] contains the following theorem which shows that the class of virtually abelian finitely generated groups coincides exactly with the class of automata presentable finitely generated groups.

THEOREM 26 (Oliver and Thomas [46]). *A finitely generated group is automatic if and only if the group is virtually abelian.*

To prove Theorem 26, it remains to prove only one direction of the classification. We will use several definitions and facts from group theory to prove this direction (cf. [22, 49, 45]).

DEFINITION 9. The *commutator* of a group G is the set

$$[G, G] = \{[g, h] = g^{-1} h^{-1} g h : g, h \in G\}.$$

The powers of a group are defined inductively as $G^0 = G$ and $G^{k+1} = [G_k, G_k]$. The group G is *solvable* if there is some n such that $G^n = \{e\}$. The maps γ_k

are defined inductively by

$$\gamma_0(G) = G, \qquad \gamma_{k+1}(G) = [\gamma(G_k), G].$$

The group G is *nilpotent* if $\gamma_n(G) = \{e\}$ for some n.

FACT 27. If G is nilpotent then G is solvable.

The following theorems of group theory relate algorithmic and growth properties of a group to the group theoretic notions defined above. These theorems are instrumental in proving Theorem 26 since we understand the algorithmic and growth properties of groups with automata presentations. Note that a finitely generated group is said to have *polynomial growth* if the size of the nth generation (G_n) of the set of generators of the group is polynomial in n (recall the definition of the generations of a structure from Subsection 2.2).

THEOREM 28 (Gromov [22]). *If a finitely generated group has polynomial growth then it is virtually nilpotent.*

THEOREM 29 (Romanovskiĭ [49]; Noskov [45]). *A virtually solvable group has a decidable first-order theory if and only if it is virtually abelian.*

PROOF OF THEOREM 26. Let G be an automatic finitely generated group. Suppose $G = \langle a_1, \dots, a_k \rangle$. By the generation lemma for monoids (Lemma 2 on page 152), for each n, $G_n(\{a_1, \dots, a_k\}) \subseteq \Sigma^{C \cdot \log(n)}$. Therefore, $|G_n(\{a_1, \dots, a_k\})| \leq n^C$, and so G has polynomial growth. By Gromov's theorem, G is virtually nilpotent, hence Fact 27 implies that it is virtually solvable. Since G is automatic, it has a decidable first-order theory. Hence, Romanovskiĭ and Noskov's theorem implies that G is virtually abelian, as required. ⊣

We have just seen a complete description of the finitely generated groups which have word automata presentations. In the Boolean algebra case, such a description led to an algorithm for the isomorphism question. However, whether such an algorithm exists in the current context is still an open question: is the isomorphism problem for automatic finitely generated groups decidable?

Many other questions can be asked about automata presentations for groups. For example, we might shift our attention away from finitely generated groups and ask whether the isomorphism problem for automatic torsion-free abelian groups is decidable. A question in this vein which has generated interest is whether $(\mathcal{Q}; +)$ has a word automata presentation; Tsankov has recently announced that $(\mathcal{Q}; +)$ is not an automatic structure. More details about the current state of the art in automata presentable groups may be found in the survey paper [44].

§3. Complicated structures.

3.1. Scott ranks of word automatic structures. In the previous lecture we saw positive characterization results and relatively low tight bounds on classes of automatic structures. In particular, we saw that ω is the tight bound on the

Cantor-Bendixson ranks of automatic partial order trees, and that ω^ω is the tight bound on the ordinal heights of automatic well-founded relations. We will now prove results at the opposite end of the spectrum: results that say that automatic structures can have arbitrarily high complexity in some sense. This will have complexity theoretic implications on the isomorphism problem for the class of automatic structures.

We begin by recalling an additional notion of rank for the class of countable structures. This *Scott rank* was introduced in [54] in the context of Scott's famous isomorphism theorem for countable structures.

DEFINITION 10. Given a countable structure $\mathcal{A}$, for tuples $\bar{a}, \bar{b} \in A^n$ we define the following equivalence relations.

- $\bar{a} \equiv^0 \bar{b}$ if $\bar{a}$ and $\bar{b}$ satisfy the same quantifier free sentences,
- for ordinal $\alpha > 0$, $\bar{a} \equiv^\alpha \bar{b}$ if for all $\beta < \alpha$, for each tuple $\bar{c}$ there is a tuple $\bar{d}$ such that $\bar{a}, \bar{c} \equiv^\beta \bar{b}, \bar{d}$ and for each tuple $\bar{d}$ there is a tuple $\bar{c}$ such that $\bar{a}, \bar{c} \equiv^\beta \bar{b}, \bar{d}$.

The Scott rank of tuple $\bar{a}$ is the least β such that for all $\bar{b} \in A^n$, $\bar{a} \equiv^\beta \bar{b}$ implies $(\mathcal{A}, \bar{a}) \cong (\mathcal{A}, \bar{b})$. The *Scott rank* of the structure $\mathcal{A}$, denoted $SR(\mathcal{A})$, is the least ordinal greater than the Scott ranks of all tuples of $\mathcal{A}$.

The Scott rank was extensively studied in the context of computable model theory. In particular, Nadel [42] and Harrison [24] showed that the tight upper bound for the Scott rank of computable structures is $\omega_1^{CK}+1$. Recall that ω_1^{CK} is the first uncomputable ordinal; that is, it is the least ordinal which is not isomorphic to a computable well-ordering of the natural numbers. However, most common examples of automatic structures have low Scott ranks. The following theorem from [30] proves that automatic structures have the same tight upper bound on the Scott rank as computable structures.

THEOREM 30 (Khoussainov and Minnes [30]). *For each $\alpha \leq \omega_1^{CK}+1$, there is an automatic structure with Scott rank at least α. Moreover, there is an automatic structure with Scott rank ω_1^{CK}.*

PROOF. We outline the idea of the proof. The main thrust of the argument is the transformation of a given computable structure to an automatic structure which has similar Scott rank. Let $\mathcal{C} = (C; R_1, \dots, R_m)$ be a computable structure. For simplicity in this proof sketch, we assume $m = 1$ and that $C = \Sigma^*$ for some finite Σ. Let $\mathcal{N}$ be a Turing machine computing the relation R. The configuration space $Conf(\mathcal{N})$ of the machine $\mathcal{N}$ is a graph whose nodes encode configurations of $\mathcal{N}$ and where there is an edge between two nodes if $\mathcal{N}$ has an instruction which takes it in one step from the configuration represented by one node to that of the other (see Example 11 on page 142). Recall that $Conf(\mathcal{N})$ is an automatic structure. We call a deterministic Turing machine *reversible* if its configuration space is well-founded; that is, it consists only of finite chains or chains of type ω. The following lemma from [5] allows us to restrict our attention to reversible Turing machines.

LEMMA 12 (Bennett [5]). *Any deterministic Turing machine may be simulated by a reversible Turing machine.*

Without loss of generality, we assume that $\mathcal{N}$ is a reversible Turing machine which halts if and only if its output is "yes". We classify the chains in $Conf(\mathcal{N})$ into three types: terminating computation chains are finite chains whose base is a valid initial configuration, nonterminating computation chains are infinite chains whose base is a valid initial configuration, and unproductive chains are chains whose base is not a valid initial configuration. We perform the following smoothing operations to $Conf(\mathcal{N})$ in order to capture the isomorphism type of $\mathcal{C}$ within the new automatic structure. Note that each of these smoothing steps preserves automaticity. First, we add infinitely many copies of ω chains and finite chains of every finite size. Also, we connect to each base of a computation chain a structure which consists of infinitely many chains of each finite length. Finally, we connect representations of each tuple $\bar{x}$ in $\mathcal{C}$ to the initial configuration of $\mathcal{N}$ given $\bar{x}$ as input. We call the resulting automatic graph $\mathcal{A}$. The following lemma can be proved using the defining equivalence relations of Scott rank and reflects the idea that the automatic graph contains witnesses to many properties of the computable structure.

LEMMA 13. $SR(\mathcal{C}) \leq SR(\mathcal{A}) \leq 2 + SR(\mathcal{C})$.

At this point, we have a tight connection between Scott ranks of automatic and computable structures. The following lemmas from [21] and [37] describe the Scott ranks that are realized by computable structures,

LEMMA 14 (Goncharov and Knight [21]). *For each computable ordinal $\alpha < \omega_1^{CK}$, there is a computable structure whose Scott rank is above α.*

LEMMA 15 (Knight and Millar [37]). *There is a computable structure with Scott rank ω_1^{CK}.*

We apply Lemma 13 to Lemmas 14 and 15 to produce automatic structures with Scott ranks above every computable ordinal and at ω_1^{CK}. To produce an automatic structure with Scott rank $\omega_1^{CK} + 1$, we apply Lemma 13 to Harrison's ordering from [24]. This concludes the proof of Theorem 30. ⊣

COROLLARY 21 (Khoussainov, Rubin, Nies, and Stephan [33]). *The isomorphism problem for automatic structures is Σ_1^1-complete.*

PROOF. In the proof of Theorem 30, the transformation of computable structures to automatic structures preserves isomorphism types. Hence, the isomorphism problem for computable structures is reduced to that for automatic structures. Since the isomorphism problem for computable structures is Σ_1^1-complete, the isomorphism problem for automatic structures is Σ_1^1-complete as well. ⊣

3.2. More high bounds for word automatic structures. The previous subsection contained an example where the behaviour of automatic structures matched that exhibited by the class of computable structures. The techniques of Theorem 30 can be used to obtain similar results in a couple of other contexts. We first revisit the idea of automatic trees, introduced in Subsection 2.4. In that subsection, we saw that all automatic partial order trees have finite Cantor-Bendixson rank (Theorem 21 on page 161). Consider now a different viewpoint of trees.

DEFINITION 11. A *successor tree* is $\mathcal{A} = (A; S)$ where, if $\leq_S$ is the transitive closure of S, then $(A; \leq_S)$ is a partial order tree.

Observe that the successor tree associated with a given partial order tree is first-order definable in the partial order tree. Hence, Corollary 3 (page 146) implies that any automatic partial order tree is an automatic successor tree. However, the inclusion is strict: the following theorem from [30] shows that the class of automatic successor trees is far richer than the class of automatic partial order trees.

THEOREM 31 (Khoussainov and Minnes [30]). *For each computable ordinal $\alpha < \omega_1^{CK}$ there is an automatic successor tree of Cantor-Bendixson rank α.*

The proof of Theorem 31 utilizes the configuration spaces of Turing machines, as in the proof of Theorem 30. Another setting in which these tools are useful is that of well-founded relations (see [30]). In Subsection 2.3, we proved that automatic well-founded partial orders have ordinal heights below ω^ω. If we relax the requirement that the relation be a partial order, we can attain much higher ordinal heights.

THEOREM 32 (Khoussainov and Minnes [30]). *For each computable ordinal $\alpha < \omega_1^{CK}$ there is an automatic well-founded relation whose ordinal height is at least α.*

Theorem 32 answers a question posed by Moshe Vardi in the context of program termination. Given a program P, we say that the program is terminating if every computation of P from an initial state is finite. If there is a computation from a state x to y then we say that y is reachable from x. Thus, the program is terminating if the collection of all states reachable from the initial state is a well-founded set. Vardi asked whether the ordinal height of the state space of a terminating program has a low bound. The connection between well-foundedness and program termination is explored further in [6].

3.3. Borel structures. Most of Lectures 2 and 3 so far dealt exclusively with word automatic structures. To conclude this lecture, we turn again to automata on infinite strings (Büchi automata) and infinite trees (Rabin automata). Recall from Subsection 2.1 that word automatic structures are a strict subset of tree automatic structures. We would like a similar separation between Büchi automatic structures and Rabin automatic structures. To arrive at such a separation, we recall a complexity hierarchy of sets from descriptive set theory (a good reference is [28]).

DEFINITION 12. A set is called *Borel* in a given topology if it is a member of the smallest class of sets which contains all open sets and closed sets and is closed under countable unions and countable intersections.

In the context of automatic structures, we have an underlying topology which depends on the objects being processed by the automata. If the input objects are infinite binary strings (as in the case of Büchi automata), the basic open sets of the topology are defined as

$$[\sigma] = \{\alpha \in \{0,1\}^\omega : \sigma \prec \alpha\}$$

for each finite string σ. Based on this topology, we have the following definition.

DEFINITION 13. A structure is *Borel* if its domain and basic relations are Borel sets.

FACT 33. Every Büchi automatic structure is Borel.

We can use Fact 33 to prove that not all Rabin automatic structures are recognizable by Büchi automata. [26] contains the following theorem about Rabin automatic structures.

THEOREM 34 (Hjorth, Khoussainov, Montalbán, and Nies [26]). *There is a Rabin automatic structure that is not Borel and, therefore, is not Büchi automatic.*

PROOF. Consider the set of labelled trees

$$V = \{(\mathcal{T}, v) : \text{each path through } \mathcal{T} \text{ has finitely many } 1s\}.$$

It is easy to see that V is Rabin recognizable. However, it is not Borel: consider the embedding of $\omega^{<\omega}$ into $\mathcal{T}$ which takes $n_1 \cdots n_k$ to $1^{n_1}01^{n_2}0 \ldots 1^{n_k}$. The pre-image of V is the set of trees in $\omega^{<\omega}$ which have no infinite path. This set is an archetypal example of a nonBorel set, and hence V is not Borel. We will now transfer this example to the setting of structures by coding V into a Rabin automatic structure. The domain of this structure is the class of $\{0,1\}$-labelled trees $(\mathcal{T}, v)$. The structure has two unary predicates: $S = \{(\mathcal{T}, v) : \text{there is a unique } x \text{ for which } v(x) = 1\}$, and V as above. In addition, the structure contains two unary functions S_0', S_1' which mimic the operations on $\mathcal{T}$. Then $(D; S, V, S_0', S_1')$ is Rabin automatic. But, if it had a Borel copy then V would be Borel, and so the structure is not Borel. Thus, we have a Rabin automatic structure that is not Büchi automatic. ⊣

We conclude these lectures by proving Theorem 6 from Subsection 1.4 (page 144). To do so, we need to exhibit a Büchi automatic structure and a Büchi recognizable equivalence relation such that their quotient is not Büchi recognizable. By Fact 33, it suffices to show that the quotient is not Borel. We will use the following well-known fact from descriptive set theory (see [25] for a survey of relevant results in this area). Recall that for $X, Y \subseteq \mathcal{N}$, $X =^\star Y$ means that X and Y agree except at finitely many points.

FACT 35. There is no Borel function $F : \mathcal{P}(\mathcal{N}) \to \{0, 1\}^\omega$ such that for all $X, Y \subseteq \mathcal{N}$, $X =^\star Y$ if and only if $F(X) = F(Y)$

PROOF OF THEOREM 6. The structure we will consider is an expansion of the disjoint union of two familiar structures. Let $\mathcal{B} = (\mathcal{P}(\mathcal{N}); \subseteq)$ and $\mathcal{B}^\star = (\mathcal{P}(\mathcal{N})/ =^\star; \leq)$. We will study the disjoint of union of $\mathcal{B}$ and $\mathcal{B}^\star$ along with a unary relation $U(x)$ satisfying $U(x) \iff x \in \mathcal{P}(\mathcal{N})$ and a binary relation $R(x, y)$ interpreted as the canonical projection of $\mathcal{P}(\mathcal{N})$ into $\mathcal{P}(\mathcal{N})/ =^\star$.

Suppose that $\mathcal{B}_0 = (B_0; \leq_0)$ and $\mathcal{B}_1 = (B_1; \leq_1)$ are two disjoint Büchi automatic presentations of $\mathcal{B} = (\mathcal{P}(\mathcal{N}); \subseteq)$. Let U^0 pick out B_0 and R^0 be the bijection from B_0 to B_1 which acts like the identity on $\mathcal{P}(\mathcal{N})$. Then $\mathcal{A} = (B_0 \sqcup B_1; \leq_0 \sqcup \leq_1, U^0, R^0)$ is a Büchi recognizable structure. We define the equivalence relation

$$E(x, y) \iff \begin{cases} x, y \in B_0 \ \& \ x = y \\ x, y \in B_1 \ \& \ x =^\star y \end{cases}.$$

It is not hard to see that $E(x, y)$ is Büchi recognizable given that $\mathcal{B}_0$ and $\mathcal{B}_1$ are. Observe that $\mathcal{A}/E \cong (\mathcal{P}(\mathcal{N}) \sqcup \mathcal{P}(\mathcal{N})/ =^\star; \leq, U, R)$.

Assume for a contradiction that $\mathcal{A}/E$ has a Büchi automata presentation. By Fact 33, this implies that $\mathcal{A}/E$ has a Borel presentation $\mathcal{C} = (C; \leq^C, U^C, R^C)$. Let $\Phi : \mathcal{A}/E \to \mathcal{C}$ be an isomorphism and $G : \mathcal{B}_0 \to \mathcal{C}$ be the restriction of Φ to $\mathcal{B}_0$. The following lemma tells us that G is a Borel function.

LEMMA 16. *If $\mathcal{S} = (S; \leq)$, $\mathcal{S}' = (S'; \leq')$ are Büchi structures and E, E' are Büchi equivalence relations such that $\mathcal{S}/E, \mathcal{S}'/E' \cong (\mathcal{P}(\mathcal{N}); \subseteq)$ then if $\Psi : \mathcal{S}/E \to \mathcal{S}'/E'$ is an isomorphism, $graph(\Psi) = \{\langle x, y\rangle \in S \times S' : \Phi([x]_E) = [y]_{E'}\}$ is Borel.*

To prove the lemma, it suffices to show that $graph(\Psi)$ is a countable intersection of Borel sets. Recall that $(\mathcal{P}(\mathcal{N}); \subseteq)$ can be viewed as a Boolean algebra whose atoms are singleton sets $\{n\}$ for $n \in \mathcal{N}$. Let $\{[a_n]_E : n \in \mathcal{N}\}$ be a list of atoms of $\mathcal{S}/E$ and let $b_n \in S'$ be such that $\Psi([a_n]_E) = [b_n]_{E'}$. Since Ψ is an isomorphism of Boolean algebras, for any $x \in S$, $y \in S'$

$$\Psi([x]_E) = [y]_{E'} \iff \forall n \big(a_n \leq x \iff b_n \leq' y\big).$$

Because $\mathcal{S}, \mathcal{S}'$ are Büchi structures, the relations $a_n \leq x$ and $b_n \leq' y$ are Borel and we have a Borel definition for $\Psi([x]_E) = [y]_{E'}$. The lemma is now proved and we return to the proof of Theorem 6.

The function $G : \mathcal{B}_0 \to \mathcal{C}$ satisfies the hypotheses of the above lemma and hence is Borel. We define the map $F = R' \circ G$ from $\mathcal{B}_0 \cong \mathcal{P}(\mathcal{N})$ to $\mathcal{C}$. Note that since F is the composition of Borels, it is Borel. Moreover, for any $X, Y \subseteq \mathcal{N}$

$$\begin{aligned} X =^\star Y \iff R^0(X) = R^0(Y) &\iff \Phi(R^0(X)) = \Phi(R^0(Y)) \\ \iff R'(G(X)) = R'(G(Y)) &\iff F(X) = F(Y), \end{aligned}$$

contradicting Fact 35. ⊣

§4. **Conclusion.** These lectures have surveyed word automatic structures, Büchi automatic structures, tree automatic structures, and Rabin structures. Our emphasis has been on the interactions between these structures, various logics, and algorithmic issues. This area of research is vibrant and growing. Some current lines of investigation include studying the isomorphism problem for mathematically meaningful classes of objects (is the isomorphism problem for word automatic linear orders decidable? for word automatic groups?) and classifications of Büchi automatic and tree automatic structures.

REFERENCES

[1] P. A. Abdulla, K. Čerāns, B. Jonsson, and Y. Tsay, *Algorithmic analysis of programs with well quasi-ordered domains*, ***Information and Computation***, vol. 160 (2000), no. 1-2, pp. 109–127, LICS 1996, Part I (New Brunswick, NJ).

[2] L. Aceto, A. Burgueno, and K. G. Larsen, *Model checking via reachability testing for timed automata*, Proc. TACAS '98, pp. 263–280, 1998.

[3] V. Bárány, ***Automatic Presentations of Infinite Structures***, Diploma Thesis, RWTH Aachen, Germany, 2007.

[4] V. Bárány, L. Kaiser, and S. Rubin, *Cardinality and counting quantifiers on omega-automatic structures*, ***Proc. STACS '08*** (Susanne Albers and Pascal Weil, editors), Leibniz International Proceedings in Informatics, vol. 08001, 2008, pp. 385–396.

[5] C. H. Bennett, *Logical reversibility of computation*, ***International Business Machines Corporation. Journal of Research and Development***, vol. 17 (1973), pp. 525–532.

[6] A. Blass and Y. Gurevich, *Program termination and well partial orderings*, ***ACM Transactions on Computational Logic***, vol. 9 (2008), no. 3, pp. Art. 18, 26.

[7] A. Blumensath, ***Automatic Structures***, Diploma Thesis, RWTH Aachen, Germany, 1999.

[8] A. Blumensath and E. Grädel, *Automatic structures*, ***15th Annual IEEE Symposium on Logic in Computer Science (Santa Barbara, CA, 2000)***, IEEE Computer Society Press, Los Alamitos, CA, 2000, pp. 51–62.

[9] A. Bouajjani, J. Esparza, and O. Maler, *Reachability analysis of pushdown automata: application to model-checking*, ***CONCUR '97: Concurrency Theory (Warsaw)***, Lecture Notes in Computer Science, vol. 1243, Springer, Berlin, 1997, pp. 135–150.

[10] J. R. Büchi, *Weak second-order arithmetic and finite automata*, ***Zeitschrift für Mathematische Logik und Grundlagen der Mathematik***, vol. 6 (1960), pp. 66–92.

[11] ———, *On a decision method in restricted second order arithmetic*, ***Logic, Methodology and Philosophy of Science (Proc. 1960 Internat. Congr.)*** (E. Nagel, P. Suppes, and A. Tarski, editors), Stanford University Press, Stanford, California, 1962, pp. 1–11.

[12] D. Cenzer and J. Remmel, *Polynomial-time versus recursive models*, ***Annals of Pure and Applied Logic***, vol. 54 (1991), no. 1, pp. 17–58.

[13] C. Delhommé, *Automaticité des ordinaux et des graphes homogènes*, ***Comptes Rendus Mathématique. Académie des Sciences. Paris***, vol. 339 (2004), no. 1, pp. 5–10.

[14] J. E. Doner, *Decidability of the weak second-order theory of two successors*, ***Notices of the American Mathematical Society***, vol. 12 (1965), p. 819.

[15] D. B. A. Epstein, James W. Cannon, Derek F. Holt, Silvio V. F. Levy, Michael S. Paterson, and William P. Thurston, ***Word Processing in Groups***, Jones and Bartlett Publishers, Boston, MA, 1992.

[16] Yu. L. Ershov, *Decidability of the elementary theory of distributive lattices with relative complements and the theory of filters*, ***Algebra i Logika***, vol. 3 (1964), pp. 17–38.

[17] Yu. L. Ershov, S. S. Goncharov, A. Nerode, J. B. Remmel, and V. W. Marek (editors), ***Handbook of Recursive Mathematics: Recursive Model Theory. Vol. 1***, Studies in Logic and the Foundations of Mathematics, vol. 138, North-Holland, Amsterdam, 1998.

[18] Yu. L. Ershov, S. S. Goncharov, A. Nerode, J. B. Remmel, and V. W. Marek (editors), ***Handbook of Recursive Mathematics: Recursive Algebra, Analysis and Combinatorics. Vol. 2***, Studies in Logic and the Foundations of Mathematics, vol. 139, North-Holland, Amsterdam, 1998.

[19] A. A. Fraenkel, ***Abstract Set Theory***, Third revised edition, North-Holland, Amsterdam, 1966.

[20] A. Fröhlich and J. C. Shepherdson, *Effective procedures in field theory*, ***Philosophical Transactions of the Royal Society of London. Series A. Mathematical and Physical Sciences***, vol. 248 (1956), pp. 407–432.

[21] S. S. Goncharov and J. F. Knight, *Computable structure and non-structure theorems*, ***Algebra and Logic***, vol. 41 (2002), pp. 351–373.

[22] M. Gromov, *Groups of polynomial growth and expanding maps*, ***Institut des Hautes Études Scientifiques. Publications Mathématiques***, (1981), no. 53, pp. 53–78.

[23] V. S. Harizanov, *Pure computable model theory*, ***Handbook of Recursive Mathematics, Vol. 1*** (Yu. L. Ershov, S. Goncharov, A. Nerode, and J. Remmel, editors), Studies in Logic and the Foundations of Mathematics, vol. 138, North-Holland, Amsterdam, 1998, pp. 3–114.

[24] J. Harrison, *Recursive pseudo-well-orderings*, ***Transactions of the American Mathematical Society***, vol. 131 (1968), pp. 526–543.

[25] G. Hjorth, *Borel equivalence relations which are highly unfree*, ***The Journal of Symbolic Logic***, vol. 73 (2008), no. 4, pp. 1271–1277.

[26] G. Hjorth, B. Khoussainov, A. Montalbán, and A. Nies, *From automatic structures to Borel structures*, ***Proc. LICS '08***, 2008, pp. 431–441.

[27] B. R. Hodgson, *On direct products of automaton decidable theories*, ***Theoretical Computer Science***, vol. 19 (1982), no. 3, pp. 331–335.

[28] A. S. Kechris, ***Classical Descriptive Set Theory***, Graduate Texts in Mathematics, vol. 156, Springer-Verlag, New York, 1995.

[29] B. Khoussainov, J. Liu, and M. Minnes, *Unary automatic graphs: an algorithmic perspective*, ***Mathematical Structures in Computer Science. A Journal in the Applications of Categorical, Algebraic and Geometric Methods in Computer Science***, vol. 19 (2009), no. 1, pp. 133–152.

[30] B. Khoussainov and M. Minnes, *Model theoretic complexity of automatic structures (extended abstract)*, ***Theory and Applications of Models of Computation***, Lecture Notes in Computer Science, vol. 4978, Springer, Berlin, 2008, pp. 514–525.

[31] B. Khoussainov and A. Nerode, *Automatic presentations of structures*, ***Logic and Computational Complexity (Indianapolis, IN, 1994)***, Lecture Notes in Computer Science, vol. 960, Springer, Berlin, 1995, pp. 367–392.

[32] ———, *Open questions in the theory of automatic structures*, ***Bulletin of the European Association for Theoretical Computer Science. EATCS***, (2008), no. 94, pp. 181–204.

[33] B. Khoussainov, A. Nies, S. Rubin, and F. Stephan, *Automatic structures: Richness and limitations*, ***Logical Methods in Computer Science***, (2004), pp. 2:2, 18 pp. (electronic), Special issue: Conference "Logic in Computer Science 2004".

[34] B. Khoussainov, S. Rubin, and F. Stephan, *On automatic partial orders*, Proc. LICS '03, pp. 168–177, 2003.

[35] ———, *Definability and regularity in automatic structures*, ***STACS 2004***, Lecture Notes in Computer Science, vol. 2996, Springer, Berlin, 2004, pp. 440–451.

[36] S. C. Kleene, *Representation of events in nerve nets and finite automata*, ***Automata Studies***, Annals of Mathematics Studies, vol. 34, Princeton University Press, Princeton, N.J., 1956, pp. 3–41.

[37] J. F. Knight and J. Millar, *Computable structures of rank* ω_1^{CK}, Submitted to ***J. Math. Logic***; posted on arXiv 25 Aug. 2005.

[38] D. Kuske and M. Lohrey, *First-order and counting theories of* ω*-automatic structures*, ***Foundations of Software Science and Computation Structures***, Lecture Notes in Computer Science, vol. 3921, Springer, Berlin, 2006, pp. 322–336.

[39] J. Liu, ***Automatic structures (provisional title)***, Ph.D. thesis, University of Auckland, Auckland, in progress.

[40] A. I. Mal'cev, *Constructive algebras. I*, ***Akademiya Nauk SSSR i Moskovskoe Matematicheskoe Obshchestvo. Uspekhi Matematicheskikh Nauk***, vol. 16 (1961), no. 3 (99), pp. 3–60.

[41] M. Minnes, ***Automatic structures (provisional title)***, Ph.D. thesis, Cornell University, Ithaca, NY, 2008.

[42] M. Nadel, $\mathcal{L}_{\omega_1\omega}$ *and admissible fragments*, ***Model-Theoretic Logics*** (K. J. Barwise and S. Feferman, editors), Perspectives in Mathematical Logic, Springer, New York, 1985, pp. 271–316.

[43] A. Nerode and J. B. Remmel, *Polynomial time equivalence types*, ***Logic and Computation (Pittsburgh, PA, 1987)***, Contemporary Mathematics, vol. 106, Amer. Math. Soc., Providence, RI, 1990, pp. 221–249.

[44] A. Nies, *Describing groups*, ***The Bulletin of Symbolic Logic***, vol. 13 (2007), no. 3, pp. 305–339.

[45] G. A. Noskov, *The elementary theory of a finitely generated almost solvable group*, ***Izvestiya Akademii Nauk SSSR. Seriya Matematicheskaya***, vol. 47 (1983), no. 3, pp. 498–517, (Russian); ***Math. USSR Izv. 22***, 465-482, 1984 (English translation).

[46] G. P. Oliver and R. M. Thomas, *Automatic presentations for finitely generated groups*, ***STACS 2005***, Lecture Notes in Computer Science, vol. 3404, Springer, Berlin, 2005, pp. 693–704.

[47] M. O. Rabin, *Computable algebra, general theory and theory of computable fields.*, ***Transactions of the American Mathematical Society***, vol. 95 (1960), pp. 341–360.

[48] ———, *Decidability of second-order theories and automata on infinite trees.*, ***Transactions of the American Mathematical Society***, vol. 141 (1969), pp. 1–35.

[49] N. S. Romanovskiĭ, *The elementary theory of an almost polycyclic group*, ***Matematicheskiĭ Sbornik. Novaya Seriya***, vol. 111(153) (1980), no. 1, pp. 135–143, 160, (Russian); ***Math. USSR Sb.***, **39**, 1981 (English translation).

[50] J. Rotman, ***An Introduction to the Theory of Groups***, fourth ed., Graduate Texts in Mathematics, vol. 148, Springer-Verlag, New York, 1995.

[51] S. Rubin, ***Automatic structures***, Ph.D. thesis, University of Auckland, Auckland, 2004.

[52] ———, *Automata presenting structures: A survey of the finite string case*, ***The Bulletin of Symbolic Logic***, vol. 14 (2008), no. 2, pp. 169–209.

[53] T. Scanlon, *Infinite finitely generated fields are biinterpretable with* $\mathbb{N}$, ***Journal of the American Mathematical Society***, vol. 21 (2008), no. 3, pp. 893–908.

[54] D. Scott, *Logic with denumerably long formulas and finite strings of quantifiers*, ***Theory of Models (Proc. 1963 Internat. Sympos. Berkeley)*** (J. Addison, L. Henkin, and A. Tarski, editors), North-Holland, Amsterdam, 1965, pp. 329–341.

[55] G. S. Shmelev, *Classification of indecomposable finite-dimensional representations of the Lie superalgebra W(0, 2)*, ***Doklady Bolgarskoĭ Akademii Nauk. Comptes Rendus de l'Académie Bulgare des Sciences***, vol. 35 (1982), no. 8, pp. 1025–1027 (Russian).

[56] A. Tarski, ***A Decision Method for Elementary Algebra and Geometry***, RAND Corporation, Santa Monica, Calif., 1948.

[57] A Tarski, *Arithmetical classes and types of boolean algebras*, ***American Mathematical Society. Bulletin***, vol. 55 (1949), p. 63.

[58] J. W. Thatcher and J. B. Wright, *Generalized finite automata theory with an application to a decision problem of second-order logic*, ***Mathematical Systems Theory. An International Journal on Mathematical Computing Theory***, vol. 2 (1968), pp. 57–81.

[59] W. Thomas, *Automata on infinite objects*, ***Handbook of Theoretical computer Science, Vol. B***, Elsevier, Amsterdam, 1990, pp. 133–191.

DEPARTMENT OF COMPUTER SCIENCE
UNIVERSITY OF AUCKLAND
AUCKLAND, NEW ZEALAND
E-mail: bmk@cs.auckland.ac.nz

MATHEMATICS DEPARTMENT
CORNELL UNIVERSITY
ITHACA, NEW YORK 14853, USA
E-mail: minnes@math.cornell.edu

PILLAY'S CONJECTURE AND ITS SOLUTION—A SURVEY

YA'ACOV PETERZIL

§1. Introduction. These notes were originally written for a tutorial I gave in a Modnet Summer meeting which took place in Oxford 2006. I later gave a similar tutorial in the Wroclaw Logic colloquium 2007. The goal was to survey recent work in model theory of o-minimal structures, centered around the solution to a beautiful conjecture of Pillay on definable groups in o-minimal structures. The conjecture (which is now a theorem in most interesting cases) suggested a connection between arbitrary definable groups in o-minimal structures and compact real Lie groups.

All the results discussed here have already appeared in print (mainly [34, 8, 28, 18]). The goal of the notes is to put the results together and to provide a direct path through the proof of the conjecture, avoiding side-tracks and generalizations which are not needed for the proof. This is especially true for the last paper in the list [18] which was often written with an eye towards generalizations far beyond o-minimality.

The last section of this paper has gone through substantial changes in the final stages of the writing. Originally, it contained several open questions and conjectures which arose during the work on Pillay's Conjecture. However, most of these questions were recently answered in a paper of Hrushovski and Pillay, [20], in which the so-called Compact Domination Conjecture has been solved. In another paper, [27], the assumptions for Pillay's Conjecture were weakened from o-minimal expansions of real closed fields to o-minimal expansions of groups. These recent results are now briefly discussed here. I also list some related work which appeared since the original conjecture was formulated.

The paper is aimed for readers who are familiar with the basic model theoretic language and with the introductory definitions of o-minimality (for more on o-minimality, see v. d. Dries' book, [38]).

§2. A motivating example and the conjecture. Before stating Pillay's conjecture, with all its technical terminology, let's consider the main motivating example.

Logic Colloquium '07
Edited by Françoise Delon, Ulrich Kohlenbach, Penelope Maddy, and Frank Stephan
Lecture Notes in Logic, 35

Consider the group:

$$G = SO(2,\mathbb{R}) = \left\{ \begin{pmatrix} a & b \\ -b & a \end{pmatrix} \in GL(2,\mathbb{R}) : a^2 + b^2 = 1 \right\}$$

G is isomorphic, as a Lie group, to the circle group. Namely,

$$G \simeq \mathbb{T}^1 = \{z \in \mathbb{C} : |z| = 1\},$$

with its complex field-multiplication. Both groups, together with their group operations and the isomorphism between them, are definable in the real field $\overline{\mathbb{R}} = \langle \mathbb{R}, <, +, \cdot, 0, 1 \rangle$, so from a model theoretic view-point they are equivalent to each other.

Consider now a κ-saturated real closed field $\mathcal{R} \succ \overline{\mathbb{R}}$ (κ large). We write $G(\mathcal{R})$ for the realization of G in $\mathcal{R}$. Namely, $G(\mathcal{R}) = SO(2,\mathcal{R})$.

Because $SO(2,\mathbb{R})$ is a compact group the *standard-part mapping*, which sends every element of $\mathcal{R}$ of "finite" size to its nearest real element, induces a group-homomorphism $st : SO(2,\mathcal{R}) \to SO(2,\mathbb{R})$, defined by:

$$st \begin{pmatrix} a & b \\ -b & a \end{pmatrix} = \begin{pmatrix} st(a) & st(b) \\ st(-b) & st(a) \end{pmatrix}.$$

We have

$$ker(st) = \mu(I) = \bigcap_{n \in \mathbb{N}} \{A : |A - I| < 1/n\},$$

the intersection of countably many definable sets in $\mathcal{R}$.

One says in this case that $ker(st)$ is *type-definable*, i.e., it can be written as the intersection of less than κ-many definable sets.

A-priori, the map $st(g)$ is just an abstract group homomorphism. The first observation of Pillay, [34], establishes a connection between definability in $G(\mathcal{R})$ and the Euclidean topology on G:

Two topologies on $G(\mathcal{R})/\mu(I)$. We identify $G(\mathcal{R})/\mu(I)$ with $SO(2,\mathbb{R})$ and denote by the *E-topology* its standard Euclidean topology. We also define another topology on this quotient, called the *Logic topology* (L-topology), by: $F \subseteq SO(2,\mathbb{R})$ is *L-closed* iff $st^{-1}(F) \subseteq G(\mathcal{R})$ is *type-definable* in the ordered field structure on $\mathcal{R}$.

Logical compactness, together with the saturation of $\mathcal{R}$ relative to the size of $SO(2,\mathcal{R})/\mu(I)$ imply (see [34]) that the L-topology is compact and Hausdorff.

FACT 2.1. *A set $F \subseteq SO(2,\mathbb{R})$ is E-closed if and only if it is L-closed.*

PROOF. Because both topologies are compact and Hausdorff it is sufficient to prove only one of the two implications.

Assume that $F \subseteq SO(2,\mathbb{R})$ is closed in the Euclidean topology. We will show that $st^{-1}(F)$ is type-definable.

Because $F \subseteq SO(2,\mathbb{R})$ is compact, for every $g \in SO(2,\mathbb{R}) \setminus F$ there is $n_g \in \mathbb{N}$ such that the distance between g and F is $> 1/n_g$.

CLAIM. $st^{-1}(F) = p(\mathcal{R})$ for the type:

$$p(x) = \{x \in SO(2, \mathcal{R})\ \&\ |x - g| \geqslant 1/n_g \ :\ g \in SO(2, \mathbb{R}) \setminus F\}.$$

Indeed, assume that $st(h) = g' \in F$. Then, for every $g \in SO(2, \mathbb{R}) \setminus F$, we have $|g' - g| > 1/n_g$. Because h is infinitesimally close to g' we have $|h - g| > 1/n_g$. Hence, $h \models p(x)$.

For the opposite inclusion, assume that $h \notin st^{-1}(F)$. It follows that $st(h) = g \in SO(2, \mathbb{R}) \setminus F$, and therefore $|h - g| < 1/n_g$, and $h \notin p(\mathcal{R})$. ⊣

REMARKS. 1. The type $p(x)$ defining $st^{-1}(F)$ is parameterized by a subset of $SO(2, \mathbb{R})$ hence uses at most $2^{\aleph_0}$-many formulas. Moreover, the type is given *uniformly*, namely there is a fixed formula $\phi(x, y)$ such that all formulas in p are of the form $\phi(x, b)$ for varying b's. As we will see later on, this plays an important role in the general theory.

2. The quotient group $G(\mathcal{R})/\mu(I)$ ($\simeq SO(2, \mathbb{R})$) is independent of the structure $\mathcal{R}$. I.e. every coset, even in elementary extensions of $\mathcal{R}$, is already represented in $\mathcal{R}$. In such a case $\mu(I)$ is said to have *bounded index* in G. An equivalent condition is that the cardinality of $G(\mathcal{R})/\mu(I)$ is smaller than κ (recall that $\mathcal{R}$ is κ-saturated). Note that if H is *a definable* subgroup of G of bounded index then the quotient is necessarily finite.

   An example of a type-definable subgroup which is *not* of bounded index is the infinitesimal subgroup $\mu(0)$ of $\langle \mathcal{R}, <, + \rangle$. The quotient in this case is *not* $\langle \mathbb{R}, + \rangle$ because as we extend $\mathcal{R}$ to elementary extensions one can realize more and more elements which are not infinitesimally close to each other.

3. The Logic topology on $G(\mathcal{R})/\mu(I)$ is *not* the quotient topology with respect to the topology of the real closed field, because $\mu(I)$ is open in this topology (so the quotient topology is discrete).

4. As was pointed out by Zil'ber, one can carry out the above process starting with any compact Hausdorff topological space X, instead of $SO(2, \mathbb{R})$, as long as a definable basis for the topology is uniformly definable. In this case, if we consider an elementary extension X^* of X then $\pi : X^* \to X$ is defined by: $\pi(x) =$ the unique $y \in X$ such that every X-definable open set containing y also contains x.

Generalizing the example. Assume now that we move in the opposite direction. Namely, we start with an arbitrary group G definable in an arbitrary (sufficiently saturated) o-minimal structure. The goal is to associate to G a compact real Lie group H and a surjective group-homomorphism $\pi : G \to H$ whose kernel is type-definable, such that the logic topology agrees with the Euclidean topology on H. Ideally, H should capture certain properties of G, such as dimension, the structure of torsion points, cohomological structure and elementary theory. This is the idea behind Pillay's Conjecture.

Before stating the conjecture in full we need to review some topological concepts in the theory of definable groups in o-minimal structures:

Assume that $\mathcal{M} = \langle M, <, \cdots \rangle$ is an o-minimal structure. $\mathcal{M}$ is an ordered structure and as such it is a topological space. The cartesian products of M admit the product topology. Now, if G is a definable group in $\mathcal{M}$ whose universe is a subset of M^n then the set G inherits the subspace topology from M^n but this might not be compatible with the group operation on G. (Consider for example, the interval $[0, 1)$ in $\mathbb{R}$, with addition *mod* 1. This is a definable group in the real field but the group operation is not continuous with respect to the real topology).

A fundamental theorem of Pillay, [33], says: Let $\langle G, \cdot \rangle$ be definable group whose underlying set G is a subset of M^n. Then there exists a topology τ on G with the properties:

(1) For all $g \in G$ outside a definable set of small dimension, if $\{U_s : s \in S\}$ is a basis for the open neighborhoods of g in M^n then $\{h \cdot U_s : s \in S, h \in G\}$ is a basis for τ.

(2) G, together with τ, is a topological group. Namely, the group operation, and the group-inverse map are continuous with respect to τ.

It is easy to see that any two topologies satisfying (1) and (2) are equal and hence we sometimes call τ *the group topology*. Actually, Pillay proves a much stronger result, as he shows that G can be covered by finitely many τ-open sets, each definably homeomorphic to an open subset of M^k for some fixed k (the o-minimal dimension of G). This implies for example, that just like definable sets in the o-minimal topology, every definable subset of G has finitely many definably τ-connected components (a set is called *definably τ-connected* if it not contained the disjoint union of two non-empty definable τ open sets).

It turns out, [38], that if $\mathcal{M}$ expands a real closed field then every definable group G is definably homeomorphic (with its τ-topology) to a definable group $H \subseteq M^r$, for some r, such that the topology on H is the subspace topology. We call such a group H *an affinely embedded* group.

Definable compactness. If one works in a sufficiently saturated o-minimal structure $\mathcal{M}$ then the underlying topology on M^n is very far from being locally compact. In fact, it is not difficult to see that *no* infinite definable subset of M is compact. Also, sequences are quite useless in this setting since the only converging sequences are those which are eventually constant.

What should be then the correct analogue of compactness? The first attempt is to restrict oneself to definable covers of open sets. However, this fails as the following example shows:

Consider the interval $[0, 1]$ in a nonstandard real closed field $\mathcal{R}$, and take $\alpha \in \mathcal{R}$ to be a positive infinitesimal. The family

$$\mathcal{U} = \{(x - \alpha, x + \alpha) : x \in [0, 1]\}$$

is a *definable* open cover but it has no finite subcover.

So, instead of using either open covers or converging sequences, we use "converging" definable curves (see [32]):

DEFINITION 2.2. A definable group G is *definably compact* if every definable continuous $f : (a, b) \to G$ has a limit point in G (with respect to the τ-topology), as t tends to either a or b in $\mathcal{M}$.

Examples of definably compact groups. 1. If $G \subseteq M^n$ is an affinely embedded group then G is definably compact if and only if it is closed and bounded. In particular, if we work over $\mathbb{R}$, the notions of definable compactness and compactness are the same for definable groups.

2. Compact real Lie groups are definably compact in any o-minimal structure in which they are definable.

3. If A is an abelian variety over a real closed field R then $A(R) =$ the set of R-points of A, is definably compact.

4. The interval $[0, a)$, in any ordered divisible abelian group, with addition *mod* a is definably compact.

As a result of the work on Pillay's conjecture, and mainly as a result of the work of Dolich, [10], one obtains an equivalent definition for definable compactness, in terms of open covers:

FACT 2.3. [28] *G is definably compact if and only if every uniformly definable open cover of G which is parameterized by a complete type, has a finite sub-cover.*

Pillay's Conjecture. We are now ready to state Pillay's conjecture in full:

Pillay's Conjecture *PC* [34]. Let G be a definable group in a κ-saturated o-minimal structure $\mathcal{M}$ (large κ). Then:

(1) G has a minimal (minimum) type-definable normal subgroup of bounded index, call it G^{00}.
(2) G/G^{00}, equipped with the Logic topology, is isomorphic, as a topological group, to a compact Real Lie group.
(3) If G is definably compact then
$$\dim_{Lie}(G/G^{00}) = \dim_{\mathcal{M}}(G).$$

The beauty of this conjecture is that it offers a surprising connection between the pure lattice of definable sets in definable groups in o-minimal structures and Real Lie groups. It implies that every definably compact group in an o-minimal (large) structure has a homomorphism onto a canonical Real Lie group that is associated to it. The pull-back under this homomorphism of every Euclidean closed set is type-definable and vice-versa. Such quotients are called in Model Theory "hyper-imaginaries" (in contrast to standard imaginaries, which are quotients of definable sets by definable equivalence relations).

Some examples.

(1) Let G be an elementary extension of a compact Lie group H (the group H can be realized, say, as a real algebraic group). Then, as in the case

of $SO(2,\mathbb{R})$, the group G^{00} is just $\mu(e)\cap G$ and $G/G^{00}\simeq H$ (where $\mu(e)$ is the infinitesimal neighborhood of the identity element $e\in G$, in the sense of the real closed field in which G is realized). If G is definably isomorphic in $\mathcal{M}$ to such a group we say that G has *very good reduction*.

In these examples the choice of G^{00} is determined by the infinitesimals of the associated saturated real closed field $\mathcal{R}$, i.e. by the valuation ring of $\mathcal{R}$. This is not the case in the next example.

(2) Consider a sufficiently saturated real closed field $\mathcal{R}$, α a positive infinitesimal element, and let $G=\langle[0,\alpha),+\bmod\alpha\rangle$. In this case the whole of G is contained in the kernel of the standard part map, so we need to use an "internal" notion of valuation:

$$G^{00}=\left\{g\in[0,\alpha):\forall n\in\mathbb{N}\ g<\alpha/n\vee\alpha-\frac{\alpha}{n}<g<\alpha\right\}$$

and G/G^{00}, as a Lie group is again $SO(2,\mathbb{R})$.

(3) $G=\langle R,+\rangle$ (R a real closed field). In this case $G^{00}=G$, so G/G^{00} is trivial.

(4) A non-elementary example: Take $A(R)$ to be the R-points of an abelian variety A defined over a real closed field R, $\dim A=n$, and let $G=A(R)^0$ be its semi-algebraic connected component. By ***PC***, there exists a homomorphism from G onto an n-dimensional real torus $\mathbb{T}^n$, whose kernel is type-definable in R, and such that the logic topology agrees with the Euclidean topology on $\mathbb{T}^n$.

The current status of *PC*. The existence of G^{00}, and the fact that G/G^{00} is a Lie group was proven in [8] without any restrictions on the structure $\mathcal{M}$. ***PC*** is now proven in full when $\mathcal{M}$ expands a real closed field (the last step in the proof is in [18]). ***PC*** was also proved in the case when $\mathcal{M}$ is an ordered vector space over an ordered division ring, [24], [15], and as a result also in o-minimal expansions of ordered groups, [27] (see Section 11 for these latest updates).

It is still unknown whether ***PC*** holds in arbitrary o-minimal structures. As we will point out, the only obstacle here is the understanding of torsion points in definably compact groups in such structures.

§3. A sketch of the proof. Because the proof of Pillay's Conjecture has several components, coming from different papers, we use this section to outline its proof. Details will be given in subsequent sections.

Assume that G is a definable group in an o-minimal structure $\mathcal{M}$.

In Section 4 it is explained why G^{00} necessarily exists. Moreover, it is shown that G has no infinite descending chain of type-definable subgroups of bounded index, and therefore, by Pillay's original paper, [34], G/G^{00} with its Logic topology is isomorphic to a compact real Lie group.

It is also established there that if G is abelian then G^{00} is divisible and furthermore every type-definable subgroup of G which is torsion-free and of bounded index must equal G^{00} (later on it will be shown that G^{00} itself is torsion-free as well). This section is based on the work with Berarducci, Otero and Pillay, [8].

It is now left to prove that for a definably compact group G, the Lie group dimension of G/G^{00} equals the o-minimal dimension of G. This is done as follows:

The abelian case

Step 1: G^{00} is torsion-free:

Given a definable $X \subseteq G$, the type-definable set $Stab_{ng}(X) \subseteq G$ is defined to be the collection of all $g \in G$ such that the symmetric difference $X \triangle gX$ is non-generic in G. The results in Section 5, based on the work with Pillay, [28] and on work of Dolich, [10], show that $Stab_{ng}(X)$ is actually a subgroup of G.

In Section 6 it is proved that $Stab_{ng}(X)$ is a subgroup of bounded index and hence contains the minimal such group G^{00}. This section uses the fact that o-minimal structures have the Non Independence Property. The notion of measure comes in as well and in particular the fact that every abelian group is amenable and thus admits a left invariant, finitely additive, real valued probability measure. The results here are based on the work with Hrushovski and Pillay, [18], which itself uses ideas of Keisler, [21].

Given $n \in \mathbb{N}$, one uses the fact that there are only finitely many n-torsion elements for every n (see Strzebonski, [37]) and Definable Choice to obtain a definable $X_n \subseteq G$ with $Stab_{ng}(X_n)$ containing no n-torsion points. Because every $Stab(X_n)$ contains G^{00} it follows that G^{00} is torsion-free.

Step 2: $\dim_{Lie}(G/G^{00}) = \dim G$.

(See Section 7). This is based on the work of Edmundo and Otero, [13], on the torsion subgroup of a definably compact, definably connected abelian group, in o-minimal expansions of real closed fields. Their result shows that if $\dim(G) = n$ then its torsion subgroup is identical to that of a real n-dimensional torus. Because G^{00} is divisible and torsion-free (see above), it follows that the torsion subgroup of G/G^{00} equals to that of G and hence to that of an n-dimensional torus. This implies that G/G^{00} is itself an n-dimensional torus, thus completing the proof of ***PC*** in the abelian case (in o-minimal expansions of real closed fields).

The semisimple case

A definable group G is called semisimple if it has no definable infinite normal abelian subgroup. By the work with Pillay and Starchenko, [29], every definably connected semisimple group can be written as the almost direct product of almost definably simple groups. Namely, each component is a noncommutative definable group which, modulo its finite center, has no definable normal subgroup. The word "almost" implies that up to a finite central subgroup the product of the groups is direct.

In addition, it is shown in [30] that every definably simple group is definably isomorphic to a semi-algebraic group definable over the real algebraic numbers. The fact that ***PC*** holds for semisimple groups follows from the work in [28].

The general case

Given a definably compact, definably connected group G in an o-minimal structure, it is shown in [31] (joint work with Starchenko) that $G/Z(G)$ is semisimple. By the above, we already know that ***PC*** holds for both $Z(G)$ (the abelian case) and for $G/Z(G)$ (the semisimple case), thus we already have on one hand,

$$\dim_{Lie}((G/Z(G))/(G/Z(G))^{00}) = \dim(G/Z(G))$$

and

$$\dim_{Lie}(Z(G)/Z(G)^{00}) = \dim(Z(G)),$$

while on the other hand

$$\dim(G) = \dim(G/Z(G)) + \dim(Z(G)).$$

It is thus left to see that

$$\dim_{Lie}(G/G^{00}) = \dim_{Lie}((G/Z(G))/(G/Z(G))^{00}) + \dim_{Lie}(Z(G)/Z(G)^{00}),$$

and, as is not hard to verify, this reduces to showing:

$$G^{00} \cap Z(G) = Z(G)^{00}.$$

Because $Z(G)$ is abelian and $G^{00} \cap Z(G)$ is type-definable of bounded index in $Z(G)$, it is sufficient, by the result in [8] mentioned above, to show that $G^{00} \cap Z(G)$ is torsion-free (see Section 8).

For that purpose, it is needed to understand better the group G^{00} in the general non-abelian case. Once again, it is needed to develop a theory for definable generic sets in definably compact groups and to prove, for example, that the definable sets which are not left generic form an ideal.

The new notion introduced here is the so-called fsg (finitely satisfied generics) property of a group G, which implies that for every definable $X \subseteq G$ the type-definable $Stab_{ng}(X)$ is a subgroup which is moreover of bounded index in G. The fact that definably compact abelian groups and definably compact semi-simple groups have the fsg property is then deduced from the previous work. This can be shown to imply that G has fsg as well. (See Section 9).

Finally (see Section 10), in order to show that $G^{00} \cap Z(G)$ is torsion-free one produces, just as in the abelian case, a set X_n, for every $n \in \mathbb{N}$, such that $Stab_{ng}(X_n) \cap Z(G)$ has no n-torsion points. It follows that $G^{00} \cap Z(G)$ is torsion-free which completes the proof of ***PC*** for o-minimal expansions of real closed fields.

We now give the details of the proof.

§4. The existence of G^{00} and some corollaries.

The material in this section is contained in [8].

In [34] Pillay shows, for a group G definable anywhere, that the existence of G^{00} and the fact that G/G^{00} with the group topology is a compact Lie group are together equivalent to the Descending Chain Condition for type-definable subgroups of bounded index.

Throughout this section G is a definable group in an arbitrary o-minimal structure.

THEOREM 4.1. [8]

(1) *G satisfies DCC for type-definable subgroups of bounded index. Namely, there is no infinite descending chain of type-definable subgroups of G of bounded index.*

(2) *If G is definably connected then G/G^{00} is connected.*

ABOUT THE PROOF. By [29], every definable group in an o-minimal structure has a definable normal solvable subgroup H such that G/H is semisimple, namely has no infinite definable normal abelian subgroup. DCC for a semisimple group follows from its decomposition into an almost direct product of definably almost simple groups (see [29]) and the fact that definably simple groups have very good reduction, [30]. By analyzing each abelian step which makes up the solvable group H, we are reduced to the abelian case, so we assume that G is abelian.

An important ingredient of the proof is the notion of a *definably connected* type-definable set X. By that we mean that there are no definable open sets $U_1, U_2 \subseteq G$ such that $U_1 \cap X$ and $U_2 \cap X$ are both nonempty and pairwise disjoint. As is proved in the paper, every type-definable, definably connected subgroup of G has a type-definable subgroup of bounded index which is definably connected. This latter subgroup can be written as the directed intersection of definably connected sets.

Assume now that DCC fails. Then there exists a descending chain of type-definable subgroups of bounded index $H_1 > \cdots H_n > \cdots$, which we may assume are all definably connected. Using standard model theoretic arguments one may assume that all groups are defined over a countable model M_0 using a countable language. Let H be the minimal type-definable subgroup of bounded index definable over M_0 (this does exist!). Most of the work now is towards proving that G/H, equipped with the Logic topology, is a compact Lie group. That is done using topological arguments, together with the fact that G has a finite number of elements of every given finite order (see [37]). Once it is established that G/H is Lie group, the sequence H_i/H is a descending chain of closed subgroups, which is impossible.

REMARK. 1. In [35], Shelah proves the existence of G^{00} (but not DCC!) for any group with NIP (see precise definition in Section 6) and therefore in particular for o-minimal structures.

(*The following discussion and example were suggested by Pillay*):

2. There are two other related notions for a group G (in a sufficiently saturated structure): Consider all definable subgroups of G of finite index. If the intersection of these groups has bounded index (equivalently, the intersection does not change when we move to an elementary extension) then it is called G^0. In o-minimal structures and in groups of finite Morley rank, G^0 itself is definable and has finite index in G.

Another notion is that of G^{000}: For $A \subseteq M$ a small subset, let G_A^{000} be the smallest subgroup of G of bounded index which is invariant under automorphisms fixing A point-wise (note that this group always exists). If G_A^{000} does not depend on A then we call this group G^{000}.

It is not difficult to see that the existence of G^0 (G^{00}) is equivalent to the fact that every descending chain of (type-) definable subgroups of G of finite (bounded) index must have bounded cardinality. Similarly, the existence of G^{000} is equivalent to the fact that every descending chain of automorphism-invariant subgroups (over small sets) has bounded cardinality. It therefore follows that the existence of G^{000} implies the existence of G^{00} and this in turn implies the existence of G^0. In stable theories they all exist and are equal to each other.

It was shown by Shelah, [36], that if G is abelian and has NIP (see definition below) then G^{000} exists. Later on this was generalized by Gismatullin, [17], to an arbitrary group with NIP. However, it is still unknown in the NIP context (and even in the o-minimal case), whether $G^{00} = G^{000}$.

EXAMPLE. Consider the group $G = \langle \mathbb{Z}^\omega, \cdot \rangle$ in the two-sorted structure $\langle G; \mathbb{N} \rangle$, with a predicate $P \subseteq \mathbb{Z}^\omega \times \mathbb{N}$ such that $(x, n) \in P$ if and only if $x_n = 0$.

Note that for every $x \in \mathbb{Z}^\omega$, if $x_n \neq 0$ then the group $\{y \in \mathbb{Z}^\omega : y_n = 0\} = P(G, n)$ has index 2 in $\mathbb{Z}^\omega$. Hence, the theory of the structure says that for every $0 \neq g \in G$ there exists an $n \in \mathbb{N}$ such that $P(G, n)$ is a subgroup of index 2 which avoids g. This is easily seen to imply that the group G^0 (and therefore also G^{00} and G^{000}) does not exist in elementary extensions.

We now return to the o-minimal context. Here are two important corollaries of Theorem 4.1;

COROLLARY 4.2. *Assume that G is abelian. Then*:

(1) *G^{00} is divisible.*
(2) *Let H be a type-definable subgroup of bounded index. If H is torsion-free then $H = G^{00}$.*

PROOF. (1) We need to see that for every n, the map $\sigma_n(x) = x^n$ sends G^{00} onto itself. It is easy to see that $\sigma_n(G^{00})$ has bounded index in $\sigma_n(G)$. However, σ_n has finite kernel, [37], and therefore $\dim \sigma_n(G) = \dim(G)$, so $\sigma_n(G)$ has finite index in G. It now follows that $\sigma_n(G^{00})$ has bounded index in G, and because it is contained in G^{00} it follows from minimality that $\sigma_n(G^{00}) = G^{00}$.

(2) The group H/G^{00} is a closed subgroup of the compact Lie group G/G^{00}, therefore either $H = G^{00}$ or H/G^{00} has torsion. If $H \neq G^{00}$ then the latter holds and hence, because G^{00} is divisible, H must have torsion as well. Contradiction. ⊣

COROLLARY 4.3. *If G is torsion-free then $G^{00} = G$.*

Notice that up until now we have not even established that in a definably compact group we have $G^{00} \neq G$. Indeed, the main remaining difficulty in proving ***PC*** is the dimension equality:

REMAINING CONJECTURE. *If G is definably compact then* $\dim_M(G) = \dim_{Lie}(G/G^{00})$.

§5. Some theory of generic sets I. *Most of the material in this section is taken from* [28].

Here G is definable in an o-minimal structure. However, some of the results work in any model theoretic setting, or at least when there is a reasonable notion of rank.

DEFINITION 5.1. (1) A definable set $X \subseteq G$ is called *left k-generic* if $G = \bigcup_{i=1}^{k} g_i X$, for some $k \in \mathbb{N}$ and $g_i \in G$. X is *left generic* if it is left k-generic for some $k \in \mathbb{N}$. The notion of right-generic is similarly defined. X is *generic* if it is both left and right generic.
(2) If $X \subseteq G$ is definable then X is called *large* if $\dim(G \setminus X) < \dim G$.

REMARK. In ω-stable connected groups the notions of "generic" and "large" are the same and both are equivalent to $RM(X) = RM(G)$. In o-minimal structures generic sets are not necessarily large and $\dim(X) = \dim(G)$ does not imply that X is generic:

1. In $\langle R, + \rangle$ (R an ordered divisible abelian group), a set is generic if and only if it is of the form $(-\infty, a) \cup (b, +\infty)$, for $a, b \in R$.
2. In elementary extensions of the circle group $\mathbb{T}^1$ a definable set is generic if and only if its length not an infinitesimal. Equivalently, one of its definably connected components contains at least two torsion elements.

Recall that for an A-definable set $X \subseteq M^n$, an element $x \in X$ is called *generic in X over A* if $\dim(x/A) = dim(X)$, where $dim(x/A)$ is taken to be the size of the maximal $acl_{\mathcal{M}}$-independent subset of the tuple x.

FACT 5.2. *If X is large in G and* $\dim(G) = n$ *then X is (both left and right)* $n+1$*-generic.*

PROOF. Without loss of generality, X is $\emptyset$-definable.

We show: If g is generic in G and $h \in G \setminus (X \cup gX)$ then

$$\dim(h/g) < \dim(h/\emptyset) < n.$$

Indeed, if the left inequality fails then $\dim(h/g) = \dim(h/\emptyset)$ and hence (by the addition formula for dimension) we have $\dim(g/h) = \dim(g/\emptyset) = n$. It follows that

$$\dim(g^{-1}h/h) = \dim(g^{-1}/h) = \dim(g/h) = n.$$

In particular, $g^{-1}h$ is generic in G and because X was large we must have $g^{-1}h \in X$ and hence $h \in gX$, contradicting the assumption on h.

The inequality $\dim(h/\emptyset) < n$ follows from the fact that $h \in G \setminus X$ and X is large.

It follows from the above dimension inequality that $\dim(G \setminus (X \cup gX)) < \dim(G \setminus X) < dim(G)$. We now replace X by $X \cup gX$ and proceed by induction. ⊣

Our goal in this section is to discuss the following result:

THEOREM 5.3. [28] *Assume that G is a definably compact affinely embedded group, $\mathcal{M}$ expands an ordered group and $X \subseteq G$ is a definable set which is not left-generic. Then $G \setminus X$ is right generic.*

We start with the following:

FACT 5.4. (i) *If $X \subseteq G$ is not left-generic then $Cl(X)$ is not left-generic.*

(ii) *If $X \subseteq G$ is generic then $Int(X)$ is also generic. (Here and below we use the τ-topology of G which was described above, and we take $Cl(X)$ and $Int(X)$ to be the topological closure and interior, respectively, with respect to that topology).*

PROOF. (i) We use the following basic fact about a definable set X in o-minimal structures: For $Fr(X) = Cl(X) \setminus X$, $\dim Fr(X) < \dim(X)$.

If $Cl(X)$ is left-generic then

$$G = \bigcup_{i=1}^{k} g_i Cl(X) = \bigcup_{i=1}^{k} g_i X \cup \bigcup_{i=1}^{k} g_i Fr(X).$$

But $\dim(Fr(X)) < \dim(G)$, hence $\dim(\bigcup_{i=1}^{k} g_i Fr(X)) < \dim(G)$, and therefore the set $\bigcup_{i=1}^{k} g_i X$ is large in G. By Fact 5.2, this last set is generic and therefore X is generic.

(ii) Use the fact that for any $X \subseteq G$, we have $\dim(X \setminus Int(X)) < \dim(G)$, and proceed as in (i). $\dashv$

The connection of generic sets to Pillay's Conjecture comes through the following result (Fact 5.5 and Fact 5.6 hold in arbitrary structures):

FACT 5.5. *If $H \subseteq G$ is a type-definable subgroup then H has bounded index in G if and only if can be written as the intersection of a directed family of left generic sets. (A "directed family" of sets here is a family for which the intersection of any two sets contains another set in the family).*

PROOF. Notice first that any type-definable group can be written as the intersection of a directed family of definable sets (by adding to the original family which defines H any finite intersection of sets in the family).

If H has bounded index and is contained in a definable set X then G can be covered by boundedly many left translates of X (namely the number of cosets of H). By compactness, finitely many left translates of X cover G.

Assume now that $H = \bigcap_{i<\lambda} X_i$ is the intersection of a directed family of left generic sets and let $A = \{g_j : j < \lambda\}$ be a set of elements such that for every X_i, we have $G = AX_i$. It is sufficient to see that every complete type over A is contained in a single coset of H (for then $[G; H] < 2^{|A|}$.)

Indeed, if g, h realize the same type over A then for every $i \in I$ there exists $a_i \in A$ such that $g, h \in a_i X_i$. It follows that for every $i \in I$, we have $g^{-1}h \in X_i^{-1}X_i$. However, since H is a subgroup (and the family $\{X_i : i \in I\}$ directed) we can also write $H = \bigcap_{i\in I} X_i^{-1}X_i$, hence $g^{-1}h \in H$. $\dashv$

FACT 5.6. *If $X \subseteq G$ is not left-generic then for any small $M_0 \subseteq M$ (where "small" means $|M_0| < \kappa$) there exists $g \in G$ such that $Xg \cap M_0 = \emptyset$.*

PROOF. By assumption, for every $h_1, \dots, h_k \in G$, $k \in \mathbb{N}$, there is $g \in G$ such that

$$\bigwedge_{i=1}^{k} g \notin h_i X,$$

or equivalently

$$\bigwedge_{i=1}^{k} h_i^{-1} \notin Xg^{-1}.$$

Clearly then, for every $h_1, \dots, h_k \in G$, $k \in \mathbb{N}$, there is $g \in G$ such that

$$\bigwedge_{i=1}^{k} h_i \notin Xg.$$

It follows that if M_0 is any small subset of M then, by the saturation of $\mathcal{M}$, there is $g \in G$ such that $M_0 \cap Xg = \emptyset$. $\dashv$

Digression: Dolich's work. In [10], Dolich examines the notion of forking and dividing in o-minimal structures. The paper contains many interesting and highly nontrivial results about types in o-minimal structures. In [28] we extract from his work the following:

THEOREM 5.7. *Let $X(a) \subseteq M^n$ be a closed and bounded a-definable set (a a finite tuple) in a sufficiently saturated o-minimal structure $\mathcal{M}$ expanding an ordered group and let $\mathcal{M}_0 \subseteq \mathcal{M}$ be a small model. Assume that the set $\{X(a') : a' \equiv_{M_0} a\}$ has the finite intersection property (namely, the intersection of every finite sub-family is nonempty).*
Then $X(a) \cap M_0 \neq \emptyset$.

The proof of this result is too long to discuss here. We only make few observations:

1. Consider the simplest case of Theorem 5.7, where $X(a_1a_2)$ is the closed interval $[a_1, a_2] \subseteq M$, and M_0 a small submodel of M. If $[a_1, a_2] \cap M_0 = \emptyset$ then $a_1 \equiv_{M_0} a_2$ (otherwise, by o-minimality, there is an interval J over M_0, containing one of the a_i's but not the other. It follows that one of the endpoints of this interval, which must be in M_0, also belongs to $[a_1, a_2]$. Contradiction). Moreover, because $a_1 \notin M_0$ there exists $a_2' < a_1$ with $a_2' \equiv_{M_0} a_1 \equiv_{M_0} a_2$. By homogeneity, there is $a_1' < a_2'$ such that $a_1'a_2' \equiv_{M_0} a_1a_2$. We now have $X(a_1'a_2') \cap X(a_1a_2) = \emptyset$, so the family

$$\{X(a_1', a_2') : a_1'a_2' \equiv_{M_0} a_1a_2\}$$

does not have the finite intersection property.

2. Theorem 5.7 is false when $X(a)$ is not closed and bounded. Consider for example the open interval $(0, \alpha)$ in a nonstandard real closed field, for an infinitesimal $\alpha > 0$, and take M_0 to be the real algebraic numbers. The family

$$\{(0, \alpha') : \alpha' \equiv_{M_0} \alpha\}$$

has the finite intersection property but $(0, \alpha) \cap M_0$ is empty.

3. In the stable case, the analogous theorem to Theorem 5.7 is true for any definable set because the assumption is equivalent to the forking of $X(a)$ over $\mathcal{M}_0$.

4. The description of a definably compact set using a type-definable open covering (see Fact 2.3) easily follows from Fact 5.7.

End of Digression.

PROOF OF THEOREM 5.3. Because G is affinely embedded it is closed and bounded in M^k. Assume $X \subseteq G$ is not left generic. By Fact 5.4 we may assume that X is closed. Fix $\mathcal{M}_0$ such that X is definable over M_0. By Fact 5.6, there exists $g \in G$ such that $M_0 \cap Xg = \emptyset$. By Theorem 5.7, there are $g_1, \ldots, g_k$ (each realizing the same type as g over M_0) such that

$$\bigcap_{i=1}^{k} Xg_i = \emptyset.$$

By taking complement we get

$$\bigcup_{i=1}^{k}(G \setminus X)g_i = G.$$

Hence, $G \setminus X$ is right-generic. ⊣

REMARKS. 1. Theorem 5.3 fails without the definable compactness assumption: The set $(a, +\infty)$ and its complement are both not generic in $\langle R, +\rangle$ (here left and right-generic are the same).

2. The analogue of Theorem 5.3 in the stable setting is true for any definable subset of the group G.
3. Recently, Eleftheriou has pointed out that the assumption that G is affinely embedded can be omitted Theorem 5.7 by working in the manifold charts of G.

Here are two easy consequences:

FACT 5.8. *Assume that G is definably compact and abelian.*

(1) *The non-generic sets form an ideal.*
(2) *Every formula defining a generic set in G belongs to a complete "generic" type p (over $\mathcal{M}$). Namely, every formula in p defines a generic set in G.*

§6. Some theory of generic sets II: Measure and the NIP. *The content of Sections 6-9 is mostly taken from* [18]. *The connection between the Non Independence Property and measure is due to Keisler* [21] *and the proof of Lemma* 6.4 *below is modeled after a proof from Keisler's paper.*

The next notion is due to Shelah. The definition we use is equivalent to the original one.

DEFINITION 6.1. A theory T is said to be *dependent* or to have the *Non Independence Property* (NIP) if for every indiscernible sequence $\langle a_i : i < \omega\rangle$ over A and $\phi(x, y)$ a formula over A the type $\{\phi(x, b_{2j})\Delta\phi(x, b_{2j+1}) : j < \omega\}$ is inconsistent. (We take $\phi\Delta\psi$ to mean $(\phi \wedge \neg\psi) \vee (\neg\phi \wedge \psi)$)

Stable theories, o-minimal theories, the theory of p-adically closed fields all have the NIP, while the theory of pseudo-finite fields fails to have it. In fact, by Shelah's work any simple unstable theory fails to have NIP.

DEFINITION 6.2. We say that G admits a left invariant *Keisler measure* if there exists a real valued finitely additive measure $\mu : Def(G) \to \mathbb{R}$ on the definable subsets of G, such that $\mu(G) = 1$ and for every definable $X \subseteq G$ and $g \in G$, we have $\mu(gX) = \mu(X)$.

In the rest of this section we make the following assumptions on the group G (equipped with the definable sets induced by the ambient structure):

- *G has NIP.*
- *The non left-generic sets form an ideal.*
- *G admits a left-invariant Keisler measure.*

As we will eventually show, every definably compact group in an o-minimal structure satisfies all of the above. For now, notice that any abelian definably compact group satisfies the above assumptions. Indeed, o-minimality implies NIP, and by Fact 5.8 the non-generic sets form an ideal. Because every abelian group is amenable, it admits a left-invariant real valued finitely additive probability measure on all subsets.

DEFINITION 6.3. For $X, Y \subseteq G$ definable, we write $X \approx_{ng} Y$ if $X \triangle Y$ is not left-generic.

Notice that since we assume that the non left-generics form an ideal $\approx_{ng}$ is an equivalence relation. The NIP assumption is crucial for the following.

LEMMA 6.4. *The equivalence relation $\approx_{ng}$ is bounded. I.e., there exists a fixed small model M_0 such that every equivalence class of $\approx_{ng}$ is already represented by a definable set over M_0.*

PROOF. Let μ denote the finitely additive left-invariant measure on $Def(G)$, the family of definable subsets of G. Note that if $X \subseteq G$ is a definable n-generic set then we have $\mu(X) \geqslant 1/n$.

Assume that $\approx_{ng}$ is unbounded. Then there exists a formula $\phi(x, y)$ over the empty set, with the variable x for elements in G, and unboundedly many b_i's, such that $\phi(G, b_i) \triangle \phi(G, b_j)$ is generic.

By standard Ramsey-type arguments, there exists a fixed n and an indiscernible sequence $\langle a_i : i < \omega \rangle$ such that for every $i \neq j$, the set $\phi(G, a_i) \triangle \phi(G, a_j)$ is n-generic.

Consider the family $\mathcal{F} = \{Y_j = \phi(G, a_{2j}) \triangle \phi(G, a_{2j+1}) : j < \omega\}$. By indiscernibility, there exists a natural number k such that every k sets in $\mathcal{F}$ have empty intersection. However, for every j, $\mu(Y_j) \geqslant 1/n$, and because $\mu(G) = 1$ it is impossible that every k sets in $\mathcal{F}$ intersect trivially. Contradiction. $\dashv$

DEFINITION 6.5. For $X \subseteq G$ definable, let

$$Stab_{ng}(X) = \{g \in G : gX \approx_{ng} X\}.$$

Under our assumptions, the set $Stab_{ng}(X)$ is a subgroup of G. It is type-definable because for every n, the set of all g such that n translates of $gX \triangle X$ *do not* cover G, is definable. The map $g \mapsto gX/\approx_{ng}$ induces an injective map from $G/Stab_{ng}(X)$ into $Def(G)/\approx_{ng}$ and therefore we proved:

THEOREM 6.6. *For any definable $X \subseteq G$, the subgroup $Stab_{ng}(X)$ has bounded index in G.*

§7. The proof of *PC* in the abelian case. *We assume in this section that $\mathcal{M}$ expands a real closed field and that G is definably compact and abelian.*

Our goal here is to prove:

THEOREM 7.1. *If G is definably compact and abelian then* $\dim_{Lie}(G/G^{00}) = \dim_{\mathcal{M}} G$.

PROOF. Because G^0 has finite index in G we may assume that G is definably connected.

The proof is based on two ingredients. The first one is a deep theorem of Edmundo and Otero on the torsion points in definably connected, definably compact abelian groups. (Presumably, this was one of the main justifications to the original conjecture of Pillay). Its proof is based on Cohomological tools and uses extensively the triangulation theorem which is true only in o-minimal expansions of real closed fields:

THEOREM 7.2. [13] *Assume that $\mathcal{M}$ expands a real closed field and that G is a definably compact, definably connected abelian group of dimension n. Then*

$$Tor(G) \simeq Tor(\mathbb{T}^n).$$

(*where $\mathbb{T}^n$ is the n-dimensional torus and $Tor(G)$ is the subgroup of torsion elements in G*).

The second ingredient, which we will prove below is:

LEMMA 7.3. *G^{00} is torsion-free.*

Let us see how the two results, taken together, imply ***PC*** in the abelian case: Lemma 7.3, together with the divisibility of G^{00} (see Corollary 4.2 (1)) imply that

$$Tor(G/G^{00}) \simeq Tor(G).$$

If $\dim(G) = n$, it follows from theorem [13], that

$$Tor(G/G^{00}) \simeq Tor(\mathbb{T}^n).$$

Because G/G^{00} is a connected (Theorem 4.1) abelian compact Lie group, it is Lie isomorphic to a direct sum of $\mathbb{T}^1$'s. The number of these $\mathbb{T}^1$'s is determined by, say, the number of 2-torsion points, therefore $G/G^{00} \simeq (\mathbb{T}^1)^n$ and so the real dimension of G/G^{00} is n.

PROOF OF LEMMA 7.3. For every $n \in \mathbb{N}$, consider the map $\sigma_n : g \mapsto g^n$ from G onto G. By definable choice, there exists a definable set $X \subseteq G$ such that $\sigma_n|X$ is a bijection of X and G (we assume that G is definably connected).

By [13] (or actually by the preceding results in [37]), $T_n = ker(\sigma_n)$ is finite. It clearly contains all n-torsion points and, as easily checked, G equals a finite disjoint union of the gX's, for $g \in T_n$. Thus X and all the gX's are generic and pair-wise disjoint, and therefore $T_n \cap Stab_{ng}(X) = \{0\}$. Because this is true for every n, the group $Stab_{ng}(X)$ is torsion-free.

By Theorem 6.6, $Stab_{ng}(X)$ has bounded index in G, and therefore $G^{00} \subseteq Stab_{ng}(X)$. It follows that G^{00} is torsion-free, ending the proof of Lemma 7.3 and thus ***PC*** in the abelian case. ⊣

§8. Proof of *PC* for arbitrary definably compact G. *We assume in this section that G is a definably compact group in an o-minimal expansion of a real closed field.*

Here are some preliminary facts about noncommutative definably compact groups:

As shown in [31], $G/Z(G)$ is semisimple, namely has no infinite definable abelian normal subgroup. If we let $N = Z(G)^0$ then the same is true for G/N. By [29], G/N can be written as an almost direct product of definably almost simple groups and every definably simple group is definably isomorphic to a semialgebraic linear group defined over the real algebraic numbers. In particular, definably simple groups have very good reduction. The proof of ***PC*** for groups with very good reduction is partly contained in the Introduction (see [28] for more details). It easily follows that ***PC*** holds for definably compact almost simple groups and therefore also for an almost direct product of such groups. Therefore, ***PC*** holds for both N (Theorem 7.1) and for G/N.

We thus have:

$$\dim_{\mathcal{M}}(G) = \dim_{\mathcal{M}}(G/N) + \dim_{\mathcal{M}}(N) =$$

$$\dim_{Lie}((G/N)/(G/N)^{00}) + \dim_{Lie}(N/N^{00}).$$

We also have

$$\dim_{Lie}(G/G^{00}) = \dim_{Lie}(G/G^{00}N) + \dim_{Lie}(G^{00}N/G^{00}).$$

It is easy to verify that $G^{00}N/N = (G/N)^{00}$. Hence,

$$(G/N)/(G/N)^{00} \backsimeq G/G^{00}N.$$

In order to show that $\dim_{\mathcal{M}}(G) = \dim_{Lie}(G/G^{00})$ it is therefore sufficient to prove:

$$G^{00}N/G^{00} \backsimeq N/N^{00}.$$

The group on the left is isomorphic to $N/(G^{00} \cap N)$, hence in order to prove ***PC*** we are left to prove:

LEMMA 8.1. *If G is definably compact then $N^{00} = G^{00} \cap N$.*

The fact that $N^{00} \subseteq (G^{00} \cap N)$ follows from the fact that the group on the right has bounded index in N. However, in order to prove the opposite inclusion (which fails for arbitrary groups, even with NIP) we need to take one more de'tour, through the general theory of generic sets.

§9. Some theory of generic sets III. *In this section we make no assumption on the group G unless otherwise stated.*

Definition 9.1. The theory of G is said to have *finitely satisfiable generics*[1] (in short fsg) if there exists a complete type p over $\mathcal{M}$ such that if $\phi(x) \in p$ then:

(i) $\phi(G)$ is left generic.
(ii) There exists a small model $M_0 \subseteq M$ such that every left translate of $\phi(G)$ intersects M_0.

Our goal is to show that every definably compact group in an o-minimal structure has fsg. This is useful because of the following properties:

Fact 9.2. *Assume that $T = Th(G)$ has fsg. Then*

(i) *There exists a small $\mathcal{M}_0 \subseteq \mathcal{M}$ such that every left generic set and every right generic set intersect M_0.*
(ii) *Given $X \subseteq G$ definable, X is left-generic if and only if it is right generic.*
(iii) *The definable non generic sets in G form an ideal.*
(iv) *G^{00} exists and $G^{00} = \bigcap\{Stab_{ng}(X) : X \in Def(G)\}$.*

Proof. Assume that p and $\mathcal{M}_0$ witness fsg.

(i) If X is a left generic set then there are $g_1, \ldots, g_k \in G$ such that the formula $x = x$ is equivalent to the finite disjunction of the formulas $x \in g_i X$. Hence, there is g_i such that "$x \in g_i X$" is in p. By assumption on p, $X \cap M_0 \neq \emptyset$. Consider the type $p(x^{-1})$. Because $x \mapsto x^{-1}$ is a $\emptyset$-definable bijection of G it easily follows that if $\phi(x) \in p(x^{-1})$ then $Y = \phi(G)$ is right generic and every right translate of Y intersects M_0. As above, it follows that every right generic set intersects M_0.

(ii) Assume X is not a left generic set. By Fact 5.6, there exists a right translate of X which does not intersect M_0, hence by (i), X is not right generic as well.

(iii) Since p is a complete generic type it must contain the complement of every nongeneric set.

(iv) For the existence of G^{00}, see [18], Corollary 4.3.

Now fix a small model $\mathcal{M}_0$ witnessing (i). Given $X \subseteq G$ definable, let $\mathcal{M}_1$ be a small model containing $\mathcal{M}_0$ over which X is definable. If $g \equiv_{\mathcal{M}_1} h$ then $gX \cap M_1 = hX \cap M_1$ and hence $(gX \triangle hX) \cap M_0 = \emptyset$. By (i), $gX \triangle hX$ is nongeneric. Thus, every coset of $Stab_{ng}(X)$ contains all the realizations of some complete type over $\mathcal{M}_1$. In particular, in a (still small) model where every complete type over $\mathcal{M}_1$ is realized, there is a representative for every coset of $Stab_{ng}(X)$, so $Stab_{ng}(X)$ has bounded index, and therefore it contains G^{00}.

For the opposite inclusion, since G^{00} has bounded index it can be obtained as the intersection of definable generic sets (Fact 5.5). If g belongs to $Stab_{ng}(X)$ for every such X then it must belong to G^{00} (otherwise, by compactness, there is X containing G^{00} such that $gX \cap X = \emptyset$, which implies that $g \notin Stab_{ng}(X)$. ⊣

[1] For simplicity, we slightly modified the definition from [18].

LEMMA 9.3. *Assume that N is a definable normal subgroup of G. If N and G/N have fsg then G has fsg.*

PROOF. See Proposition 4.5 in [18]. ⊣

We return to the o-minimal setting.

LEMMA 9.4. *If G is definably compact and abelian in an o-minimal structure then G has fsg.*

PROOF. Since we do have complete generic types in abelian groups (see Fact 5.8), it is sufficient to show that there is a small $\mathcal{M}_0$ such that every generic set intersects $\mathcal{M}_0$.

Let $\mathcal{M}_0$ be a small model such that every $\approx_{ng}$-class in $Def(G)$ has a representative definable over $\mathcal{M}_0$ (recall, Lemma 6.4, that $\approx_{ng}$ is bounded because G has NIP).

Given $X \subseteq G$ generic, there exists $X_1 \subseteq X$ such that X_1 is still generic and $Cl(X_1) \subseteq X$. Indeed, the following argument for that fact was suggested by the UIUC Logic seminar (it assumes that $\mathcal{M}$ expands an ordered group but this is unnecessary, as the argument in [18] shows):

$$G = \bigcup_{i=1}^{k} g_i X.$$

By Fact 5.4, $Int(X)$ is also generic. For every $\varepsilon > 0$ let

$$X^\varepsilon = \{g \in X : d(g, Fr(X)) > \varepsilon\}$$

(the notion of $d(g, Fr(X))$, the distance of g from the frontier of X, assumes the presence of an underlying group). We have,

$$X = \bigcup_{\varepsilon>0} X^\varepsilon.$$

It is now sufficient to take ε realizing the complete type $p(x)$ of the infinitesimals right of zero. So,

$$G = \bigcup_{\varepsilon \models p} \bigcup_{i=1}^{k} g_i X^\varepsilon.$$

We obtained a definable open covering of G parameterized by a complete type. By the equivalent definition to definable compactness, Fact 2.3, there is a finite subcover, which easily implies that some X^ε is generic. If we let $X_1 = X^\varepsilon$ then $Cl(X_1) \subseteq X$. We may therefore assume that X is closed.

Let $Y \subseteq G$ be a set definable over $\mathcal{M}_0$ such that $Y \triangle X$ is nongeneric (the existence of Y follows from our assumption on $\mathcal{M}_0$). Because X is generic so is Y. Again, by Fact 5.4, we may assume that Y is closed, so both X and Y are definably compact. We will show that $(Y \cap X) \cap M_0 \neq \emptyset$ and in particular $X \cap M_0 \neq \emptyset$. Notice that both X and Y are definably compact.

By Theorem 5.7, if $X \cap Y \cap M_0 = \emptyset$ then there are finitely many $\mathcal{M}_0$-conjugates of $X \cap Y$ whose intersection is empty. Because Y is $\mathcal{M}_0$-definable there are $X_1, \dots, X_k$ all $\mathcal{M}_0$-conjugates of X such that

$$\bigcap_{i=1}^{k} X_i \cap Y = \emptyset.$$

Since Y is generic this implies that for some X_i we must have $Y \setminus X_i$ generic. Contradiction to $Y \triangle X$ being non-generic. ⊣

LEMMA 9.5. *If H is definably compact and semisimple in an o-minimal structure then H has fsg.*

The proof of this lemma is based on the almost-decomposition into definably almost simple groups. The definably simple case is handled in [28] using measure theoretic arguments based on [7] and [1].

Using Lemmas 9.4, 9.5, and 9.3, we can conclude:

THEOREM 9.6. *Every definably compact group in an o-minimal structure has fsg.*

The above theorem, together with Fact 9.2, implies that the set of left (hence also right) generics in G form an ideal, and that for any definable set X, $Stab_{ng}(X)$ is a type-definable group of bounded index. Finally (and this is the main fact which forced us to take this de'tour through the notion of "fsg"), the group G^{00} is the intersection of all stabilizers of definable subsets of G.

§10. Completing the proof of *PC*. We can now return to the missing ingredient in the proof of ***PC***, namely the proof of Lemma 8.1. We need to show that $G^{00} \cap N = N^{00}$, where N is a definably connected normal central subgroup. By Corollary 4.2, it is sufficient to prove that $G^{00} \cap N$ is torsion-free.

Given $n \in \mathbb{N}$, let $T_n = Tor_n(N)$ and $X \subseteq N$ be a definable set such that $g \mapsto g^n$ gives a bijection of X and N. By Definable Choice, there is $D \subseteq G$ which intersects every coset of N exactly once. It is now easy to verify that G is the finite disjoint union of the translates of DX by the elements of T_n. In particular, DX is generic and

$$T_n \cap Stab_{ng}(DX) = \{e\}.$$

Because this is true for every n, we have $Stab_{ng}(DX) \cap Tor(G) \neq \emptyset$.

By the fsg property, $G^{00} \subseteq Stab_{ng}(DX)$, therefore $G^{00} \cap Tor(N) = \{e\}$.

We thus proved that $G^{00} \cap N = N^{00}$, completing the proof of ***PC*** (see the argument preceding Lemma 8.1). ⊣

10.1. Defining measure on G. As a result of the work on Pillay's Conjecture, the following theorem was established in [18].

THEOREM 10.1. *If G is definably compact in an o-minimal structure then it admits a left-invariant Keisler measure on the definable subsets of G. For a definable $X \subseteq G$, we have $\mu(X) = 0$ if and only if X is non-generic.*

PROOF. As we already pointed out, the existence of such measure is immediate when G is abelian. In the general case, we first note that G/G^{00}, as a compact Lie group, admits a left-invariant finitely additive probability measure on a boolean algebra of sets containing all Borel measurable sets- the Haar measure $\boldsymbol{m}$.

We first fix a complete generic type $p(x)$ over G. Given a definable set X, we consider the set

$$\hat{X} = \{gG^{00} \in G/G^{00} : p \models \text{“}x \in gX''\}.$$

(note that X is well defined. Namely, if $gh^{-1} \in G^{00}$ then in particular, $gX \triangle hX$ is non-generic and therefore not in p. It follows that $gX \in p$ if and only if $hX \in p$). The main part of the proof is to show that $\hat{X}$ is a Borel set in G/G^{00} (see Proposition 6.2 in [18]). We then define

$$\mu_p(X) = \boldsymbol{m}(\hat{X}).$$

Clearly, μ_p is left invariant, and it is easy to verify that it is also finitely additive (if $X_1 \cap X_2 = \emptyset$ then $\hat{X}_1 \cap \hat{X}_2 = \emptyset$).

Finally, if X is generic then finite additivity implies that $\mu_p(X) > 0$ and if X is non-generic then $\hat{X} = \emptyset$ and therefore $\mu_p(X) = 0$. ⊣

§11. Related work and some open questions. This section has gone through substantial changes in the last stages of writing. As will be explained below, most of the open questions listed here were solved in a recent paper by Hrushovski and Pillay, [20].

11.1. Omitting the real closed field assumption. As was pointed out early on, the only remaining obstacle for proving ***PC*** without the assumption that $\mathcal{M}$ expands a real closed field is the lack of an analogue to Theorem 7.2 on the number of torsion points, without the field assumption. Such a theorem was proved by Eleftheriou and Starchenko [16] when $\mathcal{M}$ was assumed to be an ordered division ring over an ordered vector space and hence ***PC*** holds in this case as well. Actually, a very clear description of definable groups in this setting is given in the paper, out of which the number of torsion points is easily read.

In order to prove the torsion points result under weaker assumptions it seems important to develop similar topological tools to the ones originally used, but this time without the triangulation theorem. Indeed, Sheaf Cohomology in expansions of ordered groups has been the subject of several papers of Edmundo, Jones and Peatfield (see [11] and [12]) and of Beraducci and Fornasiero (see [4]). In [14], Edmundo and Terzo prove Pillay's conjecture under a relatively weak assumption on $\mathcal{M}$ but with an additional assumption of "orientability" on the group G.

In a recent result, [27], I was able to prove the question about the torsion points and hence Pillay's Conjecture, in o-minimal expansions of ordered groups, follows.

The questions formulated below were written prior to the publication of the recent pre-print by Hrushovski and Pillay [20]. As I will eventually point out, most of these questions are now solved by that paper, either explicitly or implicitly. I leave them here because I find that their discussion could still be of some interest.

11.2. Uniform definability of G^{00}. An important feature of the basic example of Pillay's conjecture (where we start with a compact real Lie group and view it in an elementary extension) is the fact that the type defining G^{00} is given by a single formula, with varying parameters. Namely,

$$G^{00} = \{g \in G : |g| < a : a \in \mathbb{R}\}.$$

Consider the structure G_{ind} whose universe is G/G^{00}, with a function symbol for the group operation and a predicate for every set of the form $\pi(X)$, for $X \subseteq G^n$ definable in the o-minimal structure $\mathcal{M}$. In [18] we showed, using a theorem of Baysalov and Poizat, that if $G = \langle [0,1)^n, +mod 1\rangle$ (in an o-minimal expansion of an ordered divisible abelian group) then structure G_{ind} is definable in an o-minimal structure over the reals. Later, in [22], Marikova reproved this result without referring to [1], and provided a much finer analysis of the definable sets in this structure. The uniformity in parameters plays an important role in both works. For more recent work of Marikova and v.d. Dries see [23], [39]

CONJECTURE. *If G is definably compact then there is a formula $\phi(x, y)$, where x varies over elements of G, and a set of parameters A, such that*

$$G^{00} = \left\{ g \in G : \bigwedge_{a \in A} \phi(g, a) \right\}.$$

CONJECTURE. *The structure on the compact Lie group G_{ind} is definable in some o-minimal structure over the real numbers.*

Related to the above conjecture is the following:

QUESTION. *What is the structure which G induces on $Tor(G)$? In particular, what subsets of $Tor(G)$ are of the form $X \cap Tor(G)$ for a definable subset of G?*

Note that when G is abelian its torsion group can be realized as a definable set in the o-minimal structure $\langle \mathbb{Q}, +, < \rangle$, namely it is isomorphic as a group to $\langle [0,1)^n, +mod 1\rangle$, viewed inside of $\mathbb{Q}$. It was shown by Wilkie, [40], that there are nontrivial o-minimal expansions of this structure. Moreover, if G itself equals to the real points of $\langle [0,1)^n, +mod 1\rangle$, in the structure of the real field, then the torsion points of G inherit the ring operations and therefore the induced structure is unstable and undecidable. However, even in this case it is interesting to ask which definable subsets of $\mathbb{Q}^n$ can be obtained as the trace of a definable set in G.

11.3. The distribution of torsion points. Somewhat surprisingly for those of us who have worked on this problem, the solution of Pillay's Conjecture did not yield a much better understanding of the distribution of torsion points in a definably compact abelian G. Here are some conjectures on this matter:

CONJECTURE A. *If $X \subseteq G$ is generic then it contains a torsion point.*

CONJECTURE B. *If $G \setminus X$ is non-generic then X contains a torsion point.*

Clearly, (A) implies (B) and both imply the following result, recently proved in [26]:

THEOREM. *If* $\dim(G \setminus X) < \dim G$ *then X contains a torsion point.*

11.4. Other related work. In other work generated by Pillay's conjecture the precise relationship between G and G/G^{00} was investigated. In [2], Berarducci discusses the o-minimal spectrum $\tilde{G}$ of a definably compact G and proves that G/G^{00} is a topological quotient of $\tilde{G}$ (for more recent work on related topological issues, see [3, 5, 6], the last two are together with Mamino, and then with Otero). In [12], Edmundo, Jones and Peatfield examine the connections between the cohomology groups of G and of G/G^{00}. In her PhD thesis, [9], Conversano investigates G/G^{00} when G is not definably compact.

In [25], Onshuus and Pillay study the analogous conjecture in the p-adic setting and show cases where it fails and other cases where the conjecture holds.

In a recent result, Hrushovski, Pillay and the present author (see [19]), prove that every definably compact group G is elementarily equivalent, as a pure group, to G/G^{00}. A better understanding of the group theoretic structure of G can then be deduced, and in particular, one concludes that the commutator subgroup of G is definable and that G is the almost direct product of $Z(G)$ and $[G, G]$.

Finally, the recent paper of Hrushovski and Pillay [20] puts some of the notions which were examined in [18] in a very general context, and examines forking, stability and measure in several different abstract settings, mainly in groups under the assumptions of NIP and the existence of some measure. The machinery and results obtained there are very powerful and, as I now explain, yielded answers to most of the questions raised above.

11.5. The Compact Domination Conjecture and its recent solution. At first, it seems as if the most natural way to define measure on definable subsets of G would be directly through the map $\pi : G \to G/G^{00}$. Namely, to let $\mu(X)$

equal $m(\pi(X))$ (the Haar measure of $\pi(X)$) for any definable set $X \subseteq G$. However, a difficulty arises when one tries to prove finite additivity:

Take $X_1, X_2 \subseteq G$ two disjoint definable sets. Finite additivity should imply that the Haar measure of $\pi(X_1) \cap \pi(X_2)$ is zero (note that $\pi(X_1), \pi(X_2)$ need not be disjoint anymore). However, until very recently this remained an open question and, as we will soon see, it is equivalent to the Compact Domination Conjecture below.

We first introduce some notation: Given $X \subseteq G$, we let

$$B(X) = \{y \in G/G^{00} : \pi^{-1}(g) \cap X \neq \emptyset \,\&\, \pi^{-1}(g) \cap (G \setminus X) \neq \emptyset\}.$$

We say that G is *compactly dominated by G/G^{00}, via π, in a measure theoretic sense* if for every definable $X \subseteq G$, the Haar measure of $B(X)$ is zero. We say that G is *compactly dominated by G/G^{00}, via π, in a topological sense* if every such $B(X)$ is nowhere dense in G. (The term "compact domination" is modeled after the notion "stably dominated" referring to a situation where an unstable set is "controlled" by a stable one).

The Compact Domination Conjecture, formulated in [18] stated that every definably compact group in an o-minimal structure is compactly dominated (in both senses). In an earlier version of these notes several equivalences to the above conjectures were proved, implying for example that the measure theoretic conjecture implies the topological one. Both are known to imply Conjecture (A) above about the density of torsion points. However, the recent preprint of Hrushovski and Pillay [20] proves the Compact Domination Conjecture, and at the same time the torsion point and the uniform definability conjectures formulated above.

THEOREM 11.1. [20] *Every definably compact group is compactly dominated, in the measure theoretic* (*and hence topological*) *sense.*

Since the paper is new I will only try to very roughly sketch the ideas behind this solution:

First of all, using results from [19], the problem is reduced to abelian groups.

As in [18], the authors make use of the theorem of Baysalov and Poizat mentioned above. Namely, they consider an elementary extension $\mathcal{M}^*$ of $\mathcal{M}$, and for every $\mathcal{M}^*$-definable set X they add a predicate to $\mathcal{M}$ for the trace of X, on M^n. The main theorem in [1] (later generalized by Shelah to any theory with NIP), implies that this new structure eliminates quantifiers and in particular it is weakly o-minimal. We denote it by $\bar{\mathcal{M}}$.

The main difficulty is to prove that G^{00} is actually definable in the structure $\bar{\mathcal{M}}$. Hrushovski and Pillay do it after showing first that it can be written as the set theoretic stabilizer of any global generic type in G. It is here that they also show the uniform definability of G^{00}, which was conjectured above.

G/G^{00} is now a compact Lie group, given as a quotient of two definable sets in the weakly o-minimal structure $\bar{\mathcal{M}}$. After a fine analysis of the topological situation (and using the knowledge of the fundamental group of G), they prove that G/G^{00} is *semi o-minimal* in this weakly o-minimal structure. Namely it is in the definable closure of finitely many o-minimal structures, all definable in $\bar{\mathcal{M}}$. In particular, this settles the conjecture on G_{ind} mentioned above.

Once this machinery is established, definable subsets of G/G^{00} of Haar measure zero are just sets of smaller dimension (in the o-minimal sense). It is now not difficult to prove compact domination similarly to the simple cases handled in [18].

Acknowledgements. I thank M. Otero, E. Baro and A. Pillay for reading and commenting on an early version of this survey.

REFERENCES

[1] Y. Baisalov and B. Poizat, *Paires de structures o-minimales*, ***The Journal of Symbolic Logic***, vol. 63 (1998), no. 2, pp. 570–578.

[2] A. Berarducci, *O-minimal spectra, infinitesimal subgroups and cohomology*, ***The Journal of Symbolic Logic***, vol. 72 (2007), no. 4, pp. 1177–1193.

[3] ———, *Cohomology of groups in o-minimal structures: acyclicity of the infinitesimal subgroup*, to appear in ***The Journal of Symbolic Logic***.

[4] A. Berarducci and A. Fornasiero, *O-minimal cohomology: Finiteness and invariance results*, preprint, 2007.

[5] A. Berarducci and M. Mamino, *Equivariant homotopy of definable groups*, preprint, 2009.

[6] A. Berarducci, M. Mamino, and M. Otero, *Higher homotopy of groups definable in o-minimal structures*, to appear in ***Israel Journal of Mathematics***.

[7] A. Berarducci and M. Otero, *An additive measure in o-minimal expansions of fields*, ***The Quarterly Journal of Mathematics***, vol. 55 (2004), no. 4, pp. 411–419.

[8] A. Berarducci, M. Otero, Y. Peterzil, and A. Pillay, *A descending chain condition for groups definable in o-minimal structures*, ***Annals of Pure and Applied Logic***, vol. 134 (2005), no. 2-3, pp. 303–313.

[9] A. Conversano, ***On the Connectionss between Definable Groups in o-Minimal Structures and Real Lie Groups in the Noncompact Case***, Ph.D. thesis, U. Siena, 2009.

[10] A. Dolich, *Forking and independence in o-minimal theories*, ***The Journal of Symbolic Logic***, vol. 69 (2004), no. 1, pp. 215–240.

[11] M. Edmundo, G. Jones, and N. Peatfield, *O-minimal sheaf cohomology with supports*, preprint, 2006.

[12] ———, *Hurewicz theorems for definable groups, lie groups and their cohomologies*, preprint, 2007.

[13] M. Edmundo and M. Otero, *Definably compact abelian groups*, ***Journal of Mathematical Logic***, vol. 4 (2004), no. 2, pp. 163–180.

[14] M. Edmundo and G. Terzo, *On pillay's conjecture for orientable definable groups*, preprint, 2008.

[15] P. Eleftheriou, *Groups definable in linear o-minimal structures: the noncompact case*, to appear in ***The Journal of Symbolic Logic***.

[16] P. Eleftheriou and S. Starchenko, *Groups definable in oredered vector spaces over ordered division rings*, ***The Journal of Symbolic Logic***, vol. 72 (2007), no. 4, pp. 1108–1140.

[17] J. Gismatullin, *Model theoretic connected components of groups*, preprint, 2007.

[18] E. Hrushovski, Y. Peterzil, and A. Pillay, *Groups, measures, and the NIP*, ***Journal of the American Mathematical Society***, vol. 21 (2008), no. 2, pp. 563–596.

[19] ———, *On central extensions and definably compact groups in o-minimal structures*, preprint, 2008.

[20] E. Hrushovski and A. Pillay, *On nip and invariant measures*, preprint, 2007.

[21] H. J. Keisler, *Measures and forking*, ***Annals of Pure and Applied Logic***, vol. 34 (1987), no. 2, pp. 119–169.

[22] J. Maříková, *The structure on the real field generated by the standard part map on an O-minimal expansion of a real closed field*, ***Israel Journal of Mathematics***, vol. 171 (2009), pp. 175–195.

[23] ———, *O-minimal fields with standard part map*, preprint, 2009.

[24] A. Onshuus, *Groups definable in* $\langle \mathbb{F}, +, < \rangle$, preprint.

[25] A. Onshuus and A. Pillay, *Definable groups and compact p-adic Lie groups*, ***Journal of the London Mathematical Society. Second Series***, vol. 78 (2008), no. 1, pp. 233–247.

[26] M. Otero and Y. Peterzil, *G-linear sets and torsion points in definably compact groups*, ***Archive for Mathematical Logic***, vol. 48 (2009), no. 5, pp. 387–402.

[27] Y. Peterzil, *Returning to semibounded sets*, to appear in ***The Journal of Symbolic Logic***.

[28] Y. Peterzil and A. Pillay, *Generic sets in definably compact groups*, ***Fundamenta Mathematicae***, vol. 193 (2007), no. 2, pp. 153–170.

[29] Y. Peterzil, A. Pillay, and S. Starchenko, *Definably simple groups in o-minimal structures*, ***Transactions of the American Mathematical Society***, vol. 352 (2000), no. 10, pp. 4397–4419.

[30] ———, *Linear groups definable in o-minimal structures*, ***Journal of Algebra***, vol. 247 (2002), no. 1, pp. 1–23.

[31] Y. Peterzil and S. Starchenko, *Definable homomorphisms of abelian groups in o-minimal structures*, ***Annals of Pure and Applied Logic***, vol. 101 (2000), no. 1, pp. 1–27.

[32] Y. Peterzil and C. Steinhorn, *Definable compactness and definable subgroups of o-minimal groups*, ***Journal of the London Mathematical Society. Second Series***, vol. 59 (1999), no. 3, pp. 769–786.

[33] A. Pillay, *On groups and fields definable in o-minimal structures*, ***Journal of Pure and Applied Algebra***, vol. 53 (1988), no. 3, pp. 239–255.

[34] ———, *Type-definability, compact Lie groups, and o-minimality*, ***Journal of Mathematical Logic***, vol. 4 (2004), no. 2, pp. 147–162.

[35] S. Shelah, *Minimal bounded index subgroup for dependent theories*, ***Proceedings of the American Mathematical Society***, vol. 136 (2008), no. 3, pp. 1087–1091.

[36] ———, *Definable groups and 2-dependent theories*, prepprint.

[37] A. Strzebonski, *Euler characteristic in semialgebraic and other o-minimal groups*, ***Journal of Pure and Applied Algebra***, vol. 96 (1994), no. 2, pp. 173–201.

[38] Lou van den Dries, ***Tame Topology and o-Minimal Structures***, London Mathematical Society Lecture Note Series, vol. 248, Cambridge University Press, Cambridge, 1998.

[39] Lou van den Dries and J. Marikova, *Triangulation in o-minimal fields with standard part map*, preprint, 2009.

[40] A. J. Wilkie, *Fusing o-minimal structures*, ***The Journal of Symbolic Logic***, vol. 70 (2005), no. 1, pp. 271–281.

DEPARTMENT OF MATHEMATICS
UNIVERSITY OF HAIFA
HAIFA, ISRAEL
E-mail: kobi@math.haifa.ac.il

PROOF THEORY AND MEANING: ON THE CONTEXT OF DEDUCIBILITY

GREG RESTALL

Abstract. I examine Belnap's two criteria of *existence* and *uniqueness* for evaluating putative definitions of logical concepts in inference rules, by determining how they apply in four different examples: conjunction, the universal quantifier, the indefinite choice operator and the necessity in the modal logic S5. This illustrates the ways that definitions may be evaluated relative to a background theory of consequence, and the ways that different accounts of consequence provide us with different resources for making definitions.

The idea is compelling to many of us: The concept of conjunction is given fully in its rules of inference. Everything we need to know about the behaviour of "*and*" in the logicians' sense than is given by rules like these.

$$\frac{p \quad q}{p \wedge q}\,[\wedge I] \qquad \frac{p \wedge q}{p}\,[\wedge E_1] \qquad \frac{p \wedge q}{q}\,[\wedge E_2]$$

The virtues of this view are simple: When we learn inference rules, we learn how to *use* the connectives in deductive reasoning. One does not need to be a disciple of Wittgenstein to think that there is a connection between meaning and use, and in inference rules like these the connection with use is present to hand. The $[\wedge I]$ rule tells us how to *get* a conjunction, and the $[\wedge E]$ rules tell us what we can *do* with a conjunction when we have it. Inference rules speak to the matter of the *use* of the concept of conjunction without positing some "semantic value" to be the referent (or extension or whatever) of the concept.

Of course, there are very many issues with this way of looking at things. For example, while determinate and precise inference rules are available for many logical particles such as connectives, operators, quantifiers and the like, it is harder to see how one might present other meanings in the same fashion[1]. The issue I wish to focus on in this paper is not the general problem of giving an account of meaning in terms of rules of inference. Instead, I wish to concentrate

This paper is a draft, and comments from readers are very welcome. Please check the webpage `http://consequently.org/writing/ptm-context` for the latest version of the paper, to post comments and to read comments left by others.

[1]This does not mean that there aren't those who try. Inferentialists, such as Robert Brandom [5, 6] are committed to giving an account of meaning in terms of inferential propriety.

Logic Colloquium '07
Edited by Françoise Delon, Ulrich Kohlenbach, Penelope Maddy, and Frank Stephan
Lecture Notes in Logic, 35

on the more local issue of the propriety of defining logical concepts in this way. Can one present inference rules and take them to define a connective, an operator or a quantifier? Is such a definition always legitimate, or are potential definitions beholden to some higher court of appeal? If logical concepts are freely definable, then what are the boundaries of such freedom? More interestingly still, what are we *doing* when we define a logical concept in this way?

The most striking formulation of this issue is due to Arthur Prior, in his influential paper "The Runabout Inference Ticket" [19]. He asks us to consider the binary connective "tonk" given by these two inference rules.

$$\frac{p}{p \text{ tonk } q}\,[\text{tonk}I] \qquad \frac{p \text{ tonk } q}{q}\,[\text{tonk}E]$$

If we can freely define connectives with inference rules, we have defined something. However, using the concept tonk which we have apparently introduced in a definition, we may infer any conclusion from any premise:

$$\dfrac{\dfrac{p}{p \text{ tonk } q}\,[tonkI]}{q}\,[\text{tonk}E]$$

This is too much. Something has gone *wrong* in our definition. But what?

Responses to Prior's paper vary. One tradition is represented in J. T. Stevenson's "Roundabout the Runabout Inference-Ticket" [23]. Stevenson argues that tonk is unacceptable because the rules impose different truth-table conditions on tonk. According to [tonk$I$], p tonk q must be true when p is true. According to [tonk$E$], when q is false, p tonk q must be false. Since no connective can meet both of those conditions, there is no connective like tonk. There is more than one way of understanding Stevenson's criteria. The straightforward way is to take the possibility of a truth table definition for a connective like tonk as a criterion for its definability. This is to give the game (of *defining* the connective proof-theoretically) away. Truth tables are where the real action is.

This is not the approach I wish to consider. Instead, I wish to look again at the *other* kind of response to Prior's paper, that given by Nuel Belnap in his "Tonk, Plonk and Plink" [3]. For Belnap, criteria governing the acceptability of a definition can be given in proof theoretical terms. What makes tonk unacceptable is not some *categorical* feature of tonk, but in the interaction of tonk with the inference rules we already endorse.

> It seems to me that the key to a solution lies in observing that even on the synthetic view, we are not defining our connectives *ab initio*, but rather in terms of an *antecedently given context of deducibility*, concerning which we have some definite notions. By that I mean that before arriving at the problem of characterizing

> connectives, we have already made some assumptions about the nature of deducibility [3, page 131].

The point is that rules for tonk are given relative to a background set of commitments concerning inference. (Which may involve, for example, the commitment that the argument from p to q is invalid.) If the addition of the rules for tonk change our commitments concerning inference (we now have a proof from p to q), then we are forced with a choice. Either reject those rules, or modify our original assumptions. Both are options. If we *endorse* the antecedent theory — or what Belnap calls our *antecedently given context of deducibility*, then this gives us a standard by which we can judge putative definitions.

> We may now state the demand for the consistency of the definition of the new connective, *plonk*, as follows: the extension must be *conservative*; i.e., although the extension may well have new deducibility-statements, these new statements will all involve *plonk*. The extension will not have any new deducibility-statements which do not involve *plonk* itself. It will not lead to any deducibility-statement not containing *plonk*, unless that statement is already provable in the absence of the *plonk*-axioms and *plonk*-rules. The justification for unpacking the demand for consistency in terms of conservativeness is precisely our antecedent assumption that we already had all the universally valid deducibility-statements not involving any special connectives [3, page 132].

On these grounds, if the background account of deducibility included the commitment that some p did not entail some q, then, relative to that background, tonk fails the demand of consistency. This is *one* of the tests Belnap considers in the paper. In the case of a natural deduction proof theory or a sequent calculus, we can demonstrate that this criterion is met by means of a *normalisation proof* or a *cut elimination argument*, which usually has as a consequence a *subformula property*. These proof-theoretical results show that if one has a proof for some argument (or a derivable sequent) then we can find a special *normal* proof for that argument (or a *cut-free* derivation of that sequent) in which all formulas used in that proof are *sub-formulas* of the formulas in the premises or conclusion (or are subformulas of the formulas in the target sequent). This means that if the proof (or sequent) does not use our newly defined connective, then it has a normal proof (or cut-free derivation) not using the *rules* for our newly defined connective, for the rules for this connective must involve it somewhere, and the normal proof (cut-free derivation) does not feature this connective at all. It follows that if the argument is proved (sequent is derived) in the system *using* the new concept, then it can still be proved (derived) in the system *before* its introduction.

Belnap considered another criterion to add to the criterion of conservative extension.

> The mathematical analogy leads us to ask if we ought not also to add *uniqueness* as a requirement for connectives introduced by definitions in terms of deducibility (although clearly this requirement is not as essential as the first, or at least not in the same way). Suppose, for example, that I propose to define a connective *plonk* by specifying that $B \vdash A\text{-}plonk\text{-}B$. The extension is easily shown to be conservative, and we may, therefore, say "There is a connective having these properties". But is there only one? It seems rather odd to say we have defined *plonk* unless we can show that *A-plonk-B* is a function of A and B, i.e., given A and B there is only one proposition *A-plonk-B*. But what do we mean by uniqueness when operating from a synthetic, contextualist point of view? Clearly that at most one inferential role is permitted by the characterization of *plonk*; i.e., that there cannot be two connectives which share the characterization given to plonk but which otherwise sometimes play different roles. Formally put, uniqueness means that if exactly the same properties are ascribed to some other connective, say *plink*, then *A-plink-B* will play exactly the same role in inference as *A-plonk-B* both as premiss and as conclusion. [3, page 133]

The idea here is straightforward. Inference rules only truly *define* a connective if they go beyond *describing* some of its inferential properties, and go on to determine its behaviour in valid argument. This is the criterion of UNIQUENESS. Keeping on using Belnap's mathematical analogy, we will call his first criterion the EXISTENCE criterion. The antecedently given context of deducibility provides the space for the definition of some concept (it allows for its *existence*), if it may be conservatively extended to include it. The concept is defined *uniquely* if whenever we attempt to define it *twice*, the two concepts defined agree.

In the rest of this paper, I will use Belnap's criteria of existence and uniqueness to examine four different issues in the philosophy of logic, concerning a connective ($\wedge$), a quantifier ($\forall$), a variable binding term operator (ε) and a modal operator ($\Box$). Along the way we will learn more about connectives, quantifiers, variable binders and modal operators, and we will also hopefully learn more about what is involved in using inference rules to *define*.

§1. $\wedge$. If you were forced to pick an unproblematic binary connective, *conjunction* would be a smart choice. For the conditional, there is no critical consensus: some defend of the material conditional, others propose some kind of modal or intensional analysis, and yet others think that the conditional is not really a connective at all, but a complicated speech-act more like conditional assertion than anything else. Even *disjunction* has its own disputes—how are we to mark the difference (if there is a real difference) between inclusive and exclusive disjunction? If I use "or" in an exclusive way, does this *imply* or merely *presuppose* that the disjuncts are not both true? There are no disputes like *these* over conjunction. The rules $[\wedge I]$ and $[\wedge E]$ given on page 204 are not that complicated and are not in that much dispute. However, there is *some* dispute over the meaning of "*and*" and Belnap's two criteria give us a way to understand the kinds of discussions concerning it.

Existence. First, consider the *existence* problem. Do the rules $[\wedge I]$ and $[\wedge E]$ conservatively extend a proof system to which they are added? The answer would *seem* to be affirmative, since Prawitz has taught us that natural deduction systems for intuitionistic logic or for classical logic, featuring just these rules, have the *normalisation* property [18]. Any non-normal proof (a proof with adjacent steps where a particle is introduced only to be eliminated straight after) may be converted into a normal proof. This means, in the case of intuitionistic logic and many other logics, that the proof system has the *subformula* property for normal proofs. If there is a normal proof from premises X to conclusion A, then the proof will use only subformulas of the formulas in X or A. This means if we have an argument from X to A, where X and A do not feature "$\wedge$", then a *normal* proof from X to A will not feature the "$\wedge$" rules. If for every proof from X to A we can find some normal proof, this tells us that adding the rules for "$\wedge$" do not let us prove any inferences in the old vocabulary that we couldn't prove before. Conjunction is a conservative extension. It passes the existence test. Or does it?

Results in proof theory like these are notoriously sensitive to being formulated precisely. Normalisation results are no exception. As a matter of fact, the rules $[\wedge I]$ and $[\wedge E]$ do *not* conservatively extend all proof systems. Consider a very basic proof system, in the style of Gentzen [11, 12] or Prawitz [18]. Proofs are trees of formulas, with premises as *leaves* and conclusion as the *root*. The *identity* proof for a formula A is the simple tree

$$A$$

whose root is identical to its only leaf—the formula A itself. The rules for conjunction are as given on page 204. We can, for example, construct a proof whose premises are instances of $p \wedge (q \wedge r)$ and whose conclusion is $(p \wedge q) \wedge r$ by composing many $[\wedge I]$ and $[\wedge E]$ steps

$$\dfrac{\dfrac{\dfrac{p \wedge (q \wedge r)}{p}[\wedge E] \quad \dfrac{\dfrac{p \wedge (q \wedge r)}{q \wedge r}[\wedge E]}{q}[\wedge E]}{p \wedge q}[\wedge I] \quad \dfrac{\dfrac{p \wedge (q \wedge r)}{q \wedge r}[\wedge E]}{r}[\wedge E]}{(p \wedge q) \wedge r}[\wedge I]$$

The result is a proof system for lattice conjunction. However, in this little system, the *primitive* system (with identity as its only rule) is *not* conservatively extended by the conjunction rule, at least when you look at matters closely. The reasoning is straightforward. In the system with conjunction rules, we have a proof whose premises are p and q and whose conclusion is q.

$$\dfrac{\dfrac{p \quad q}{p \wedge q}[\wedge I]}{q}[\wedge E]$$

There is no such proof that avoids using the conjunction rules. In this system, the only proof ending in q that uses no connectives is the identity proof in which q is the sole premise and conclusion.

So, we have seen a way in which even rules such as $[\wedge I]$ and $[\wedge E]$ may fail Belnap's existence test. The case may seem *ad hoc*, but it is not as cooked up as it seems. If, for example, you have concerns for *relevance* [1, 2, 8, 21] in a consequence relation, then you may think that you do not *have* a proof from p together with q to the conclusion q. There may be a proof from q to q (the identity proof we have already mentioned), but bringing the extra premise p *alongside* the other premise does not necessarily involve it in the proof. If our antecedently given context of deducibility respects this strict criterion of relevance in this way, then the conjunction rules $[\wedge I]$ and $[\wedge E]$ are to be rejected as definitions — they fail Belnap's existence criterion[2].

Uniqueness. We may say something about uniqueness as an illustration of the other criterion. If we are happy with $[\wedge I]$ and $[\wedge E]$ as rules, and our system admits them because we were already committed to the validity of the argument from p, q to p, then we may ask if the rules merely *describe* conjunction or truly *define* it. So suppose that we introduce *two* conjunction connectives $\wedge$ and &, with the same form of rules. Are they equivalent?

We have the following proofs:

$$\dfrac{\dfrac{A \,\&\, B}{A}[\&E] \quad \dfrac{A \,\&\, B}{B}[\&E]}{A \wedge B}[\wedge I] \qquad \dfrac{\dfrac{A \wedge B}{A}[\wedge E] \quad \dfrac{A \wedge B}{B}[\wedge E]}{A \,\&\, B}[\&I]$$

So $\wedge$ is interchangeable with & as a premise or a conclusion in any argument. They are *equivalent*. If we have a point at which we require $A \wedge B$ as a premise, we may justify it on the basis of the assumption of $A \,\&\, B$, and vice versa.

[2]This happens in real-life natural deduction systems. In Lemmon's *Beginning Logic* [16], the only way to derive the classically valid inference from p to $q \supset p$ is *via* a conjunction such as $p \wedge q$. There is no way to *vacuously* discharge a hypothesis q that is not actually present as a premise, and the only way to involve it as a premise is to use another connective such as conjunction. So, either Lemmon's rules for $\supset$ are incomplete, or, in his system, the conjunction rules fail Belnap's test.

Similarly, if we wish to derive $A \wedge B$, we may derive $A \,\&\, B$ and deduce $A \wedge B$ from it, using these proofs[3].

So, we have uniqueness. Does this uniqueness result mean that any two connectives introduced with these rule have the same *meaning*? To decide this would take us too far away into general considerations in the theory of meaning. All we have seen so far is that $\wedge$ and & would play the same role in deductive inference. Perhaps this does not mean that they are fully interchangeable in every context in every respect. For example, you might think of $\wedge$ as "*and*" and & as "*but*." Perhaps these have the same *deductive* power. Perhaps it is always legitimate to infer *p butq* from the premises p and q, even if it is odd to *say* it in cases where q is not *surprising* given the assumption of p. If this is the case (and some theories of "but" distinguish it from "and" not in what it *says* but in what uses of it *presupposes*), then it may well be that "and" and "but" agree when it comes to inferential power and derivability, but they do not agree in every aspect of meaning. If this is the case, then inferential equivalence can define a connective like conjunction up to a point, but it cannot determine its meaning in any sense more precise than that.

§**2.** $\forall$. Now let us consider the universal quantifier. It has two relatively straightforward inference rules

$$\frac{A[c]}{(\forall x)A[x]}\,[\forall I] \qquad \frac{(\forall x)A[x]}{A[t]}\,[\forall E]$$

These rules are *relatively* straightforward, but they have complexities beyond the rules of a propositional logic. These rules appeal to the syntax of names and variables. In [$\forall E$], we may derive $A[t]$ (a sentence in which the singular term t may appear some number of times), from the premise $(\forall x)A[x]$, where $A[x]$ is found by replacing the instances of t in $A[t]$ by the variable x. The [$\forall E$] rule is simple. We infer from a universally quantified claim any of its *instances*.

The more complicated rule is the introduction rule [$\forall I$]. Here, we don't derive $(\forall x)A[x]$ from $A[c]$ by itself. We may only derive $(\forall x)A[x]$ if we have derived $A[c]$ in a special way. We can make this inference when we have derived $A[c]$ from premises *in which the name c does not occur*. The idea is straightforward. If we had any other object (call it d), then we could derive $A[d]$ in just the same way from just the same premises. If we presumed nothing "about c," then the proof of $A[c]$ is completely general and could have served

[3]There is, however, a little subtlety in doing this. In the Gentzen–Prawitz-style system I am using, we trade in the use of the premise $A \wedge B$ by *two* appeals to the premise $A \,\&\, B$. If our antecedent theory of deducibility distinguishes multiplicities of premises, then $A \,\&\, B$ and $A \wedge B$ need not be equivalent. For those keeping score of structural rules appealed to [21], for *existence* we used WEAKENING, and for *uniqueness*, we need CONTRACTION.

as well as a proof of $A[d]$, or of $A[x]$ for *any* object x. In other words, this gives us grounds to conclude $(\forall x)A[x]$.

Existence. The standard normalisation argument in natural deduction or cut elimination argument sequent systems show that the addition of these rules in well-behaved proof systems is conservative. Adding these quantifier rules in sensible proof theories is always *conservative*, we pass Belnap's existence test. The crucial step in a normalisation proof shows us that if we have a proof in which a universal quantifier is introduced only to be eliminated immediately at the next step.

$$\dfrac{\dfrac{\begin{array}{c}X\\ \vdots\,\pi\\ A[c]\end{array}}{(\forall x)A[x]}\,[\forall I]}{A[d]}\,[\forall E]$$

Here, to justify the appeal to $[\forall I]$, we know that c does not appear in the proof π in the premises X. This means that we may replace c by d in the proof π to get

$$\begin{array}{c}X\\ \vdots\,\pi^{c}_{d}\\ A[d]\end{array}$$

which is a proof of the same conclusion $A[d]$ from the same premises X, without going through the detour formula $(\forall x)A[x]$. The side condition on [∀I] is just what we need to prove existence.

Uniqueness. The curious feature here is uniqueness. Think of the kind of argument that would show that two universal quantifiers — let's say, $(\forall x)$ and (x) — are equivalent. The simplest such argument would go like this:

$$\dfrac{\dfrac{(x)A[x]}{A[c]}\,[(\,)E]}{(\forall x)A[x]}\,[\forall I] \qquad \dfrac{\dfrac{(\forall x)A[x]}{A[d]}\,[\forall E]}{(x)A[x]}\,[(\,)I]$$

These arguments succeed only under certain conditions. The first proof, from $(x)A$ to $(\forall x)A$, works only if the name c substituted for x in $[(\,)E]$ is also allowed to count as a term appropriate for substitution in the rule $[\forall I]$. The second proof, from $(\forall x)A$ to $(x)A$, works only if the name d substituted for x in $[\forall E]$ is also allowed to count as a term appropriate for substitution in the rule $[(\,)I]$. Only when the terms used in the rules for one quantifier agree with the terms used in the other can we manage to prove equivalence.

This restriction is important. Many sorted predicate logics may have different universal quantifiers corresponding to different domains. *Each* quantifier will satisfy rules like [$\forall I/\forall E$] without the quantifiers coinciding. The inferences from one quantifier to another would be blocked, because we cannot step from $(\forall x)A[x]$ to $A[c]$ to $(x)A[x]$. The problem is not that the inference steps from truth to falsity at any point — it is that the result is syntactically ill-formed. To make matters precise, suppose we have a language with two kinds of singular terms (for simplicity, let us say that they are disjoint — no singular term is of both kinds), a-terms and b-terms, and we correspondingly have a-variables and b-variables. The syntax of each n-place predicate is determined not only by its arity, but also by the kind of each place. There may be 1-place predicates which take a-terms and 1-place predicates which take b-terms. Call these $\langle a\rangle$-predicates and $\langle b\rangle$-predicates respectively. Similarly, an 2-place predicate may be an $\langle a, a\rangle$ predicate, an $\langle a, b\rangle$-predicate, a $\langle b, a\rangle$-predicate or a $\langle b, b\rangle$-predicate, and so on. The rules for forming complex formulas are as usual, but we have the following restrictions on quantifiers. If $A[c]$ is a sentence, in which an a-term c occurs, then $(\forall_a x)A[x]$ is a sentence, where $A[x]$ is the result of replacing the a-term c in $A[c]$ by the a-variable x. Similarly, if d is a b-term, then $(\forall_b y)A[d]$ is a sentence, in which we quantify into b-term position. Now the inference from $(\forall_a x)Fx$ to $(\forall_b y)Fy$ will fail, not because the premise is true and the conclusion is false, but if the premise $(\forall_a x)Fx$ is syntactically well-formed, then the conclusion $(\forall_b y)Fy$ is not. For $(\forall_a x)Fx$ to make sense, F must be an $\langle a\rangle$-predicate. For $(\forall_b y)Fy$ to make sense, F must be a $\langle b\rangle$-predicate. In the language under discussion, nothing is both.

So what, then, of Belnap's criteria? Do quantifiers pass the uniqueness test? Does the possibility of having two different universal quantifiers mean that we should banish $\forall$ and $\exists$ from the canon of logical constants? Surely such a conclusion is too extreme. However, we must acknowledge that in the face of considerations like this, the choice of quantifiers as logical constants is *relative* to the syntax of the language under consideration. Once we identify a category of singular terms (of names or variables or whatever), then *relative to this choice*, the quantifiers (for that category) are logical constants. Existence and uniqueness proofs work, and the meanings of the quantifiers are fixed.

This means, of course, that the choice of syntax has a burden to carry. In some cases — such as mathematical reasoning — the task does not seem too onerous to discharge. In the language of set theory, for example, it is not too hard to discern what singular terms are in use[4].

For inferentialists, who wish to use rules like these to determine meanings, the identification of the category of singular terms is a matter of pressing importance. For example, Brandom devotes many pages of *Making it Explicit* [5]

[4]Though it is not without its share of controversy. Are class terms able to be paraphrased away without appeal to singular terms for classes? If so, well and good, ZFC and theories without classes seem sufficient. If not, then perhaps NBG and theories with classes, or more radical views are appropriate [9, 17].

to the issue of determining the characteristics of singular terms. Figuring out *what* these are, and what reason we might have for a single category of them—rather than multiple disjoint classes—is a pressing project for those who take first-order quantification to tell us something about *ontology* [20]. Getting the category right tells us something about the significance of taking $\forall$ as a logical constant and something of the significance of taking it to be defined by its inference rules. This attempt at a definition only succeeds *given* the analysis of singular terms of the language.

§3. ε. A more controversial concept in the vicinity of first-order predicate logic is Hilbert's "ε", the variable-binding term operator of indefinite choice. The syntax is relatively straightforward. If $A[t]$ is a sentence in which term t appears, then $\varepsilon x A[x]$ is a singular term, where $A[x]$ is given by replacing the term t in $A[t]$ by x. The intended interpretation is that if there is some x such that $A[x]$, then $\varepsilon x A[x]$ denotes some such object. (If there is no such x, then the denotation of $\varepsilon x A[x]$ is not so important.) With this understanding of ε, the universal and existential quantifiers may be *defined*. We have

$$(\exists x)A[x] \;=_{\mathrm{df}}\; A[\varepsilon x A[x]]$$
$$(\forall x)A[x] \;=_{\mathrm{df}}\; A[\varepsilon x \neg A[x]]$$

Choice is useful for a number of different reasons. Hilbert's original motivation was to do without the quantifiers. You may also be interested in choice in order to more straightforwardly model natural deduction reasoning—it is quite tempting to be able to infer from $(\exists x)A[x]$ the conclusion $A[t]$ for some term t (many students prefer to make a step like this than to use the Official $[\exists E]$ Rule from their natural deduction system). Choosing $\varepsilon x A[x]$ for t makes this inference valid. From $(\exists x)A[x]$ we *may* infer $A[\varepsilon x A[x]]$[5] Another use for ε may be found in formulating and defending choice principles in mathematical theories. If the *syntax* allows us to formulate arbitrary choices, then we have scope for "proving" the axiom of choice in some set theory on the basis of more fundamental axioms or principles [7]. The general strategy is clear: if we need to show that!

$$(\forall x)(\exists y)Fxy \supset (\exists f)(\forall x)Fxf(x)$$

— if f is to be a choice function, where $f(x)$ is a choice of a representative y where Fxy, then taking $f(x)$ to be $\varepsilon y Fxy$ has the desired behaviour, since we may prove

$$(\forall x)(\exists y)Fxy \supset (\forall x)Fx(\varepsilon y Fxy)$$

[5]However, this matter is subtle [13]. It is especially difficult in the presence of intensional operators such as $\Box$ and $\Diamond$. If $\Diamond(\exists x)A[x]$, we may infer $\Diamond A[\varepsilon x A[x]]$. It looks to all the world that we may infer that there is an object c (the denotation of the term $\varepsilon x A[x]$ in $\Diamond A[\varepsilon x A[x]]$) such that $\Diamond A[c]$, in other words, $(\exists x)\Diamond A[x]$. But isn't the inference from $\Diamond(\exists x)A[x]$ to $(\exists x)\Diamond A[x]$ invalid?

So, with these thoughts in mind, it is appropriate to ask the question of what kind of concept ε turns out to be, to better understand what is involved in admitting it into our vocabulary. As with quantification, issues arise with *uniqueness* more than with *existence*. There is no doubt that ε may be conservatively extended to the language of first-order logic — at least if we admit the axiom of choice into our metatheory. Any model of the language of first order logic may be extended with a choice function selecting from the extension of a formula $A[x]$ with one free variable, an object to be the denotation of $\varepsilon x A[x]$. This kind of model not only validates the conditions on ε we have already seen, but also stronger conditions such as "left uniqueness"

$$(\forall x)(A[x] \equiv B[x]) \supset \varepsilon x A[x] = \varepsilon x B[x]$$

to the effect that if A and B are coextensive predicates, then the choice for A is identical to the choice for B[6]. So, adding choice, with quite strong principles, is conservative over a base logic like classical predicate logic. Any troubles with ε do not arise over questions of *existence*. Uniqueness is another matter.

Uniqueness. Suppose that we add two ε operators to the language. We do not need to fix on a particular collection of inference rules for ε, but a plausible pair of rules for ε are these:

$$\frac{A[c]}{A[\varepsilon x A[x]]}\,[\varepsilon I] \qquad \frac{A[\varepsilon x A[x]] \quad \begin{array}{c}[A[c]] \\ \vdots\,\pi \\ C\end{array}}{C}\,[\varepsilon E] \qquad (c \text{ occurs in no other premise in } \pi.)$$

These rules together ensure that ε makes $\exists x A[x]$ equivalent to $A[\varepsilon x A[x]]$. However, they are not enough to assure uniqueness. They may both be satisfied by wildly different choice functions. For example, if we have a model of predicate logic, we may interpret a choice function by selecting a member of every non-empty subset of the domain. Such choices may be made in very many different ways and, and so, we may have very many different interpretations of the choice function. For example, take a model of the language of arithmetic modulo 3, with the domain $\{0, 1, 2\}$, the arithmetic vocabulary of 0, 1, $+$, $\times$ with the interpretation of of arithmetic mod three. Take two choice functions, interpreting the two choice operators ε and ε'. For $\varepsilon x A[x]$, we choose the *smallest*[7] x satisfying the property $A[x]$, if there is one, and 0 if there is not. For $\varepsilon' x A[x]$, we choose the *largest*[8] x satisfying $A[x]$, if there is one, and 2 if there is not. We can see under this interpretation that the rules $[\varepsilon I]$ and $[\varepsilon E]$ are satisfied for both ε and ε', and we have left uniqueness too. Are ε and ε' in any way equivalent? Can we substitute ε for ε'?

---

[6]For a recent discussion of left uniqueness and other conditions on ε, see Claus-Peter Wirth's interesting recent paper [24].

[7]Smallest under the ordering $0 < 1 < 2$, that is.

[8]Largest under the same ordering: $0 < 1 < 2$.

No, we cannot. In this interpretation, $\varepsilon x(x = x)$ denotes 0, and $\varepsilon' x(x = x)$ denotes 2. So, in this interpretation, we have

$$\varepsilon x(x = x) \neq \varepsilon' x(x = x)$$

but we *do* have

$$\varepsilon x(x = x) = \varepsilon x(x = x) \quad \text{and} \quad \varepsilon' x(x = x) = \varepsilon' x(x = x)$$

So, ε is in no way equivalent to ε'. If *I* learn to use ε by way of inference rules like $[\varepsilon I]$ and $[\varepsilon E]$ (and left uniqueness and perhaps other conditions also satisfied in our models), and you do too, then there is no guarantee that our use of the concept will agree. I could "mean" ε and you could "mean" ε'. Our usage will not converge on the one interpretation. Even given the choices we made to fix the logical vocabulary of singular terms, in order to give uniqueness for the quantifiers, we cannot assume that we have uniqueness for choice. Choice does not pass the uniqueness test, and it cannot be thought to be a logical constant in the same manner as the connectives and the quantifiers, unless we either set aside the criterion of uniqueness (so our inference principles do not need to determine the interpretation of the concept), or we find some other concept upon which to ground the interpretation of ε.

Our last example will be another case in which uniqueness is at risk, and in which we may perhaps find the resources to restore uniqueness.

§4. $\Box$. Consider modality: in particular, consider a normal modal operator $\Box$ — necessity — interpreted as satisfying the simple modal logic S5 [4, 14, 15]. As is well known, there are many different models for the modal logic S5. If we think of Kripke models in which formulas are evaluated at "worlds", then given any equivalence relation $\approx$, if we take $\Box A$ to hold at world w iff A holds at every world $v \approx w$, then the result is a model of the logic S5. A *universal* model is one in which the equivalence relation $\approx$ on the domain W of worlds is the total relation: $w \approx v$ always. So, in a universal model, $\Box A$ is true at some world if and only if A is true at *every* world.

Now, it appears that these simple facts show that $\Box$ is not a true logical constant in our strict sense. We may interpret $\Box_1$ and $\Box_2$, both satisfying the rules of the logic S5, with different equivalence relations $\approx_1$ and $\approx_2$ on the one model. The *logic* S5 does not force a unique interpretation on our two operators, and there is no way to force $\Box_1$ and $\Box_2$ to be equivalent, from the assumption that they are both S5 necessity operators. To put it starkly, the inference rules for $\Box$ according to S5 *describe* the behaviour of $\Box$ but they do not define it.

We could leave matters there and feel satisfied that we have a neat answer, which puts $\wedge$ and $\forall$ on one side of the fence (logical constants, fully defined by their inference rules) and ε and $\Box$ on the other (requiring more to define them that their inferential properties alone). Such a position is consonant with the main tradition in contemporary modal logic, where we admit that there are *many* operators satisfying the rules of this or that modal logic, and there are many ways that different modal operators might interact[9]. However, ending here would obscure one important point. The case of $\Box_1$ and $\Box_2$ is analogous to the case of the universal quantifier in a multi-sorted vocabulary. We could only prove uniqueness for a universal quantifier relative to a particular choice of syntax — a choice of singular terms and variables. If the analogy with quantifiers holds, perhaps we can make a similar choice in the case of modality. In this modal case, the choice is not so much in the syntax of formulas themselves (in propositional modal logic we have no terms denoting worlds or any other modal "ontology" or "ideology") it must be elsewhere. One possible place to find such a choice of syntax is found in a *proof theory* for the modal logic S5 I introduced at *Logic Colloquium* in 2005 [22]. Here, we extend Gentzen's sequent calculus to deal with *hypersequents* (multisets of sequents), where just as the sequent

$$X \vdash Y$$

tells us that asserting each sentence in X and denying each sentence in Y is ruled out as inconsistent on logical grounds, in the *hypersequent* appropriate for modal reasoning considering hypothetical situations, or reasoning with different dialogical *contexts*,

$$X_1 \vdash Y_1 \mid X_2 \vdash Y_2 \mid \cdots \mid X_n \vdash Y_n$$

tells us that asserting X_1 and denying Y_1 in one context, together with asserting X_2 and denying Y_2 in another context ... and asserting X_n and denying Y_n in another is inconsistent on logical grounds. The general principle is best illustrated by a derivation of a principle of S5, to infer $\Box\neg\Box\neg A$ from A.

$$\dfrac{\dfrac{\dfrac{\dfrac{\dfrac{A \vdash A}{A, \neg A \vdash}[\neg L]}{A \vdash \mid \Box\neg A \vdash}[\Box L]}{A \vdash \mid \vdash \neg\Box\neg A}[\neg R]}{A \vdash \mid \vdash \Box\neg\Box\neg A}[\Box R]}{A \vdash \Box\neg\Box\neg A}[merge]$$

In the second line we have $A, \neg A \vdash$ — asserting both A and $\neg A$ is ruled out. At the third line, we move to $A \vdash \mid \Box\neg A \vdash$, which tells us that if we assert A on one context, and in another assert $\Box\neg A$, this is also ruled out. This is what makes $\Box$ *modal*. Taking $\neg A$ to be *necessary* is inconsistent with supposing (in some other "possibility") that it is the case that A. The next line moves

[9]The canonical text on mutimodal logics is Gabbay, Kurucz, Wolter and Zakharyashev's *Many-Dimensional modal logics* [10].

us from the assertion of $\Box\neg A$ to the denial of $\neg\Box\neg A$, and then after that, if asserting A (in some context) is inconsistent with denying $\neg\Box\neg A$ (in another context), then it is also inconsistent with denying $\Box\neg\Box\neg A$, because *denying* $\Box B$ in general has the same import as denying B in some other arbitrary context. Finally, if it is inconsistent to assert A in some context and deny $\Box\neg\Box\neg A$ in another, it is just as bad to assert A and deny $\Box\neg\Box\neg A$ in the very same context, and hence, we have $A \vdash \Box\neg\Box\neg A$, the conclusion.

The rules for $\Box$ used here take this form:

$$\frac{X, A \vdash Y \mid \Delta}{\Box A \vdash \mid X \vdash Y \mid \Delta}[\Box L] \qquad \frac{\vdash A \mid \Delta}{\vdash \Box A \mid \Delta}[\Box R]$$

In the earlier paper I show that this hypersequent calculus admits of a cut elimination proof (and hence, we can show *existence*), and that it is indeed a system sound and complete for S5. What is more interesting from our point of view is the fact that we can prove *uniqueness* too.

Uniqueness. If $\Box_1$ and $\Box_2$ use the same sequent rules, we may prove equivalence of $\Box_1$ and $\Box_2$ as follows:

$$\cfrac{\cfrac{\cfrac{A \vdash A}{\Box_1 A \vdash \mid \vdash A}[\Box_1 L]}{\Box_1 A \vdash \mid \vdash \Box_2 A}[\Box_2 R]}{\Box_1 A \vdash \Box_2 A}[merge] \qquad \cfrac{\cfrac{\cfrac{A \vdash A}{\Box_2 A \vdash \mid \vdash A}[\Box_2 L]}{\Box_2 A \vdash \mid \vdash \Box_1 A}[\Box_1 R]}{\Box_2 A \vdash \Box_1 A}[merge]$$

This is not a difficult result. The technique is identical to that used when we proved that two universal quantifiers are equivalent. In that case, we relied upon the fact that the quantifiers trade on the one syntactic category of names and variables. In this case, we rely upon the fact that the two modal operators trade upon the one hypersequent structure. In the case of the quantifiers, uniqueness may be defended on the basis of the uniqueness of syntactic category of names and variables in the language under consideration. If the analogy is to be made out, the same thing may be said for modality, except (if the interpretation of the hypersequent proof theory I hastily sketched is accepted) the defence is not made on the basis of any syntactic feature of the *sentences* of the language in question, but rather, on the way we combine them in dialogue. The modal operator $\Box$ gains its interpretation, on this view, on the basis of the dialogical shifts from context to context made in modal reasoning. On this view, any stability we have in our understanding of the concept of necessity and possibility may be found not in the shared access we have to the one domain of possible worlds (how could we have access to *that*, except *via* or modal reasoning?) but in the way we govern shifts in dialogue when we together consider *what would happen if* ...

Belnap's criteria of existence and uniqueness were offered as a way to draw a principled boundary that would keep logical operators on one side of the fence, while keeping anomalous definitions like that of tonk on the other side. These criteria are fruitful not only in the case of barring monsters like tonk, but they also have something to teach us in the more prosaic cases of $\wedge$, $\forall$, ε and $\Box$.

Acknowledgments. Thanks to Allen Hazen, Lloyd Humberstone, Penelope Maddy, Albert Visser, Alasdair Urquhart, Heinrich Wansing and audiences at the University of Melbourne Logic Group, Logica 2007, Logic Colloquium 2007 for discussion on these topics. This research is supported by the Australian Research Council, through grant DP0343388, and Ethel's self-titled album.

REFERENCES

[1] ALAN ROSS ANDERSON and NUEL D. BELNAP, JR., ***Entailment: The Logic of Relevance and Necessity, Volume I***, Princeton University Press, Princeton, N. J., 1975, With contributions by J. Michael Dunn and Robert K. Meyer, and further contributions by John R. Chidgey, J. Alberto Coffa, Dorothy L. Grover, Bas van Fraassen, Hugues LeBlanc, Storrs McCall, Zane Parks, Garrel Pottinger, Richard Routley, Alasdair Urquhart and Robert G. Wolf.

[2] ALAN ROSS ANDERSON, NUEL D. BELNAP, JR., and J. MICHAEL DUNN, ***Entailment. The Logic of Relevance and Necessity. Volume II***, Princeton University Press, Princeton, NJ, 1992, With contributions by Kit Fine, Alasdair Urquhart et al, Includes a bibliography of entailment by Robert G. Wolf.

[3] NUEL D. BELNAP, *Tonk, Plonk and Plink*, ***Analysis***, vol. 22 (1962), pp. 130–134.

[4] PATRICK BLACKBURN, MAARTEN DE RIJKE, and YDE VENEMA, ***Modal Logic***, Cambridge Tracts in Theoretical Computer Science, vol. 53, Cambridge University Press, Cambridge, 2001.

[5] ROBERT B. BRANDOM, ***Making it Explicit***, Harvard University Press, Cambridge, Massachusetts, 1994.

[6] ———, ***Articulating Reasons: An Introduction to Inferentialism***, Harvard University Press, Cambridge, Massachusetts, 2000.

[7] JOHN P. BURGESS, *E pluribus unum: plural logic and set theory*, ***Philosophia Mathematica. Philosophy of Mathematics, its Learning, and its Application. Series III***, vol. 12 (2004), no. 3, pp. 193–221.

[8] J. MICHAEL DUNN and GREG RESTALL, *Relevance logic*, ***Handbook of Philosophical Logic*** (Dov M. Gabbay, editor), vol. 6, Kluwer Academic Publishers, London, UK, second ed., 2002, pp. 1–136.

[9] T. E. FORSTER, ***Set Theory with a Universal Set***, second ed., Oxford Logic Guides, vol. 31, The Clarendon Press Oxford University Press, New York, 1995, Exploring an untyped universe, Oxford Science Publications.

[10] D. M. GABBAY, A. KURUCZ, F. WOLTER, and M. ZAKHARYASCHEV, ***Many-Dimensional Modal Logics: Theory and Applications***, Studies in Logic and the Foundations of Mathematics, vol. 148, North-Holland, Amsterdam, 2003.

[11] GERHARD GENTZEN, *Untersuchungen über das logische Schließen. I*, ***Mathematische Zeitschrift***, vol. 39 (1935), no. 1, pp. 176–210.

[12] ———, *Untersuchungen über das logische Schließen. II*, ***Mathematische Zeitschrift***, vol. 39 (1935), no. 1, pp. 405–431.

[13] ALLEN HAZEN, *Natural deduction and Hilbert's ε-operator*, ***Journal of Philosophical Logic***, vol. 16 (1987), no. 4, pp. 411–421.

[14] G. E. HUGHES and M. J. CRESSWELL, ***An Introduction to Modal Logic***, Methuen, London, 1968.

[15] ———, ***A New Introduction to Modal Logic***, Routledge, London, 1996.

[16] E. J. LEMMON, ***Beginning Logic***, revised ed., Hackett, Indianapolis, 1978, Edited by George W. D. Berry.

[17] PENELOPE MADDY, *Proper classes*, ***The Journal of Symbolic Logic***, vol. 48 (1983), no. 1, pp. 113–139.

[18] DAG PRAWITZ, ***Natural Deduction. A Proof-Theoretical Study***, Acta Universitatis Stockholmiensis. Stockholm Studies in Philosophy, vol. 3, Almqvist and Wiksell, Stockholm, 1965.

[19] ARTHUR N. PRIOR, *The runabout inference-ticket*, ***Analysis***, vol. 21 (1960), no. 2, pp. 38–39.

[20] W. V. O. QUINE, *On What There Is*, ***Review of Metaphysics***, vol. 2 (1948), no. 5, pp. 21–38.

[21] GREG RESTALL, ***An Introduction to Substructural Logics***, Routledge, London, UK, 2000.

[22] ———, *Proofnets for S5: sequents and circuits for modal logic*, ***Logic Colloquium 2005*** (Costas Dimitracopoulos, Ludomir Newelski, and Dag Normann, editors), Lecture Notes in Logic, vol. 28, ASL, Urbana, IL, 2008, pp. 151–172.

[23] J. T. STEVENSON, *Roundabout the Runabout Inference-Ticket*, ***Analysis***, vol. 21 (1961), no. 6, pp. 124–128.

[24] CLAUS-PETER WIRTH, *Hilbert's epsilon as an operator of indefinite committed choice*, ***Journal of Applied Logic***, vol. 6 (2008), no. 3, pp. 287–317.

SCHOOL OF PHILOSOPHY, ANTHROPOLOGY AND SOCIAL INQUIRY
THE UNIVERSITY OF MELBOURNE
PARKVILLE 3010
AUSTRALIA
E-mail: restall@unimelb.edu.au

BOUNDED SUPER REAL CLOSED RINGS

MARCUS TRESSL

§1. Introduction. This note is a complement to the paper [Tr2], where super real closed rings are introduced and studied. A super real closed ring A is a commutative unital ring A together with an operation $F_A : A^n \longrightarrow A$ for every continuous map $F : \mathbb{R}^n \longrightarrow \mathbb{R}$, $n \in \mathbb{N}$, so that all term equalities between the F's remain valid for the F_A's. For example if $C(X)$ is the ring of real valued continuous functions on a topological space X, then $C(X)$ carries a natural super real closed ring structure, where $F_{C(X)}$ is composition with F. Super real closed rings provide a natural framework for the algebra and model theory of rings of continuous functions.

A *bounded* super real closed ring A is a commutative unital ring A together with an operation $F_A : A^n \longrightarrow A$ for every *bounded* continuous map $F : \mathbb{R}^n \longrightarrow \mathbb{R}$, $n \in \mathbb{N}$, so that all term equalities between the F's remain valid for the F_A's (cf. 2.7 below).

In particular every super real closed ring is a bounded super real closed ring by forgetting the operation of the unbounded functions. An example of a bounded super real closed ring, which is not a super real closed ring, is the ring $C^{\text{pol}}(\mathbb{R}^n)$ of all polynomially bounded continuous functions $\mathbb{R}^n \longrightarrow \mathbb{R}$.

We show that

- bounded super real closed rings are precisely the classical localizations of super real closed rings (cf. 3.6).
- bounded super real closed rings are precisely the convex subrings of super real closed rings (cf. 4.6).
- there is an idempotent mono-reflector $A \mapsto \hat{A}$ from the category of bounded super real closed rings to the category of super real closed rings (cf. 5.12). This means that every bounded super real closed ring A has a super real closed hull $\hat{A}$, $\hat{A}$ is minimal and uniquely determined up to a unique A-isomorphism. For example $\widehat{C^{\text{pol}}(\mathbb{R}^n)} = C(\mathbb{R}^n)$
- Inside every bounded super real closed ring A there is a largest super real closed ring A^{Υ} (cf. 6.2). For example $(C^{\text{pol}}(\mathbb{R}^n))^{\Upsilon} = C^*(\mathbb{R}^n)$ (=the ring of bounded continuous functions $\mathbb{R}^n \longrightarrow \mathbb{R}$).

2000 *Mathematics Subject Classification.* Primary: 03C60, Secondary: 46E25, 54C05, 03E15.
Key words and phrases. real closed rings, rings of continuous functions, model theory.

Logic Colloquium '07
Edited by Françoise Delon, Ulrich Kohlenbach, Penelope Maddy, and Frank Stephan
Lecture Notes in Logic, 35

The motivation for writing this note is as follows: If A is a ring of continuous functions say $A = C(\mathbb{R}^n)$, then the Zariski sheaf Spec A is in general not a sheaf of rings of continuous functions. It is indeed *not* a sheaf of super real closed rings in a natural way. On the other hand by the localization theorem 3.6 below, Spec A is always a sheaf of bounded super real closed rings and this is also true if we start with a bounded super real closed ring A.

Hence it is desirable to extend the commutative algebra of super real closed rings to the bounded case. In order not to repeat arguments, we develop tools which allow the use of the reflector $A \mapsto \hat{A}$ to explain what's going on in A. For example, forming residue rings and classical localizations behave well with respect to the reflector, cf. 5.11 and 5.13. The close relation of the ideals of A and $\hat{A}$ is worked out in 5.8 and in 5.9.

The results mentioned above can be used to transfer most of the commutative algebra, developed in [Tr2] Sections 11, 12 and 13 (with the appropriate adaptions) to bounded super real closed rings. It would be tedious to elaborate this here, instead we present instruments which allow such a transfer easily, whenever it is needed in subsequent work.

The results in Section 6 are not of this instrumental style. As stated above, we prove the existence of a largest super real closed ring *inside* every bounded super real closed ring and we state two explicit descriptions of this ring.

§2. Definition of bounded super real closed rings. We shall make use of the theory of *real closed ring* introduced by N. Schwartz (cf. [Schw]). However we will use it in the way explained in [Tr2], Section 2. We recall this briefly. Let C_n be the set of all continuous maps $\mathbb{R}^n \to \mathbb{R}$ which are 0-definable in the field $\mathbb{R}$; in other words whose graph is a boolean combination of subsets of $\mathbb{R}^n \times \mathbb{R}$ defined by polynomial inequalities $P(\bar{x}, y) \geq 0$ with $P(\bar{x}, y) \in \mathbb{Z}[\bar{x}, y]$. A ring A is real closed if there is a collection of functions $(f_A : A^n \longrightarrow A \mid n \in \mathbb{N}, f \in C_n)$, such that

(1) If $f \in C_n$ is constant 0 or constant 1, then f_A is constant 0 or constant 1; if $f : \mathbb{R} \longrightarrow \mathbb{R}$ is the identity, then $f_A : A \longrightarrow A$ is the identity; if $f : \mathbb{R}^2 \longrightarrow \mathbb{R}$ is addition or multiplication in $\mathbb{R}$, respectively, then $f_A : A^2 \longrightarrow A$ is addition or multiplication in A, respectively.
(2) If $f \in C_n$, $k \in \mathbb{N}$ and $f_i \in C_k$ $(1 \leq i \leq n)$, then

$$[f \circ (f_1, \ldots, f_n)]_A = f_A \circ (f_{1,A}, \ldots, f_{n,A}).$$

2.1. FACT. *Every real closed ring is reduced* ([Tr2], (2.2)) *and*

(i) *For every real closed ring there is a unique collection of functions as in the definition above. This is* [Tr2, (2.13)], *where the functions f_A are explicitly constructed from the pure ring A.*
(ii) *Every ring-homomorphism $A \to B$ between real closed rings respects the new functions f_A and f_B* ([Tr2], (2.16)).

⊣

Because of 2.1 we may identify a real closed ring with its underlying pure ring.

2.2. Fact. *Let A be a real closed ring. The relation $f \leq g \iff \exists h \in A : g - f = h^2$ defines a partial order on A and A together with $\leq$ is a lattice ordered ring. The supremum of f and $-f$ is denoted by $|f|$. A subring B of A is convex if $f \leq h \leq g$ and $f, g \in B$ implies $h \in B$. By* [Tr2, (10.5)], *we have*

(i) *There is a smallest convex subring* $\mathrm{Hol}(A)$ *of A, called the holomorphy ring, namely*

$$\mathrm{Hol}(A) = \{f \in A \mid |f| \leq N \textit{ for some } N \in \mathbb{N}\}.$$

(ii) *The convex subrings of A are precisely the subrings of A containing* $\mathrm{Hol}\, A$ *and all these subrings are real closed.*

(iii) *If B is a convex subring of A, then $A = S^{-1} \cdot B$ is the localization of B at $S = A^{\times} \cap B$.*

(iv) *There is a largest real closed ring C having A as a convex subring. C is called the convex closure of A (cf.* [Tr2, (11.2)]*). If $C = A$, then A is called convexly closed. For example, real closed fields are convexly closed.*

⊣

For a topological space X let $C(X)$, $C^*(X)$ denote the continuous functions, the bounded continuous functions $X \longrightarrow \mathbb{R}$, respectively.

2.3. Definition. A function $F : \mathbb{R}^n \longrightarrow \mathbb{R}$ is called *polynomially bounded* if there is some polynomial $P \in \mathbb{R}[X_1, \dots, X_n]$ with

$$|F(x)| \leq P(x) \ (x \in \mathbb{R}^n).$$

Let $C^{\mathrm{pol}}(\mathbb{R}^n)$ be the ring of polynomially bounded continuous functions $\mathbb{R}^n \longrightarrow \mathbb{R}$.

2.4. Observation. *Since every polynomial from $\mathbb{R}[X_1, \dots, X_n]$ is bounded by a power of the polynomial $2 + X_1^2 + \dots + X_n^2$, a function $F : \mathbb{R}^n \to \mathbb{R}$ is polynomially bounded if and only if there is some $p \in \mathbb{N}$ such that*

$$\frac{|F(x)|}{(2 + x_1^2 + \dots + x_n^2)^p} \leq 1 \ (x \in \mathbb{R}^n).$$

2.5. Proposition. (i) *$C^{\mathrm{pol}}(\mathbb{R}^n)$ is a convex subring of $C(\mathbb{R}^n)$*

(ii) $C^{\mathrm{pol}}(\mathbb{R}^n) = C^*(\mathbb{R}^n)[x_1, \dots, x_n] = C^*(\mathbb{R}^n)[2 + x_1^2 + \dots + x_n^2]$

(iii) $C^{\mathrm{pol}}(\mathbb{R}^n) = S^{-1} \cdot C^*(\mathbb{R}^n)$, *where*

$$S = \{F \in C^*(\mathbb{R}^n) \mid \textit{there is } Q \in \mathbb{R}[X_1, \dots, X_n] \textit{ with } F \cdot Q \geq 1 \textit{ on } \mathbb{R}^n\}.$$

Proof. This is obvious from 2.4. ⊣

2.6. DEFINITION. Recall from [Tr1, 5.1] the following notation:

$$\Upsilon := \{s : \mathbb{R} \longrightarrow \mathbb{R} \mid s \text{ is continuous and } s^{-1}(0) = \{0\}\}.$$

We define

$$\Upsilon^{\text{pol}} := \Upsilon \cap C^{\text{pol}}(\mathbb{R}).$$

2.7. DEFINITION. (a) Let $\mathcal{L}_{\Upsilon^{\text{pol}}}$ be the first order language extending the language $\{+, -, \cdot, 0, 1\}$ of rings, which has in addition an n-ary function symbol $\underline{F}$ for every polynomially bounded continuous function $F : \mathbb{R}^n \longrightarrow \mathbb{R}$ and every $n \in \mathbb{N}_0$.

(b) Let $T_{\Upsilon^{\text{pol}}}$ be the $\mathcal{L}_{\Upsilon^{\text{pol}}}$-theory which extends the theory of real closed rings and which has the following additional axioms:

1. The axioms of a commutative unital ring (with 1) in the language $\{+, -, \cdot, 0, 1\}$.
2. The axiom $\forall xy\ (\underline{+}(x, y) = x + y \wedge \underline{\cdot}(x, y) = x \cdot y \wedge \underline{\text{id}}(x) = x \wedge \underline{-}(x) = -x \wedge \underline{1}(x) = 1 \wedge \underline{0}(x) = 1)$. Hence the symbols from the language of rings have the same meaning as the corresponding symbols when reintroduced in $\mathcal{L}_{\Upsilon^{\text{pol}}}$ as symbols, naming continuous functions.
3. All the sentences

$$\forall \bar{x}\ \underline{F}(\underline{f}_1(\bar{x}), \ldots, \underline{f}_n(\bar{x})) = \underline{F \circ (f_1, \ldots, f_n)}(\bar{x}),$$

where $F \in C^{\text{pol}}(\mathbb{R}^n), f_1, \ldots, f_n \in C^{\text{pol}}(\bar{R}^{\bar{x}})$.

The models of $T_{\Upsilon^{\text{pol}}}$ are called *bounded super real closed rings.*

Observe that the null ring is also considered as a bounded super real closed ring. Moreover, since all semi-algebraic functions $\mathbb{R}^n \longrightarrow \mathbb{R}$ are polynomially bounded, it is clear that every bounded super real closed ring is real closed.

2.8. DEFINITION. A homomorphism between $\mathcal{L}_{\Upsilon^{\text{pol}}}$-structures is called a *bounded super homomorphism.* An $\mathcal{L}_{\Upsilon^{\text{pol}}}$-substructure of an $\mathcal{L}_{\Upsilon^{\text{pol}}}$-structure is called a *bounded super substructure.*

2.9. REMINDER. *If we drop the super script " pol" everywhere in* 2.7 *and* 2.8 *we get the definition of the language* $\mathcal{L}_\Upsilon$, *the definition of a super real closed ring* (*cf.* [*Tr*2, (5.1)]) *and the definition of a super homomorphism.*

An Υ-radical ideal of a super real closed ring A is an ideal I of A, which is closed under Υ (by which we mean closed under all the functions s_A, $s \in \Upsilon$). Those are precisely the kernels of super homomorphism (cf. [Tr2, (6.3)]).

If A is super real closed, then $\text{Hol}\, A$ is a super real closed subring of A, as follows immediately from [Tr2, (9.2)](i).

Bounded super real closed rings arise naturally from super real closed rings as convex subrings:

2.10. LEMMA. *If B is bounded super real closed (e.g. if B is super real closed) and A is a convex subring of B, then A is a bounded super real closed subring of B.*

PROOF. Take $F \in C^{\mathrm{pol}}(\mathbb{R}^n)$ and $a_1, \dots, a_n \in A$. We have to show $F_B(a_1, \dots, a_n) \in A$. Since F is polynomially bounded, there is some $P \in \mathbb{R}[X_1, \dots, X_n]$ with $|F| \le P$ on $\mathbb{R}^n$. Let $\chi : \mathbb{R} \longrightarrow \mathbb{R}$ be defined by $\chi(x) = -x$, if $x \le 0$ and $\chi(x) = 0$ if $x \ge 0$. Then $|F| \le P$ reads as $\chi \circ (P - |F|) = 0$ on $\mathbb{R}^n$. Since B is bounded super real closed, also $(\chi \circ (P - |F|))_B = 0$. By definition, this means $\chi_B \circ (P - |F|)_B = 0$. Since B is real closed $\chi_B(b) = 0$ is equivalent to $b \ge 0$ in B ($b \in B$). Hence we have $(P - |F|)_B(a_1, \dots, a_n) \ge 0$. In the bounded super real closed ring B, this means $|F_B(a_1, \dots, a_n)| \le P(a_1, \dots, a_n) \in A$. Since A is convex in B, $F_B(a_1, \dots, a_n) \in A$. ⊣

§3. Localization of bounded super real closed rings.

First recall how we can localize super real closed rings:

3.1. THEOREM (cf. [Tr2, (7.4)]). *Let A be a super real closed ring and let $1 \in S \subseteq A$ be closed under multiplication and Υ. Then there is a unique expansion of the localization $S^{-1}A$ to a super real closed ring such that the localization map $A \longrightarrow S^{-1}A$ is a super homomorphism.*

The operation of $F \in C(\mathbb{R}^n)$ on $(S^{-1}A)^n$ is given as follows: There are $t \in \Upsilon$ and a continuous function $G \in C(\mathbb{R}^n \times \mathbb{R})$ with

$$F(x_1, \dots, x_n) \cdot t(y) = G(x_1 \cdot y, \dots, x_n \cdot y, y) \ ((\bar{x}, y) \in \mathbb{R}^n \times \mathbb{R}).$$

Then for $f_1, \dots, f_n \in A$ and $g \in S$

$$F_{S^{-1}A}\left(\frac{f_1}{g}, \dots, \frac{f_n}{g}\right) := \frac{G_A(f_1, \dots, f_n, g)}{t_A(g)}.$$

3.2. LEMMA. *Let $F, G \in C(\mathbb{R}^n)$ such that $\{G = 0\} \subseteq \operatorname{int}\{F = 0\}$, the interior of the zero set of F. Then there is a unique $H \in C(\mathbb{R}^n)$ with $F = H \cdot G$ such that $H = 0$ on $\{G = 0\}$.*

If there are a bounded subset B of $\mathbb{R}^n$ and some $\varepsilon \in \mathbb{R}$, $\varepsilon > 0$ such that $|G|_{\mathbb{R}^n \setminus B}| \ge \varepsilon$, then $|H| \le c \cdot |F|$ for some $c \in \mathbb{R}$, $c > 0$.

PROOF. Existence and uniqueness of H is clear. Assume there are B, ε as stated. Then $K := \overline{B} \setminus \operatorname{int}\{F = 0\}$ is compact and G does not have zeroes on K. Let $c \in \mathbb{R}$, such that $c \ge \frac{1}{\varepsilon}$ and $\frac{1}{c} \le |G|_K|$. Then for every $x \in \mathbb{R}^n$ we have $|H(x)| \le c \cdot |F(x)|$: this holds true if $F(x) = 0$, since $F = H \cdot G$ and H vanishes on $\{G = 0\}$. If $F(x) \ne 0$, then $x \notin \overline{B}$ or $x \in \overline{B} \setminus \operatorname{int}\{F = 0\} = K$. In both cases we get the assertion by the choice of c. ⊣

3.3. COROLLARY. *Let A be bounded super real closed. Let $r \in \mathbb{R}$, $F_1, F_2 \in C^{\mathrm{pol}}(\mathbb{R}^n)$ and $a_1, \dots, a_n \in A$ be such that $|a_i| \le r$ ($1 \le i \le n$) and such that*

$F_1(x) = F_2(x)$ *for all* x *in an open set containing* $[-r, r]^n \subseteq \mathbb{R}^n$. *Then*

$$F_{1,A}(a_1, \ldots, a_n) = F_{2,A}(a_1, \ldots, a_n).$$

PROOF. Define $G : \mathbb{R}^n \longrightarrow \mathbb{R}$ by $G(x) = \sum_{i=1}^n \sup\{0, |x_i| - r_i\}$. Then $G \in C(\mathbb{R}^n)$ and $\{G = 0\} = [-r, r]^n \subseteq \operatorname{int}\{F_1 - F_2 = 0\}$ by assumption. By 3.2, there is some $H \in C(\mathbb{R}^n)$ with $F_1 - F_2 = H \cdot G$ and since $G \geq 1$ outside $[-r-1, r+1]^n$ we know that $|H| \leq c \cdot |F_1 - F_2|$ for some $c \in \mathbb{R}$. Since F_1, F_2 are polynomially bounded, also $H \in C^{\text{pol}}(\mathbb{R}^n)$. Thus

$$F_{1,A}(a_1, \ldots, a_n) - F_{2,A}(a_1, \ldots, a_n) = H_A(a_1, \ldots, a_n) \cdot G_A(a_1, \ldots, a_n).$$

Since $|a_i| \leq r$ for each i and A is a real closed ring we know that $G_A(a_1, \ldots, a_n) = \sum_{i=1}^n \sup\{0, |a_i| - r_i\} = 0$, which implies the corollary. ⊣

3.4. PROPOSITION AND DEFINITION. *Let* A *be a bounded super real closed ring. The holomorphy ring* $\operatorname{Hol} A$ *is a bounded super real closed subring of* A *and there is a unique super real closed ring structure on* $\operatorname{Hol} A$, *which expands the bounded super real closed ring structure.*

For $F \in C(\mathbb{R}^n)$ *and* $a_1, \ldots, a_n \in \operatorname{Hol}(A)$ *we have*

$$(\dagger) \qquad F_{\operatorname{Hol} A}(a_1, \ldots, a_n) = G_{\operatorname{Hol} A}(a_1, \ldots, a_n)$$

whenever $G \in C^{\text{pol}}(\mathbb{R}^n)$ *is such that for some* $r \in \mathbb{N}$ *with* $|a_i| \leq r$ *we have* $F(x) = G(x)$ $(x \in \mathbb{R}^n,\ |x| \leq r+1)$.

PROOF. $\operatorname{Hol} A$ is a bounded super real closed subring of A, since for all $a_1, \ldots, a_n \in \operatorname{Hol} A$ and each $F \in C^{\text{pol}}(\mathbb{R}^n)$, there are $r \in \mathbb{R}$ with $|a_i| \leq r$ and a bounded $F^* \in C^*(\mathbb{R}^n)$ such that $F(x) = F^*(x)$ $(|x| \leq r+1)$; hence by 3.3, $F_A(a_1, \ldots, a_n) = F^*_A(a_1, \ldots, a_n) \in \operatorname{Hol} A$.

By 3.3, we may use (†) to define an $\mathcal{L}_\Upsilon$-structure on $\operatorname{Hol}(A)$ which by definition expands the bounded super real closed ring structure on $\operatorname{Hol} A$. It is straightforward (using 3.3) to check that this defines the unique super real closed ring structure on $\operatorname{Hol} A$ which expands the bounded super real closed ring structure. ⊣

3.5. THEOREM. *Let* $F \in C^{\text{pol}}(\mathbb{R}^n)$ *and* $P(T) \in \mathbb{R}[T]$, $T = (T_1, \ldots, T_n)$ *of total degree* d *with* $|F| \leq |P|$ *on* $\mathbb{R}^n$. *Then there is a unique continuous function* $G \in C(\mathbb{R}^n \times \mathbb{R})$ *with*

$$(*) \qquad F(x_1, \ldots, x_n) \cdot y^{d+1} = G(x_1 \cdot y, \ldots, x_n \cdot y, y)\ ((\bar{x}, y) \in \mathbb{R}^n \times \mathbb{R}).$$

Moreover G *is polynomially bounded.*

PROOF. Existence is given by [Tr2, (7.2)](ii). Uniqueness holds, since $G(x_1, \ldots, x_n, y)$ is uniquely determined by $(*)$ for all $(\bar{x}, y) \in \mathbb{R}^n \times \mathbb{R}$ with $y \neq 0$. Moreover the proof of [Tr2, (7.2)](ii) shows that G is again polynomially bounded (by a polynomial of total degree d. ⊣

3.6. THEOREM. *Let* A *be a bounded super real closed ring and let* $1 \in S \subseteq A$ *be multiplicatively closed. Then there is a unique expansion of the localization*

$S^{-1}A$ to a bounded super real closed ring such that the localization map $A \longrightarrow S^{-1}A$ is a bounded super homomorphism.

The operation of $F \in C^{\text{pol}}(\mathbb{R}^n)$ on $(S^{-1}A)^n$ is given as follows: Pick $d \in \mathbb{N}_0$ such that F is bounded by a polynomial of total degree d and take a polynomially bounded continuous function $G \in C(\mathbb{R}^n \times \mathbb{R})$ with

$$F(x_1, \dots, x_n) \cdot y^{d+1} = G(x_1 \cdot y, \dots, x_n \cdot y, y) \; ((\bar{x}, y) \in \mathbb{R}^n \times \mathbb{R}).$$

Such functions exist by 3.5. Then for $f_1, \dots, f_n \in A$ and $g \in S$

$$F_{S^{-1}A}\left(\frac{f_1}{g}, \dots, \frac{f_n}{g}\right) := \frac{G_A(f_1, \dots, f_n, g)}{g^{d+1}} \in S^{-1}A.$$

Proof. The proof is parallel to the proof of the localization theorem [Tr2, (7.4)], using 3.5 instead of [Tr2, (7.2)](i). ⊣

3.7. Corollary. *Let $\varphi : A \longrightarrow B$ be a super homomorphism between bounded super real closed rings and let $1 \in S \subseteq A$ be multiplicatively closed such that $\varphi(S) \subseteq B^\times$. Then the natural map $S^{-1}A \longrightarrow B$ is a super homomorphism, too.*

Proof. This follows immediately from the explicit definition of the bounded super real closed structure on $S^{-1} \cdot A$ in 3.6. ⊣

§4. The super real closed hull. For a bounded super real closed ring A, we shall now define the smallest super real closed ring containing A as a bounded super real closed subring.

4.1. Theorem and Definition. *Let A be a bounded super real closed ring. Let*

$$\hat{A} = S^{-1} \cdot \text{Hol}\, A,$$

where S is the closure of $A^\times \cap \text{Hol}\, A$ under multiplication and Υ (recall: this means "closed under all the functions $s_{\text{Hol}\, A}$, $s \in \Upsilon$"); here we consider Hol(A) *equipped with the super real closed ring structure defined in* 3.4. *Then there is a unique $\mathcal{L}_\Upsilon$-structure on $\hat{A}$ such that $\hat{A}$ is a super real closed ring having A as a bounded super real closed subring. $\hat{A}$ is called the super real closed hull of A.*

Proof. Firstly, as $A^\times \cap \text{Hol}\, A \subseteq S$ we have $A = (A^\times \cap \text{Hol}\, A)^{-1} \cdot \text{Hol}\, A \subseteq S^{-1} \cdot \text{Hol}\, A = \hat{A}$. By 3.4, Hol$(A)$ is a bounded super real closed subring of A and there is a unique expansion of this structure to a super real closed ring. By definition, S is closed under multiplication and Υ. By 3.1, there is a unique $\mathcal{L}_\Upsilon$-structure on $\hat{A}$ such that $\hat{A}$ is a super real closed ring having Hol A as a super real closed subring. Since $\hat{A}$ is also the localization of A at S, 3.6 implies that A is a bounded super real closed subring of $\hat{A}$.

It remains to show that $\hat{A}$ with the $\mathcal{L}_\Upsilon$-structure defined above is the unique super real closed ring structure on $\hat{A}$ having A as a bounded super real closed subring. However, any other super real closed ring B expanding the pure ring $\hat{A}$ having A as a bounded super real closed subring, has $\mathrm{Hol}\, A$ as a super real closed subring (cf. [Tr2, (9.2)](i)) and the underlying bounded super real closed ring structure is the one induced from A. By 3.4, the super real closed ring structures of B and $\hat{A}$ induced on $\mathrm{Hol}\, A$ are equal. From the uniqueness property in 3.1 we know that B is the super real closed ring $\hat{A}$. ⊣

4.2. Corollary. *Let $F, G \in C^{\mathrm{pol}}(\mathbb{R}^n)$.*

(i) *If $\{F = 0\} \subseteq \{G = 0\}$, then $T_{\Upsilon^{\mathrm{pol}}} \vdash \forall \bar{x}\ \underline{F}(\bar{x}) = 0 \rightarrow \underline{G}(\bar{x}) = 0$.*
(ii) *If $\{F \geq 0\} \subseteq \{G \geq 0\}$, then $T_{\Upsilon^{\mathrm{pol}}} \vdash \forall \bar{x}\ \underline{F}(\bar{x}) \geq 0 \rightarrow \underline{G}(\bar{x}) \geq 0$.*

Proof. (i). Let $A \models T_{\Upsilon^{\mathrm{pol}}}$. By [Tr2, (5.5)](iv) the super real closed ring $\hat{A}$ is a model of

$$\forall \bar{x}\ \underline{F}(\bar{x}) = 0 \rightarrow \underline{G}(\bar{x}) = 0.$$

Since A is a bounded super real closed subring of $\hat{A}$ (by 4.1), also A is a model of this sentence.

(ii) follows from (i), since in every real closed ring A, the formula $x \geq 0$ is equivalent to $p_A(x) = 0$, where $p : \mathbb{R} \longrightarrow \mathbb{R}$ is the infimum of the identity function and the constant function 0. ⊣

4.3. Lemma. *Let A be a bounded super real closed subring of the super real closed ring B. There is a unique A-algebra homomorphism $\hat{A} \longrightarrow B$ and this homomorphism is an embedding of super real closed rings.*

Proof. We have $S_0 := A^\times \cap \mathrm{Hol}\, A \subseteq T := B^\times \cap \mathrm{Hol}\, B$. Since B is super real closed, T is closed under Υ: this follows from [Tr2, (6.12)], which says that all maximal ideal of B are Υ-radical.

Since $\mathrm{Hol}\, A$ is a super real closed subring of $\mathrm{Hol}\, B$ by 3.4, $T \cap \mathrm{Hol}\, A$ is Υ-closed as well. Thus the closure S of S_0 under Υ and multiplication is contained in T, too. Hence we get a unique A-algebra homomorphism $\varphi : \hat{A} = S^{-1} \cdot \mathrm{Hol}(A) \longrightarrow T^{-1} \cdot \mathrm{Hol}(B) = B$ and this map is injective. It remains to show that φ is a super homomorphism. This follows immediately from the definition of the $\mathcal{L}_\Upsilon$-structure on both rings in 3.1. ⊣

4.4. Corollary. *If A is a super real closed ring, then $\hat{A}$ (defined for the underlying bounded super real closed ring) is equal to A. In particular, the $\mathcal{L}_\Upsilon$-structure of A is uniquely determined by the $\mathcal{L}_{\Upsilon^{\mathrm{pol}}}$-structure.* ⊣

4.5. Corollary. *Let B be a super real closed ring and let A be a bounded super real closed subring of A. Then $B \cong_A \hat{A}$ as (bounded) super real closed rings if and only if B is generated by A as a super real closed ring.*

Proof. Let $C \subseteq \hat{A}$ be the super real closed subring generated by A. By 4.3 there is a super real A-algebra monomorphism $\hat{A} \longrightarrow C$. Composing this map with the inclusion $C \longrightarrow \hat{A}$ and using uniqueness shows that $C = \hat{A}$. Hence $\hat{A}$ is generated by A as a super real closed ring.

Conversely suppose B is generated by A as a super real closed ring. By 4.3, we may view $\hat{A}$ as a super real closed subring of B. Since B is generated by A we get $B = \hat{A}$. ⊣

4.6. PROPOSITION. *Let A be a bounded super real closed ring. Then A is convex in $\hat{A}$, in other words $\hat{A}$ is a subring of the convex closure B of A. There is a unique super real closed ring-structure on B extending the bounded super real closed ring structure on A. In particular, every bounded super real closed ring which is convexly closed (e.g. a field) has a unique expansion to a super real closed ring.*

PROOF. Since $\mathrm{Hol}(A)$ is convex in A, the convex closure B of $\mathrm{Hol}\, A$ contains A. By [Tr2, (11.2)](iii) we know that B is the localization of $\mathrm{Hol}\, A$ at the set T of all non zero-divisors t of $\mathrm{Hol}\, A$ with the property that $\mathrm{Hol}\, A$ is convex in $(\mathrm{Hol}\, A)_t$. It follows $A^\times \cap \mathrm{Hol}\, A \subseteq T$. Since T is closed and closed under Υ by [Tr2, (11.11)], the closure S of $A^\times \cap \mathrm{Hol}\, A$ is contained in T. Hence $\hat{A} = S^{-1} \cdot \mathrm{Hol}(A) \subseteq T^{-1} \cdot \mathrm{Hol}(A) = B$, in other words A is convex in $\hat{A}$.

By [Tr2, (11.12)], there is a (unique) expansion of B to a super real closed ring having $\hat{A}$ as super real closed subring. Since B is a localization of $\hat{A}$ we get the uniqueness statement of the proposition from the uniqueness statement in 3.6 together with 4.4. ⊣

4.7. COROLLARY. *Let A be a bounded super real closed subring of a super real closed ring B. Then A is convex in the super real closed ring generated by A in B.*

PROOF. By 4.5 and 4.6. ⊣

§5. Super real ideals.

5.1. DEFINITION. An ideal I of a bounded super real closed ring A is called *super real* if $s_A(I) \subseteq I$ for every $s \in \Upsilon^{\mathrm{pol}}$. Observe that in this case I is a radical ideal, in particular I is convex and satisfies $a \in I \iff |a| \in I$ $(a \in A)$.

Certainly, every ideal I of A is contained in a smallest super real ideal of A, denoted by $\sqrt[\Upsilon]{I}$.

If A is a super real closed ring, then by [Tr2, (6.10)], the super real ideals are precisely the Υ-radical ideals (clearly Υ^{pol} is a set of generalized root functions as defined in [Tr2, (3.2)]).

5.2. EXAMPLES. Let A be a bounded super real closed ring.

(i) If $F \in C^{\mathrm{pol}}(\mathbb{R})$ is strictly positive everywhere, then in general $F_A(a)$ is not a unit for every unit $a \in A$. For example if A is the bounded super real closed ring $C^{\mathrm{pol}}(\mathbb{R})$, $F = \exp(-x^2)$ and $a = 1 + x^2 \in A$.

(ii) If $a \in A$ is a unit, then in general, there is some $s \in \Upsilon^{\mathrm{pol}}$, which is bounded away from 0 outside $[-1, 1]$ such that $s_A(a)$ is not a unit. For example if A is the bounded super real closed ring $C^{\mathrm{pol}}(\mathbb{R})$, $s = \exp(-\frac{1}{|x|})$ and $a = \frac{1}{1+x^2} \in A$.

Hence in this example, the ideal $I = (s(a))$ of A is proper, but the super real radical of I is not proper. In particular, maximal ideals of bounded super real closed rings are not super real in general. The example also shows that this is not resolved if we replace Υ^{pol} by the set of all $s \in \Upsilon^{\mathrm{pol}}$, which are bounded away from 0 outside a neighborhood of 0: or to replace Υ^{pol} by the set of all bounded $s \in \Upsilon$ such that βs does not have zeroes different from 0 in the Stone-Cech compactification $\beta\mathbb{R}$ of $\mathbb{R}$.

5.3. Remark. If $A \subseteq B$ is an extension of rings and I is an ideal of A, then $I \cdot B$ denotes the ideal generated by I in B. Recall that for a convex subring A of a real closed ring B and every radical ideal I of A we have $I \cdot B = \{a \cdot b \mid a \in I,\ b \in B\}$ and this ideal is again radical.

Our first goal in this section is to show that 5.3 remains valid in the bounded super real closed context. That is, whenever $A \subseteq B$ is a convex extension of bounded super real closed rings and I is a super real ideal of A, then $I \cdot B$ is a super real ideal of B (cf. 5.7). In order to prove this we show that for every $s \in \Upsilon^{\mathrm{pol}}$, there are $t \in \Upsilon^{\mathrm{pol}}$ and $F \in C^{\mathrm{pol}}(\mathbb{R}^2)$ with $s(x \cdot y) = t(x) \cdot F(x, y)$. This is achieved in 5.6 below. First a preparational lemma: First two preparational lemmas from elementary analysis:

5.4. Lemma. *Let $A \subseteq \mathbb{R}^2$ be compact with projection $[a, b]$ onto the first coordinate. Let C be the convex hull of A. Then C is again compact and the function $f : [a, b] \longrightarrow \mathbb{R}$ defined by $f(x) = \max C_x$ is continuous, concave (i.e. $f(\lambda x + (1-\lambda)y) \geq \lambda f(x) + (1-\lambda) f(y)$ for all $0 \leq \lambda \leq 1$) and satisfies $f(a) = \max A_a$, $f(b) = \max A_b$. Here C_x denotes the set $\{y \in \mathbb{R} \mid (x, y) \in C\}$ and similarly for A_a, A_b.*

Moreover, if A is the graph of a strictly increasing function $[a, b] \longrightarrow \mathbb{R}$, then also f is strictly increasing.

Proof. This is clear. ⊣

5.5. Lemma. *Let $s_n \in \Upsilon$ $(n \in \mathbb{N})$. Then there is some $t \in \Upsilon$, $0 \leq t \leq 1$, symmetric (i.e. $t(-x) = t(x)$), non-decreasing and concave in $[0, \infty)$ such that for every $n \in \mathbb{N}$ there is some $\delta > 0$ with*

$$t(x) \geq |s_n(x)| \ (|x| < \delta).$$

Proof. We may assume that all s_n are symmetric, non-decreasing on $[0, \infty)$ and $0 \leq s_n \leq \frac{1}{2}$. Define

$$\tau(x) = \min\left(\frac{|x|}{2}, \frac{1}{2}\right) + \sup_n \min\left(\frac{1}{n}, s_n(x)\right)$$

which is continuous on $\mathbb{R}$, strictly increasing in $[0,1]$ with values in $[0,1]$, $\tau(0) = 0$ such that for every n, there exists $\delta > 0$ with $\tau(x) \geq s_n(x)$ for $0 \leq x \leq \delta$.

We define a function $t : [0,1] \longrightarrow [0,1]$ as follows: Let C be the convex hull of the graph of $\tau \restriction [0,1]$ and let

$$t(x) := \sup C_x \; (0 \leq x \leq 1),$$

where $C_x = \{y \in \mathbb{R} \mid (x,y) \in C\}$. By 5.4, t is a strictly increasing and concave homeomorphism $[0,1] \longrightarrow [0,\tau(1)]$. We extend t to $\mathbb{R}$ via $t(x) = t(1)$ if $x \geq 1$ and $t(x) = t(-x)$ if $x < 0$. Then t is symmetric, $0 \leq t \leq 1$ and it is clear that t is still concave in $[0,\infty)$.

Since $t(x) \geq \tau(x)$ for all $x \in [0,1]$ it is clear that for all $n \in \mathbb{N}$, there is some $\delta > 0$ with $t(x) \geq s_n(x)$ $(|x| < \delta)$. ⊣

5.6. Proposition. *Let $s \in \Upsilon$. There are $t \in \Upsilon$ with $0 \leq t \leq 1$, $c \in \mathbb{R}$ and $F \in C(\mathbb{R}^2)$ such that*

$$s(x \cdot y) = t(x) \cdot F(x,y)$$
$$\textit{and } |F(x,y)| \leq c \cdot \Big(1 + (1 + |y|) \cdot |s(x \cdot y)|\Big) \; ((x,y) \in \mathbb{R}^2)$$

Proof. Let $s_0(x) = x$ and for $n > 0$, $s_n(x) := \max_{|y| \leq n} n \cdot |s(y \cdot x)|$. Then $s_n \in \Upsilon$ and from 5.5 we get some $t \in \Upsilon$, symmetric with $0 \leq t \leq 1$, non-decreasing and concave in $[0,\infty)$ such that for every $n \in \mathbb{N}_0$ there is some $\delta > 0$ with

$$t(x) \geq |s_n(x)| \; (|x| < \delta).$$

By definition of s_n for $n \geq 1$ this means

$$(*) \qquad t(x) \geq n \cdot |s(yx)| \; (|x| < \delta, |y| \leq n).$$

We first show that the function $\frac{s(x \cdot y)}{t(x)}$, defined on $(\mathbb{R} \setminus \{0\}) \times \mathbb{R}$ has a continuous extension F through 0 on $\mathbb{R} \times \mathbb{R}$:

Pick $b \in \mathbb{R}$. For $n \in \mathbb{N}$ we have to find some $\delta > 0$ with $|\frac{s(x \cdot y)}{t(x)}| < \frac{1}{n}$ for all $x \in (-\delta, \delta)$, $x \neq 0$ and all y with $|b - y| < \delta$. Enlarge n if necessary such that $|b| < n$ and take $\delta > 0$ with $|b| + \delta < n$ such that $(*)$ holds. Let $0 < |x| < \delta$ and $|b - y| < \delta$. Then $|y| < |b| + \delta < n$. Thus $|s(xy)| \leq \frac{1}{n} t(x)$, as desired.

It remains to find $c \in \mathbb{R}$ such that for all $(x,y) \in \mathbb{R}^2$, $x \neq 0$ we have

$$(\dagger) \qquad \left|\frac{s(x \cdot y)}{t(x)}\right| \leq c \cdot (1 + (1 + |y|) \cdot |s(x \cdot y)|).$$

By choice of t there is some $\delta > 0$ such that $t(x) \geq |s(x)|$ and $t(x) \geq |x|$ for all x with $|x| < \delta$. It is enough to find an element c satisfying $(\dagger)$ separately on each of the following four subsets of $\mathbb{R}^2$, covering $\mathbb{R}^2$:

<u>Case 1.</u> $|x| \geq \delta$.

Then $t(x) = t(|x|) \geq t(\delta) > 0$ since t is symmetric and increasing in $[0, \infty)$. Hence $|\frac{s(x \cdot y)}{t(x)}| \leq |\frac{s(x \cdot y)}{t(\delta)}|$ and we may choose $c := \frac{1}{t(\delta)}$.

Case 2. $|x| < \delta$ and $|y| \leq 1$.

As F is continuous we may choose c as the maximum of $|F|$ on the rectangle $[-\delta, \delta] \times [-1, 1]$.

Case 3. $|x| < \delta$ and $|y| \geq 1$ and $|x \cdot y| \geq \delta$.

Then by the choice of δ we have $t(x) = t(|x|) \geq |x|$, hence $|\frac{s(x \cdot y)}{t(x)}| \leq |\frac{s(x \cdot y)}{x}| \leq |y \cdot \frac{s(x \cdot y)}{\delta}|$, since $\frac{1}{|x|} \leq |\frac{y}{\delta}|$ by assumption in case 3. Hence we may choose $c = \frac{1}{\delta}$.

Case 4. $|x| < \delta$ and $|y| \geq 1$ and $|x \cdot y| < \delta$.

Since t is concave in $[0, \infty)$ and $\frac{1}{|y|} \leq 1$ we have $t(x) = t(|x|) = t(\frac{1}{|y|} \cdot |x \cdot y|) \geq \frac{1}{|y|} \cdot t(|x \cdot y|) = \frac{1}{|y|} \cdot t(x \cdot y)$. Since $|x \cdot y| < \delta$ we have $t(x \cdot y) \geq |s(x \cdot y)|$ by the choice of δ. Hence

$$\left| \frac{s(x \cdot y)}{t(x)} \right| \leq |y| \cdot \frac{t(x \cdot y)}{t(x \cdot y)} = |y|$$

and we may choose $c = 1$. ⊣

5.7. Proposition. *Let A be a convex subring of a bounded super real closed ring B. If I is a super real ideal of A then $I \cdot B$ is super real, too.*

Proof. For $a \in I, b \in B$ and $s \in \Upsilon^{\mathrm{pol}}$ we have to show that $s_B(a \cdot b) \in I \cdot B$. By 5.6 there are $t \in \Upsilon^{\mathrm{pol}}$, $c > 0$ and $F \in C(\mathbb{R}^2)$ with

$$s(x \cdot y) = t(x) \cdot F(x, y) \; ((x, y) \in \mathbb{R}^2)$$

such that $|F(x, y)| \leq c \cdot (1 + (1 + |y|) \cdot |s(x \cdot y)|)$ everywhere. Since s is polynomially bounded also F is polynomially bounded. Hence $s_B(a \cdot b) = t_B(a) \cdot F_B(a, b)$. Since $t_B(a) = t_A(a) \in I$ we get the claim. ⊣

If A is a super real closed ring and I is an ideal of A, then there is a largest super real ideal I^Υ of A contained in I and $I^\Upsilon = \{a \in I \mid s_A(a) \in I \text{ for all } s \in \Upsilon\}$. (cf. [Tr2, (6.7)]).

With the aid of 5.7, this can be extended to bounded super real closed rings:

5.8. Proposition and Definition. *Let A be a bounded super real closed ring. If I is an ideal of A, then there is a largest super real ideal I^Υ contained in I. We have*

$$I^\Upsilon = \{a \in I \mid s_A(a) \in I \textit{ for all } s \in \Upsilon^{\mathrm{pol}}\} = (I \cap \operatorname{Hol} A)^\Upsilon \cdot A.$$

Proof. Let $J := (I \cap \operatorname{Hol} A)^\Upsilon$. By 5.7 we know that $J \cdot A$ is super real. Moreover it is clear that every super real ideal of A contained in I has to be contained in $K := \{a \in I \mid s_A(a) \in I \text{ for all } s \in \Upsilon^{\mathrm{pol}}\}$. In particular $J \cdot A \subseteq K$ and it remains to show that $K \subseteq J \cdot A$.

Pick $a \in K$. Since $\frac{1}{1+a^2}, \frac{a}{1+a^2} \in \operatorname{Hol} A$ we have $\frac{a}{1+a^2} \in I \cap \operatorname{Hol} A$ and it remains to show that $\frac{a}{1+a^2} \in (I \cap \operatorname{Hol} A)^{\Upsilon}$. It suffices to show $s_A(|\frac{a}{1+a^2}|) \in I$ for every strictly increasing $s \in \Upsilon$ and indeed by [Tr2, (6.7)] it suffices to take $s \in \Upsilon^{\mathrm{pol}}$. Since $|\frac{a}{1+a^2}| \leq |a|$ we have $\sqrt{s}_A(|\frac{a}{1+a^2}|) \leq \sqrt{s}_A(|a|) \in I$ by our choice of a in K. Now the convexity condition for real closed rings[1] implies that $\sqrt{s}_A(|a|)$ divides $s_A(|\frac{a}{1+a^2}|)$ in A. Hence $s_A(|\frac{a}{1+a^2}|) \in I$ as desired. ⊣

5.9. Theorem. *Let I be an ideal of a bounded super real closed ring A. Then*

$$\sqrt[\Upsilon]{I} \cdot \hat{A} = \sqrt[\Upsilon]{I \cdot \hat{A}} \text{ and } \sqrt[\Upsilon]{I} = \sqrt[\Upsilon]{I \cdot \hat{A}} \cap A.$$

Proof. The inclusion $\sqrt[\Upsilon]{I} \cdot \hat{A} \subseteq \sqrt[\Upsilon]{I \cdot \hat{A}}$ follows from $\sqrt[\Upsilon]{I} \subseteq \sqrt[\Upsilon]{I \cdot \hat{A}}$ and the inclusion $\sqrt[\Upsilon]{I} \cdot \hat{A} \supseteq \sqrt[\Upsilon]{I \cdot \hat{A}}$ holds, since by 5.7, $\sqrt[\Upsilon]{I} \cdot \hat{A}$ is super real.

Clearly $(\sqrt[\Upsilon]{I} \cdot \hat{A}) \cap A$ contains $\sqrt[\Upsilon]{I}$ and it remains to show that

$$(\sqrt[\Upsilon]{I} \cdot \hat{A}) \cap A \subseteq \sqrt[\Upsilon]{I}.$$

We may assume that $I = \sqrt[\Upsilon]{I}$. Take $b \in (I \cdot \hat{A}) \cap A$. In order to show $b \in I$ we may replace b by b^2, hence we may assume that $b \geq 0$. Since $1 + b^2$ is a unit in A we have $\frac{b}{1+b^2} \in (I \cdot \hat{A}) \cap A$. Since $b = \frac{b}{1+b^2} \cdot (1 + b^2)$ we may replace b with $\frac{b}{1+b^2}$ and we may assume that $0 \leq b \leq 1$. Since $b \in I \cdot \hat{A}$, there are $a \in I$ and $c \in \hat{A}$ with $b = a \cdot c$. As $b \geq 0$, $b = |b| = |c| \cdot |a|$ and we may assume that $a, c \geq 0$, too (observe that I is radical, hence $|a| \in I$). By 4.5, $\hat{A}$ is generated by A as a super real closed ring. Thus there are $F \in C(\mathbb{R}^n)$ and $a_1, \ldots, a_n \in A$ with $c = F_{\hat{A}}(a_1, \ldots, a_n)$.

Pick $\varphi : [0, \infty) \longrightarrow [1, \infty)$ continuous and strictly increasing with $|F(\bar{x})| \leq \varphi(|\bar{x}|)$ $(\bar{x} \in \mathbb{R}^n)$. Define $t : \mathbb{R} \longrightarrow \mathbb{R}$ by

$$t(y) = \begin{cases} \dfrac{|y|}{\varphi(\frac{1}{y^2})} & \text{if } y \neq 0 \\ 0 & \text{if } y = 0. \end{cases}$$

Using [Tr2, (7.2)](i) with $s(x) = x$ we get $t \in \Upsilon$ and some $G \in C(\mathbb{R}^n \times \mathbb{R})$ with

$$(*) \qquad F(x_1, \ldots, x_n) \cdot t(y) = G(x_1 y, \ldots, x_n y, y) \text{ on } \mathbb{R}^n \times \mathbb{R}.$$

Since φ is strictly increasing and ≥ 1 everywhere it is straightforward to see that $t|_{[0.\infty)} : [0.\infty) \longrightarrow [0.\infty)$ is an homeomorphism which is polynomially bounded and whose compositional inverse is polynomially bounded, too. Hence $t \in \Upsilon^{\mathrm{pol}}$ and there is some $t_1 \in \Upsilon^{\mathrm{pol}}$ with $t \circ t_1(y) = t_1 \circ t(y) = y$ for all $y \geq 0$. As $a \geq 0$ we get $a = t_A(t_{1,A}(a))$ from 4.2(ii).

Since I is a super real ideal, also $a_0 := t_{1,A}(a) \in I$. From $(*)$ we then get

$$b = c \cdot a = F_{\hat{A}}(a_1, \ldots, a_n) \cdot t_{\hat{A}}(a_0) = G_{\hat{A}}(a_1 \cdot a_0, \ldots, a_n \cdot a_0, a_0).$$

[1]The convexity condition says: $0 \leq a \leq b \Rightarrow b | a^2$.

Let $H := (G \vee 0) \wedge 1$. Then in $\mathbb{R}$ we have

$$\forall(\bar{x}, y) : 0 \leq G(x_1 y, \dots, x_n y, y) \leq 1 \Rightarrow$$
$$G(x_1 y, \dots, x_n y, y) = H(x_1 y, \dots, x_n y, y).$$

Since this sentence is also valid in $\hat{A}$ and $0 \leq b \leq 1$ we get

$$b = H_{\hat{A}}(a_1 \cdot a_0, \dots, a_n \cdot a_0, a_0).$$

Since H is bounded it follows $b = H_A(a_1 \cdot a_0, \dots, a_n \cdot a_0, a_0)$. As $H(0) = G(0) = 0$, there is some $s \in \Upsilon^{\text{pol}}$ with $H(z_1, \dots, z_{n+1}) \leq s(z_1^2 + \dots + z_{n+1}^2)$ for all $z_1, \dots, z_{n+1} \in \mathbb{R}$: choose s so that $s(t) \geq \max\{H(z) \mid \sum z_i^2 \leq t\}$ $(t \geq 0)$.

It follows $0 \leq b = H_A(a_1 \cdot a_0, \dots, a_n \cdot a_0, a_0) \leq s_A((a_1 a_0)^2 + \dots + (a_n a_0)^2 + a_0^2)$. Since $a_0 \in I$ and I is super real, we get $b \in I$ as desired. ⊣

Note that in general for a proper ideal I of a bounded super real closed ring A, the ideal $I^{\Upsilon} \cdot \hat{A}$ is properly contained in $(I \cdot \hat{A})^{\Upsilon}$ (e.g. if $I \cdot \hat{A} = \hat{A}$, cf. 5.2(ii))

5.10. SCHOLIUM. *Let A be bounded super real closed ring. An ideal of A is super real if and only if I is the kernel of a bounded super homomorphism $A \longrightarrow B$ into a bounded super real closed ring.*

PROOF. If $\varphi : A \longrightarrow B$ is such a homomorphism and $a \in I$, then $s_A(a) \in I$, since $\varphi(s_A(a)) = s_B(\varphi(a)) = s_B(0) = 0$.

Conversely suppose I is super real. By 5.7, $I \cdot \hat{A}$ is super real, too. Together with 5.9 it follows that $I \cdot \hat{A}$ is a super real ideal of $\hat{A}$ lying over I. By [Tr2, (6.3)], super real ideals of $\hat{A}$ are kernels of super homomorphisms. Hence we can compose $A \longrightarrow \hat{A}$ with $\hat{A} \longrightarrow \hat{A}/I \cdot \hat{A}$ and we get that I is the kernel of a bounded super homomorphism. ⊣

5.11. COROLLARY. *Let A be bounded super real closed and let $I \subseteq A$ be a super real ideal. There is a unique $\mathcal{L}_{\Upsilon^{\text{pol}}}$-structure on A/I such that A/I is a bounded super real closed ring and the residue map $A \longrightarrow A/I$ is a bounded super real homomorphism.*

Moreover, there is a unique A-algebra homomorphism $\hat{A} \longrightarrow \widehat{A/I}$ and this homomorphism is super real with kernel $I \cdot \hat{A}$. In particular, there is a unique A-algebra isomorphism of super real closed rings

$$\hat{A}/(I \cdot \hat{A}) \xrightarrow{\cong} \widehat{A/I}$$

PROOF. By 5.7 we know that $I \cdot \hat{A}$ is a super real ideal of $\hat{A}$ lying over I. Since super real ideals are kernels of super homomorphisms by 5.10, we can compose $A \longrightarrow \hat{A}$ with $\hat{A} \longrightarrow \hat{A}/I \cdot \hat{A}$ and get that I is the kernel of a bounded super homomorphism. The image is A/I and it is clear that the $\mathcal{L}_{\Upsilon^{\text{pol}}}$-structure on A/I is uniquely determined by saying that the residue map $A \longrightarrow A/I$ is a bounded super real homomorphism.

We get an embedding of rings $A/I \longrightarrow \hat{A}/I \cdot \hat{A}$, which is a bounded super real homomorphism. By 4.3, we may view $\widehat{A/I}$ as a super real closed subring of $\hat{A}/I \cdot \hat{A}$. Since $\hat{A}$ is generated by A as a super real closed ring, also $\hat{A}/I \cdot \hat{A}$ is generated by A/I as a super real closed ring, thus $\hat{A}/I \cdot \hat{A} = \widehat{A/I}$. Hence we have a super real homomorphism $\varphi : \hat{A} \longrightarrow \widehat{A/I}$ with kernel $I \cdot \hat{A}$. There can only be one such A-algebra homomorphism, since $\hat{A}$ is the localization of A at $(\hat{A})^\times \cap A$. ⊣

5.12. THEOREM. *If $\varphi : A \longrightarrow B$ is a bounded super homomorphism between bounded super real closed rings A and B, then there is a unique extension of φ to a ring homomorphism $\hat{\varphi} : \hat{A} \longrightarrow \hat{B}$ and this extension is super real.*

The functor F from bounded super real closed rings to super real closed rings, which maps A to $\hat{A}$ and φ to $\hat{\varphi}$ is an idempotent mono-reflector. This means: F is left adjoint to the inclusion from the category of super real closed rings into the category of bounded super real closed rings, $F \circ F = F$ and the adjoint morphism $A \longrightarrow \hat{A}$ is a monomorphism.

PROOF. First we prove the assertion about φ. Uniqueness again follows from the fact that $\hat{A}$ is the localization of A at $(\hat{A})^\times \cap A$. Existence of $\hat{\varphi}$ follows from 5.11 and 4.3.

Hence the functor F is well defined. By 4.4, $F \circ F = F$, which also shows that F is a reflector. F is a mono-reflector, since $A \longrightarrow \hat{A}$ is a monomorphism. ⊣

We conclude this section by showing that the reflector $A \mapsto \hat{A}$ is also well-behaved with respect to localization:

5.13. PROPOSITION. *Let A be a bounded super real closed ring, let $1 \in S \subseteq A$ be multiplicatively closed and let T be the closure of S in $\hat{A}$ under multiplication and Υ. Recall from 3.1 that there is a unique super real closed ring structure on $T^{-1} \cdot \hat{A}$ such that the localization map $\hat{A} \longrightarrow T^{-1} \cdot \hat{A}$ is a super homomorphism.*

The natural morphism $\hat{\varphi} : \hat{A} \longrightarrow \widehat{S^{-1} \cdot A}$ induced by the localization map $\varphi : A \longrightarrow S^{-1} \cdot A$, sends T into $(\widehat{S^{-1} \cdot A})^\times$ and the induced map

$$T^{-1} \cdot \hat{A} \longrightarrow \widehat{S^{-1} \cdot A}$$

is an A-algebra isomorphism of super real closed rings.

PROOF. Since $\varphi(S)$ consists of units of $S^{-1} \cdot A$ also $\hat{\varphi}(S)$ consists of units of $\widehat{S^{-1} \cdot A}$.

Since T is the closure of S under multiplication and Υ, $\hat{\varphi}(T)$ is the closure of $\hat{\varphi}(S)$ under multiplication and Υ. Since $\widehat{S^{-1} \cdot A}$ is super real closed, every maximal ideal of $\widehat{S^{-1} \cdot A}$ is super real (cf. [Tr2, (6.12)]), hence for every for every $s \in \Upsilon$ and each element $b \in \widehat{S^{-1} \cdot A}$, b is a unit in $\widehat{S^{-1} \cdot A}$ if and only if $s(b)$ is a unit in $\widehat{S^{-1} \cdot A}$. This proves that indeed $\hat{\varphi}(T) \subseteq (\widehat{S^{-1} \cdot A})^\times$.

In order to show that the induced map $T^{-1} \cdot \hat{A} \longrightarrow \widehat{S^{-1} \cdot A}$ is an isomorphism it now suffices to verify the universal condition defining $T^{-1} \cdot \hat{A}$ in the category of super real closed rings for $\widehat{S^{-1}\cdot A}$, more precisely for the morphism $\hat{A} \longrightarrow \widehat{S^{-1}\cdot A}$. Let $\psi : \hat{A} \longrightarrow B$ be a super homomorphism into a super real closed ring B with $\psi(T) \subseteq B^{\times}$. Then $\psi \restriction A : A \longrightarrow B$ is a super homomorphism with $\psi|_A(S) \subseteq B^{\times}$ and by 3.7 there is a unique super homomorphism $h : S^{-1}\cdot A \longrightarrow B$ such that $\psi|_A = h \circ \varphi$. By 5.12, $\hat{h} : \widehat{S^{-1}\cdot A} \longrightarrow B$ is the unique super homomorphism extending h with $\psi = \hat{h} \circ \hat{\varphi}$. ⊣

§6. The super real core.

6.1. Proposition. *Let A_0 be a convex subring of the super real closed ring A. Then there is a largest super real closed subring of A that is contained in A_0.*

Proof. By [Tr2, (9.2)](i), the convex hull of a super real closed subring of A is itself a super real closed subring of A. Hence, by using Zorn, it is enough to show for convex super real closed subrings B, C of A, that the ring D generated by B and C in A is again a super real closed subring of A. By [Tr2, (10.5)] we know that D is a convex subring of A and by [Tr2, (9.2)](i), it is enough to show that D is closed under Υ: Let $b_1, \dots, b_n \in B$ and $c_1, \dots, c_n \in C$. Pick $s \in \Upsilon$. It is enough to show $|s_A(b_1c_1 + \dots + b_nc_n)| \leq d$ for some $d \in D$. We may certainly assume that s is symmetric (i.e. $s(-x) = s(x)$) and strictly increasing on $(0, \infty)$. The Cauchy-Schwarz inequality implies $s(x_1y_1 + \dots + x_ny_n) \leq s(|x||y|)$ for all $x = (x_1, \dots, x_n), y = (y_1, \dots, y_n) \in \mathbb{R}^n$, where $|x|, |y|$ denote the euclidean norm of x, y respectively. Since $s(|x||y|) \leq s(|x|^2) + s(|y|^2)$ we get $s(x_1y_1 + \dots + x_ny_n) \leq s(|x|^2) + s(|y|^2)$ on $\mathbb{R}^n \times \mathbb{R}^n$. Thus $s_A(b_1c_1 + \dots + b_nc_n) \leq s_A(b_1^2 + \dots + b_n^2) + s_A(c_1^2 + \dots + c_n^2) \in D$ as desired. ⊣

6.2. Corollary and Definition. *For any bounded super real closed ring A there is a largest bounded super real closed subring, denoted by A^{Υ} with the property $\widehat{A^{\Upsilon}} = A^{\Upsilon}$. We call A^{Υ} the super real core of A.*

Proof. By 6.1, A^{Υ} is the largest super real closed subring of $\hat{A}$, which is contained in A. ⊣

Observe that A^{Υ} is convex in A, since $\operatorname{Hol} A \subseteq A^{\Upsilon}$. For a proper ideal I of A we know $I^{\Upsilon} = (I \cap \operatorname{Hol} A)^{\Upsilon} \cdot A$ from 5.8. Hence $I^{\Upsilon} = (I \cap A^{\Upsilon})^{\Upsilon} \cdot A$ as well.

On the other hand $\sqrt[\Upsilon]{I} \cap A^{\Upsilon}$ in general properly contains $\sqrt[\Upsilon]{I \cap A^{\Upsilon}}$ (e.g. if $\sqrt[\Upsilon]{I} = A$).

6.3. Corollary. *For any bounded super real closed ring A we have*

$$A^{\Upsilon} = \{a \in A \mid s_{\hat{A}}(a) \in A \textit{ for all } s \in \Upsilon\}.$$

Proof. Since A^{Υ} is a super real closed subring of $\hat{A}$ we have "$\subseteq$". Conversely take $a \in A$ with $s_{\hat{A}}(a) \in A$ for all $s \in \Upsilon$. Let B be the super real closed subring generated by a in $\hat{A}$. Thus $B = \{F_{\hat{A}}(a) \mid F \in C(\mathbb{R})\}$. Certainly every element of B is bounded in absolute value by some $s_{\hat{A}}(a)$ for some $s \in \Upsilon$.

Hence by choice of a, the convex hull C of B in $\hat{A}$ is contained in A. C is a super real closed subring of B by [Tr2, (9.2)](i). Hence $a \in C \subseteq A^{\Upsilon}$. $\dashv$

6.4. OBSERVATION. *If B is a real closed ring and $A \subseteq B$ is a convex subring, then A is a domain if and only if B is a domain, and A is local if and only if B is local (as follows from the Gelfand-Kolmogorov Theorem). In particular for every bounded super real closed ring A we have*

(i) *A is a domain $\iff$ $\hat{A}$ is a domain $\iff$ A^{Υ} is a domain $\iff$ $\operatorname{Hol} A$ is a domain.*
(ii) *A is local $\iff$ $\hat{A}$ is local $\iff$ A^{Υ} is local $\iff$ $\operatorname{Hol} A$ is local.*

6.5. EXAMPLES.

Let

$$A := \{f \in C(\mathbb{R}^2) \mid f \text{ is polynomially bounded in the second coordinate}\}.$$

Hence $f \in A$ if and only if for every $x \in \mathbb{R}$, the function $f(x, \_) : \mathbb{R} \longrightarrow \mathbb{R}$ is polynomially bounded. Clearly A is a convex subring of $C(\mathbb{R}^2)$, hence A is a bounded super real closed subring of $C(\mathbb{R}^2)$. We have $\hat{A} = C(\mathbb{R}^2)$ and

$$A^{\Upsilon} = \{f \in C(\mathbb{R}^2) \mid f \text{ is bounded in the second coordinate}\}.$$

Here a super real closed ring properly between $C^*(\mathbb{R})$ and $C(\mathbb{R})$: Take

$$A = \{f \in C(\mathbb{R}) \mid f \text{ is bounded on } (0, \infty)\}.$$

Also note that there are many super real closed ring properly between $C^*([0, \infty))$ and $C([0, \infty))$, e.g.

$$A = \{f \in C(\mathbb{R}) \mid f \text{ is bounded on } \mathbb{N}\}$$

has this property since $x \cdot \operatorname{dist}_{\mathbb{N}}(x) \in A \setminus C^*(\mathbb{R})$.

The formation of the super real core is functorial: If $\varphi : A \longrightarrow B$ is a bounded super homomorphism between bounded super real closed rings, then $\varphi|_{A^{\Upsilon}}$ is a super homomorphism $A^{\Upsilon} \longrightarrow B^{\Upsilon}$: since $\hat{\varphi}$ respects the $\mathcal{L}_{\Upsilon}$-structure on $\hat{A}$ by 5.12, $\varphi(A^{\Upsilon})$ is a super real closed subring of $\hat{B}$ contained in B, i.e. $\varphi(A^{\Upsilon}) \subseteq B^{\Upsilon}$. Hence the assignment $A \longrightarrow A^{\Upsilon}$ is functorial, by sending φ to $\varphi|_{A^{\Upsilon}}$. We shall not make use of this here. Instead, we state another description of the super real core.

Since $\operatorname{Hol} A \subseteq A^{\Upsilon} \subseteq A$, there are subsets S of $\operatorname{Hol} A$ with $A^{\Upsilon} = S^{-1} \cdot \operatorname{Hol} A$. We can compute the largest such set upon input A:

6.6. PROPOSITION. *For any bounded super real closed ring A, the largest multiplicatively closed subset S of $\operatorname{Hol} A$ satisfying $A^{\Upsilon} = S^{-1} \cdot \operatorname{Hol} A$ is*

$$S = \{a \in \operatorname{Hol} A \mid s_{\operatorname{Hol} A}(a) \in A^{\times} \text{ for all } s \in \Upsilon\}.$$

PROOF. The super real closed subrings of $\hat{A}$ contained in A are all of the form $T^{-1} \cdot \operatorname{Hol} A$, where $T \subseteq A^{\times} \cap \operatorname{Hol} A$. Since A^{Υ} is the largest super real closed subring of $\hat{A}$ contained in A, the set $T := (A^{\Upsilon})^{\times} \cap \operatorname{Hol} A$ is the largest among all of them. It remains to show $T = S$.

If $a \in T$, then a is a unit in A^{Υ} and since A^{Υ} is super real closed, all elements $s_{A^{\Upsilon}}(a)$ are units of A^{Υ} as well. Since $(A^{\Upsilon})^{\times} \subseteq A^{\times}$ we get $a \in S$.

Conversely let $a \in S$. The set $T_0 := \{s_{\operatorname{Hol} A}(a) \mid s \in \Upsilon\}$ is closed under multiplication and Υ (note that $s_1, s_2 \in \Upsilon$ implies $s_1(x) \cdot s_2(x) \in \Upsilon$ and $s_1 \circ s_2 \in \Upsilon$). Therefore $T_0^{-1} \cdot \operatorname{Hol} A$ has a unique super real closed ring structure (induced from $\hat{A}$). Since $a \in S$ we know $T_0 \subseteq A^{\times}$ and therefore $T_0^{-1} \cdot \operatorname{Hol} A \subseteq A$. So by the choice of T we obtain $T_0 \subseteq T$. Thus $a \in T_0 \subseteq T$. ⊣

REFERENCES

[Schw] N. Schwartz, *The basic theory of real closed spaces*, ***Memoirs of the American Mathematical Society***, vol. 77 (1989), no. 397, pp. viii+122.

[Tr1] M. Tressl, *Computation of the z-radical in $C(X)$*, ***Advances in Geometry***, vol. 6 (2006), no. 1, pp. 139–175.

[Tr2] ———, *Super real closed rings*, ***Fundamenta Mathematicae***, vol. 194 (2007), no. 2, pp. 121–177.

UNIVERSITY OF MANCHESTER
SCHOOL OF MATHEMATICS
OXFORD ROAD, MANCHESTER M13 9PL, UK
E-mail: marcus.tressl@manchester.ac.uk

ANALYTIC COMBINATORICS OF THE TRANSFINITE: A UNIFYING TAUBERIAN PERSPECTIVE

ANDREAS WEIERMANN

Abstract. From a Tauberian perspective we prove and survey several results about the analytic combinatorics of (transfinite) proof-theoretic ordinals. In particular we show how certain theorems of Petrogradsky, Karamata, Kohlbecker, Parameswaran, and Wagner can be used to give a unified treatment of asymptotics for count functions for ordinals. This uniform approach indicates that (Tauberian theorems for) Laplace transforms provide a general tool to establish connections between additive and multiplicative results and may therefore be seen as a contribution to Problem 12.21 in Burris's book on number theoretic density and logical limit laws. In the last section we give applications and prove in some detail phase transitions related to Friedman style combinatorial well-orderdness principles for fragments of first order Peano arithmetic.

§1. Introduction to the transfinite combinatorics of the transfinite. Some years ago the author had a discussion with another logician about which fields in mathematics are surely have no connection with each other. A suggestion was: The theory of transfinite ordinal numbers and complex analysis. Of course this seems a safe guess since it is difficult to imagine that Cauchy's integral formula has something to say about the ordinals below ω^ω or below ε_0.

It has therefore been surprising that those connections exist and give rise to interesting cross-disciplinary applications. Perhaps even more interestingly we apply in this paper results from a paper in Lie-algebra theory to questions in logic. When the author gave a talk about these developments he has moreover been asked whether there are possible connections between ordinals and Tauberian theory, in particular Karamata's theorem. The questioner was of course convinced that no natural connections exist. But the opposite is true. Karamata's theorem is of considerable importance for studying Ackermannian functions (when resulting from $< \omega^\omega$-descent recursive functions). Moreover in this article we show how classical Tauberian theorems can be used for studying the provably recursive functions for the fragments of first order Peano arithmetic.

Logic Colloquium '07
Edited by Françoise Delon, Ulrich Kohlenbach, Penelope Maddy, and Frank Stephan
Lecture Notes in Logic, 35

Historically the subject analytic combinatorics of the transfinite emerged from investigations of the author about Friedman style independence results for PA [35]. For these investigations asymptotic results on the counting of ordinals of a prescribed complexity turned out to be of crucial importance.

Quite a few logicians would be content with the insight that these complexities are bounded exponentially but for the purpose of classifying phase transitions for Gödel incompleteness results better bounds are needed. This is exactly the place where complex analysis and Tauberian theorems come into play. Cauchy's formula provides bounds for the coefficients of the generating function for the objects in question. (Similarly Tauberian theory, for example, via Karamata's or Ingham's theorem can be used for related purposes.) To be more specific let for $\beta < \varepsilon_0$ the number $N\beta$ denote the number of occurrences of ω in the Cantor normal form for β. Let

$$c_\alpha(n) := |\{\beta < \alpha : N\beta = n\}|.$$

For small values of α the function c_α can be evaluated by hands. For $\alpha = \omega^d$ already some calculation (see, for example, Theorem 2.48 in [8]) is needed to verify that

$$c_{\omega^d}(n) \sim \frac{1}{d!(d-1)!} n^{d-1}.$$

If one goes to $\alpha = \omega^\omega$ the result is even more exciting, since we obtain

$$c_{\omega^\omega}(n) \sim \frac{e^{\pi\sqrt{\frac{2n}{3}}}}{4\sqrt{3}n}. \tag{1}$$

(Here and in the sequel we denote by $\sim$ asymptotic equivalence. Thus $a_n \sim b_n$ means as usual $\lim_{n\to\infty} \frac{a_n}{b_n} = 1$.)

Now one can make a principal decision. Either one does not like such formulae at all or one get's excited and wants to know how to obtain and how to apply them. (A simple proof of the formula (1) by induction seems hopeless at least at first. But still Erdös managed to provide an elementary proof of (1) in [13]. We would also would like to mention that a Tauberian proof of (1) has been given in [19].)

Mathematical experience from the last century (and even before) suggests that attacking difficult counting problems is best done by studying related generating functions. Odlyzko commented for example in [24]:"*Analytic methods are extremely powerful and when they apply, they often yield estimates of unparalleled precision.*"

Therefore let

$$C_\alpha(z) := \sum_{n=0}^{\infty} c_\alpha(n) z^n.$$

Then under some standard conditions for $r < 1$ it holds that

$$c_\alpha(n) = \frac{1}{2\pi i} \oint_{|z|=r} \frac{C_\alpha(z)}{z^{n+1}} dz. \tag{2}$$

By comparing the coefficients of the involved power series we see that

$$C_{\omega^\omega}(z) = \prod_{i=1}^{\infty} \frac{1}{1-z^i} \tag{3}$$

where $C_{\omega^\omega}(z)$ is (uniformly compact) converging for $|z| < 1$. By exploiting (2) and (3) Hardy and Ramanujan (already in 1917) proved the formula (1) by introducing the circle method. In fact they proved a much stronger result which later was brought into perfection by Rademacher (see, for example, [3] for an accessible presentation) who provided a closed expression for $c_{\omega^\omega}(n)$.

$$c_{\omega^\omega}(n) = \frac{1}{\pi\sqrt{2}} \sum_{k=0}^{\infty} \sqrt{k} A_k(n) \frac{d}{d\tau} \left(\frac{\sinh\left(\frac{1}{k}\pi\sqrt{\frac{2}{3}}\sqrt{\tau - \frac{1}{24}}\right)}{\sqrt{\tau - \frac{1}{24}}}\Bigg|_{\tau=n} \right)$$

where

$$A_k(n) = \sum_{\substack{0<h\leq k \\ (h,k)=1}} \omega_{h,k} e^{-\frac{2\pi h n}{k}}$$

and where $\omega_{h,k}$ are certain roots of unity. Explicit details are written in full e.g. in Ostmann [26, page 58]. Hardy called these results a crowning achievement of analytic number theory.

If one looks at Hardy's and Ramanujan's paper one recognizes that there is no mentioning of ordinals below ω^ω and so one might wonder how (1) follows from their work. In fact this is easy to see, but the argument gives rise to deeper connections between the theory of partitions and the analytic combinatorics of the transfinite. So let us explain this in more detail. A partition of a natural number n is a multiset $\{i_1, \ldots, i_k\}$ of positive integers such that $\sum_{l=1}^k i_l = n$. Alternatively a partition is a sequence of positive integers $\langle i_1, \ldots, i_k \rangle$ such that $i_1 \geq \cdots \geq i_k \geq 1$ and $\sum_{l=1}^k i_l = n$. Let $p(n)$ be the number of partitions of n. What Hardy and Ramanujan showed is that

$$p(n) \sim \frac{e^{\pi\sqrt{\frac{2n}{3}}}}{4\sqrt{3}n}$$

as $n \to \infty$. Now every ordinal β below ω^ω can be written in a unique way as $\omega^{i_1-1} + \cdots + \omega^{i_k-1}$ with $i_1 \geq \cdots \geq i_k \geq 1$. Moreover in this case $N\beta = \sum_{l=1}^k i_l$. This simple argument shows that $p(n) = c_{\omega^\omega}(n)$. Moreover this argument bridges a gap between ordinal counting and partition theory (cf. e.g. [2]) and in principle the whole mathematical experience in partition theory is now ready for being applied to ordinal counting.

The asymptotic theory of partitions forms a part of general analytic combinatorics and, by this connection, even the whole cutting edge technology in analytic combinatorics (see, for example, [14]) can be applied to counting transfinite ordinals. This allows for a plethora of applications, which to a great extent have not yet been fully explored.

But this connection is not only a one way route. The counting problems emerging from ordinal counting lead to new interesting problems in analytic combinatorics itself and for solving them the abstract theory of analytic combinatorics has to be pushed further. This provides interesting material for research (as we think for several years).

In a more general context analytic combinatorics of the transfinite concerns the classification of functions of the form

$$count_\alpha(n) := |\{\beta < \alpha : complexity(\beta) = n\}|$$

or

$$Count_\alpha(n) := |\{\beta < \alpha : complexity(\beta) \leq n\}|$$

Here $complexity(\beta)$ denotes a complexity assignment to β such that

$$|\{\beta : complexity(\beta) \leq n\}|$$

is always finite for the ordinal segment in question and $|X|$ denotes the cardinality of a given set X.

It then turns out that (for natural choices of complexity measures) a rich family of partition functions reappears in the analytic combinatorics of the transfinite. It seems to the author that ordinals provide the right metaframework to define what a general partition function in the context of analytic number theory is. (Note that partition functions also show up in statistical mechanics, but we do not investigate possible connections in this direction further in this paper.)

One might still wonder whether it is worth to study count functions for ordinals as a subject in its own. In any case one can argue that there is intrinsic interest in these questions as those questions have been studied by many different leading mathematicians in related contexts. Ordinal counting is closely related to partition theory, as already mentioned but also to graphical enumeration and tree enumeration. An appropriate machinery for attacking such problems is complex analysis with its fundamental tools like Cauchy's integral formula.

In this paper we nevertheless stick to another approach via Tauberian theory. The reason is that there is already a wealth of Tauberian results available from the literature. Moreover Tauberian results have the advantage that they can be applied without too much difficulty by logicians/mathematicians/ computer scientists which are not analysts by profession.

One has only to recognize that these results apply after reformulating the count function problem as problem about suitable Laplace transforms. Let us try to explain how this can be achieved and how the Tauberian results then can be applied to yield elegant solutions to the problems in question. Moreover we want to stress that we consider the approach via Laplace transforms as a partial contribution to Problem 12.21 of Burris book [8]. A systematic investigation whether the additive and multiplicative results in Burris book can be obtained uniformly via Laplace transforms is part of current research by the author.

For a very nice introduction into Tauberian theory the reader might consult Korevaar's recent book [22]. It turns out counting problems for ordinals supply very convincing fields of applications for a plethora of Tauberian theorems proved several (sometimes more than 50) years ago. With respect to this let us mention fundamental theorems by Karamata [22], Kohlbecker [21], Parameswaran [27], Wagner [34], and moreover by Geluk, de Haan, Stadtmüller [15], and Schwarz [32].

For some of these results the first natural application in the context of partitions will be given in this paper.

A further reason to study the analytic combinatorics of the transfinite is the emergence of new fields of applications within logic itself. One concerns phase transition thresholds for Gödel incompleteness and another logical limit laws for ordinals. The latter subject (which is part of an ongoing joint research project with A. Woods) will not be treated in this paper. But let us mention that a joint result with A. Woods is that ordinals above ω^ω and below ε_0 come equipped with natural $0-1$-laws for first order (and other) logics. For a more general outline of logical limit laws we refer to the book of Burris [8].

In this paper we will give applications in the last chapter but we concentrate on applications to phase transitions in logic only. We still expect that there are more applications of analytic combinatorics of the transfinite in logic and we are looking forward for the things to come.

§2. Analytic combinatorics of the transfinite via Tauberian methods. We develop in this section the asymptotic theory for count functions related with the ordinals below ε_0. These ordinals come equipped with a canonical length norm which is defined recursively as follows

Definition 1. 1. $N0 := 0$.
2. $N\alpha := n + N\alpha_1 + \cdots + N\alpha_n$ if $\alpha = \omega^{\alpha_1} + \cdots + \omega^{\alpha_n} \geq \alpha_1 \geq \cdots \geq \alpha_n$.

This norm function has a natural interpretation in terms of non planar rooted trees which are associated to ordinals below ε_0. Let $T(0)$ be the singleton tree consisting of a root. If $\alpha = \omega^{\alpha_1} + \cdots + \omega^{\alpha_n} \geq \alpha_1 \geq \cdots \geq \alpha_n$

then $T(\alpha)$ is the tree consisting of a root and the (recursively defined) direct subtrees $T(\alpha_1), \ldots, T(\alpha_n)$. In this situation $N\alpha$ is the number of edges in $T(\alpha)$ (or equivalently the number of vertices in $T(\alpha)$ minus one). This bijective correspondence is very useful for our purposes since the whole (and by now well developed) machinery of tree enumeration can now be applied to the context of ordinals below ε_0.

The corresponding local count function (in the sequel we use standard terminology as introduced in [8]) is given (in accordance with the previous section) by

$$c_\beta(n) := |\{\alpha < \beta : N\alpha = n\}|.$$

Obviously the count function for additively decomposable ordinals can be computed via the count functions for the components.

If $\omega^\alpha + \beta$ is written in Cantor normal form, and $n \geq N\alpha - 1$, then

$$c_{\omega^\alpha+\beta}(n) = c_{\omega^\alpha}(n) + c_\beta(n - N\alpha - 1).$$

This reduces the counting problem to the count problem for additive indecomposable ordinals. It is now easy to observe that an ordinal ω^α gives rise to an additive number system in the sense of Burris [8] with prime elements ω^β for $\beta < \alpha$. (The norm for the number system is just the norm on ordinals and the sum for the number system is provided by the natural sum on ordinals.)

Therefore we have the following partition identity (fundamental identity) for the resulting generating function of the sequence $c_{\omega^\alpha}(n)$

$$\sum_{n=0}^{\infty} c_{\omega^\alpha}(n) z^n = \prod_{i=1}^{\infty} \frac{1}{(1 - z^i)^{c_\alpha(i)}}.$$

In particular

$$\sum_{n=0}^{\infty} c_{\omega^\omega}(n) z^n = \prod_{i=1}^{\infty} \frac{1}{(1 - z^i)}$$

yields the classical identity for the integer partitions.

Let $C_\alpha(z) = \sum_{n=0}^{\infty} c_\alpha(n) z^n$. By work of Rademacher it is known that C_{ω^ω} is admissible in the sense of Hayman. (See, for example, [25].)

By extending this calculation all functions $C_{\omega^{\omega^d}}$ are Hayman admissible too (The case $d = 2$ is essentially contained in [23]). Let $\omega_1(d) := \omega^d$ and $\omega_{k+1}(d) := \omega^{\omega_k(d)}$. Then by a general theorem of Odlyzko and Richmond [25] all functions $C_{\omega_k(d)}$ are Harris Shoenfeld admissible and hence Hayman admissible. Therefore we can calculate the asymptotic of $c_\beta(n)$ in principle for all $\beta < \varepsilon_0$ to a high degree of accuracy. Previous results in this direction have been obtained by Yamashita in [38].

A very readable and elementary exposition about asymptotics concerning partition functions equivalent to $c_\beta(n)$ has been provided by Petrogradsky [29] by elementary calculations with the partition identity of the generating functions. (His main concern was still to study enveloping algebras for Lie algebra's.) After translating his results back to our context we obtain the following result. (Here and in the sequel $\log_{(k)}(n)$ denotes the k-times iterated application of the natural logarithm function to n. Moreover ζ denotes the Riemann zeta function.)

THEOREM 1. *Let*

$$\gamma := \frac{1}{d!(d-1)!},$$

$$\eta := \left(1+\frac{1}{d}\right)\cdot\gamma^{\frac{1}{d+1}}\cdot\zeta(d+1)\cdot(d!)^{\frac{1}{d+1}},$$

and

$$\kappa := d\left(\frac{\eta}{d+1}\right)^{1+\frac{1}{d}}.$$

Then

1. $c_{\omega_1(d)}(n) \sim \gamma\cdot n^{d-1}$.
2. $\log(c_{\omega_2(d)}(n)) \sim \eta\cdot n^{\frac{d}{d+1}}$.
3. $\log(c_{\omega_3(d)}(n)) \sim \kappa\cdot\frac{n}{\log(n)}$.
4. *For* $k\geq 3$ *we have* $\log(c_{\omega_k(d)}(n)) \sim \kappa\cdot\frac{n}{\log_{(k-2)}(n)}$.

PROOF. See Petrogradsky [29]. Also [25] yields the same results, but some additional calculations are needed. Furthermore, Yamashita's paper contains similar formulas [38] but the exposition might be harder to follow. ⊣

Let $c_\alpha(\leq n) := |\{\beta<\alpha : N\beta\leq n\}|$.

COROLLARY 1. *Let*

$$\gamma' := \frac{1}{d!d!},$$

$$\eta = \left(1+\frac{1}{d}\right)\cdot\gamma^{\frac{1}{1+d}}\cdot\zeta(d+1)\cdot(d!)^{\frac{1}{d+1}},$$

and

$$\kappa = d\left(\frac{\eta}{d+1}\right)^{1+\frac{1}{d}}.$$

Then

1. $c_{\omega_1(d)}(\leq n) \sim \gamma\cdot n^{d}$.
2. $\log(c_{\omega_2(d)}(\leq n)) \sim \eta\cdot n^{\frac{d}{d+1}}$.
3. $\log(c_{\omega_3(d)}(\leq n)) \sim \kappa\cdot\frac{n}{\log(n)}$.
4. *For* $k\geq 3$ *we have* $\log(c_{\omega_k(d)}(\leq n)) \sim \kappa\cdot\frac{n}{\log_{(k-2)}(n)}$.

PROOF. This follows from the previous theorem by the equation

$$\sum_{n=0}^{\infty} c_\alpha(\leq n)z^n = \frac{1}{1-z}\sum_{n=0}^{\infty} c_\alpha(n)z^n.$$

Anyway, we now give an alternative proof for Corollary 1 and assertions 2.-4. in Theorem 1. We base this proof completely on Tauberian theory and the proof can be seen as a model proof.

Ad 1.

We assume that the reader is familiar with Karamata's theorem (as for example presented in [6] Theorem 1.7.1). In a first step we have to translate the count problem into a problem of Laplace transforms. This is can be considered as routine (but we still give the details for the reader's convenience). Let

$$a_{k,n} := \left|\left\{\omega^k \cdot l : N(\omega^k \cdot l) = n\right\}\right|.$$

Then $U_k(x) := \sum_{n\leq x} a_{k,n} \sim \frac{x}{k+1}$ and $\hat{U}_k(s) = \int e^{-sx} d(U_k(x)) = \sum_{n=0}^{\infty} a_{k,n}e^{-sn}$. Thus by Karamata's theorem applied for the first time

$$\sum_{n=0}^{\infty} a_{k,n}e^{-sn} \sim \frac{s}{k+1}$$

as $s \to 0$. This yields

$$\prod_{k=0}^{d-1}\sum_{n=0}^{\infty} a_{k,n}e^{-sn} \sim \frac{s^d}{d!}. \tag{4}$$

Let $U(x) := \sum_{n\leq x} c_{\omega^d}(n)$ and $\hat{U}_k(s) := \int e^{-sx} d(U(x)) = \sum_{n=0}^{\infty} c_{\omega^d}(n)e^{-sn}$. Equation (4) yields $\sum_{n=0}^{\infty} c_{\omega^d}(n)e^{-sn} \sim \frac{s^d}{d!}$. Karamata's theorem now applied for the second time yields

$$U(x) \sim \frac{x^d}{d!d!}$$

as $x \to \infty$. The first assertion follows since $U(n) = c_{\omega^d}(\leq n)$.

Ad 2.

We prove the second assertion by applying Kohlbecker's Tauberian theorem [21]. The fundamental identity for additive number systems (see, for example, Theorem 2.20 in [8]) yields

$$\prod_{k=1}^{\infty}(1-e^{-sk})^{c_{\omega^d}(k)} = \sum_{m=0}^{\infty} c_{\omega^{\omega^d}}(m)e^{-sm}.$$

Let

$$q(n) := \sum_{m_1\cdot 1+m_2\cdot 2+\cdots=n} \prod_{k=1}^{n} \binom{c_{\omega^d}(k)+m_k-1}{m_k}.$$

Then $q(n) = c_{\omega^{\omega^d}}(n)$ according to equation (2) on page 386 in [20]. This means that the assumptions of Kohlbecker's theorem apply for the system of generators $\lambda_i := i$. Let $n(u) := \sum_{k\leq u} c_{\omega^d}(k)$. Then by the first assertion we have

$$n(u) \sim \frac{1}{d!d!}u^d.$$

Let

$$P(u) := \sum_{i\leq u} q(i).$$

Then $P(n) = c_{\omega^{\omega^d}}(\leq n)$. Thus Kohlbecker's theorem yields

$$\log(P(u)) \sim \eta \cdot u^{\frac{d}{d+1}}$$

and this yields the second assertion.

Ad 3. The proof follows similar patterns. Let $U(u) := \sum_{i\leq u} c_{\omega^{\omega^d}}(i)$. Kohlbecker's result yields

$$U(u) = \exp\left((1+o(1))\eta \cdot u^{\frac{d}{d+1}}\right)$$

and by partial summation (or the abelian part of Kohlbecker's result)

$$\sum_{n=0}^{\infty} c_{\omega^{\omega^d}}(n)e^{-sn} \leq \left(\exp((1+o(1)))\kappa\frac{1}{s^d}\right).$$

Using this estimate we obtain after a short verification

$$\sum_{n=0}^{\infty} c_{\omega^{\omega^{\omega^d}}}(n)e^{-sn} \leq \exp\left(\left(\exp((1+o(1)))\kappa\frac{1}{s^d}\right)\right).$$

This is more or less standard and follows from the fundamental identity. To this end we might e.g. use estimates on geometric series as in [35] or directly Lemma 4.5 from [29] or also the classical method of Laplace (cf. exercise 208 on page 80 in [30].)

Let $U'(u) := \sum_{i\leq u} c_{\omega^{\omega^{\omega^d}}}(i)$. By the Tauberian theorem of Wagner [34] (cf. equation (10) in [34]) we then obtain that

$$\log(U'(u)) \leq (1+o(1))\kappa \cdot \frac{u}{\log(u)^{\frac{1}{d}}}.$$

The opposite inequality is somewhat harder to prove. Fix a small $\varepsilon > 0$. Let $f(s) = (1-\varepsilon)\kappa\frac{1}{s^d}$ and $h(v) = \sqrt[d+1]{\frac{(1-\varepsilon)\kappa d}{v}}$. Since

$$U(u) = \exp\left((1+o(1))\eta \cdot u^{\frac{d}{d+1}}\right),$$

we can choose a subset of $S \subseteq \{\alpha < \omega^{\omega^d}\}$ such that

$$c_S(\leq v) \sim \frac{\exp(vh(v) + f(h(v)))}{h(v)\sqrt{2\pi f''(h(v))}}$$

as $v \to \infty$ (where f'' denotes the second derivative of f and $c_S(\leq v)$ the number of elements in S with norm not exceeding v. Note that the term on the right hand side is exactly of the form of equation (21.7) in [22].) Then Laplace's method (cf. exercise 208 on page 80 in [30]) yields

$$(5) \qquad s \cdot \int_0^\infty e^{-su} c_S(\leq u)du = \sum_{n=0}^\infty c_S(n)e^{-sn} \sim \exp\left((1-\varepsilon)\kappa\frac{1}{s^d}\right)$$

as $s \to 0$ (where $c_S(\leq v)$ denotes the number of elements in S with norm equal to v).

Let $M(S)$ be the set of finite (normal form) sums (multisets) of elements from S. Then $M(S)$ is contained in $\{\alpha : \alpha < \omega^{\omega^{\omega^d}}\}$ so that $c_{\omega^{\omega^{\omega^d}}}(n) \geq c_{M(S)}(n)$. So it suffices to get a good lower bound on $c_{M(S)}(n)$. Let

$$C_S(s) := \sum_{n=0}^\infty c_S(n)e^{-sn},$$

$$H(s) := \sum_{m=1}^\infty \frac{1}{m} \sum_{n=0}^\infty C_S(sm)$$

and assume that $H(s) = \sum_{n=0}^\infty c'(n)e^{-sn}$. Then coefficientwise $c'(n) \geq c_S(n)$. Therefore the n-th coefficient in $\exp(H(s))$ is not smaller than the n-th coefficient, say $c''(n)$, in $\exp(C_S(s))$. Hence $C_{M(S)}(n) \geq c''(n)$. Let $U''(u) := \sum_{i \leq u} c''(i)$. We now have by (5)

$$\begin{aligned} & s \cdot \int_0^\infty e^{-su} U''(u)du \\ & = \sum_{n=0}^\infty c''(n)e^{-sn} \\ & = \exp\left((1+o(1))\exp((1-\varepsilon))\kappa\frac{1}{s^d}\right). \end{aligned}$$

Then we obtain by the Tauberian theorem of Wagner [34] that

$$\log(U''(u)) \sim (1-\varepsilon)\kappa \cdot \frac{u}{(\log(u))^{\frac{1}{d}}},$$

and hence the assertion, if a corresponding Tauberian condition is satisfied. This is indeed so since for $g(s) = \exp((1+\varepsilon)\kappa\frac{1}{s^d})$ we have $0 < g(s) \to \infty$ and for some fixed δ

$$0 < \delta \leq \frac{(g'(s))^2}{g(s)g''(s)} \leq 1$$

as $s \to 0$.

Ad 4. Again we first obtain

$$\sum_{n=0}^{\infty} c_{\omega_k(d)}(n)e^{-sn} \leq \exp_{k-1}\left(\kappa(1+o(1))\frac{1}{s^d}\right)$$

as $s \to 0$. This follows by induction on k using Proposition 2.33 in [8]. As before let $U(u) := \sum_{i \leq u} c_{\omega_k(d)}(i)$. Then $\int_0^\infty e^{-sx} d(U(x)) = \sum_{n=0}^\infty c_{\omega_k(d)}(n)e^{-sn}$. We obtain by the easy part of the Tauberian theorem of Wagner [34] (cf. equation (10) in [34]) that

$$\log(U(u)) \leq (1+o(1))\kappa \cdot \frac{u}{\log_{k-2}(u)^{\frac{1}{d}}}.$$

The opposite inequality is again somewhat harder to prove.

Fix a small $\varepsilon > 0$. Let

$$f(s) = \int_1^{\frac{1}{s^d}} \frac{d \exp_{k-1}((1-\varepsilon)\kappa t)}{t^{d+1}} dt$$

and

$$h(v) = \sqrt[d]{\frac{\kappa(1-\varepsilon)}{\log_{k-1}(v)}}.$$

(Note that again this term is exactly of the form of equation (21.7) in [22].) Since by induction hypothesis

$$U(u) \geq \exp_{k-1}\left((1-o(1))\kappa\frac{u}{\log_{k-3}(u)}\right)$$

we can choose a subset of $T \supseteq \{\alpha < \omega_{k-1}\}$ such that

$$c_T^S(\leq v) \sim \frac{\exp(vh(v) + f(h(v)))}{h(v)\sqrt{2\pi f''(h(v))}}.$$

Then Laplace's method yields

$$s \cdot \int_0^\infty e^{-su} c_S(\leq u)du = \sum_{n=0}^\infty c_S(n)e^{-sn} \sim \exp_{k-1}\left((1-\varepsilon)\kappa\frac{1}{s^d}\right) \tag{6}$$

as $s \to 0$.

Let $M(S)$ be the set of finite (normal form) sums (multisets) of elements from S. Then $M(S)$ is contained in $\{\alpha : \alpha < \omega_k(d)\}$ so that $c_{\omega_k(d)}(n) \geq c_{M(S)}(n)$. So it suffices to get a good lower bound on $c_{M(S)}(n)$. As before let $C_S(s) := \sum_{n=0}^{\infty} c_S(n)e^{-sn}$, let $H(s) := \sum_{m=0}^{\infty} \frac{1}{m} C_S(sm)$ and assume that $H(s) = \sum_{n=0}^{\infty} c'(n)e^{-sn}$. Then coefficientwise $c'(n) \geq c_S(n)$. Therefore the n-th coefficient in $\exp(H(s))$ is not smaller than the n-th coefficient, say $c''(n)$, in $\exp(C_S(s))$. Hence $C_{M(S)}(n) \geq c''(n)$. Let $U''(u) := \sum_{i \leq u} c''(i)$. We now have by (6)

$$\begin{aligned}
& s \cdot \int_0^{\infty} e^{-su} c_{M(S)}(\leq u)du \\
&= \sum_{n=0}^{\infty} c_{M(S)}(n)e^{-sn} = \exp(H(s)) \geq \exp(C_S(s)) \\
&= \exp\left(\sum_{n=0}^{\infty} c_S(n)e^{-sn}\right) = \sum_{n=0}^{\infty} c''(n)e^{-sn} \\
&= \exp((1+o(1))) \cdot \exp_{k-1}\left(\kappa\left((1-\varepsilon)\frac{1}{s^d}\right)\right).
\end{aligned}$$

The Tauberian theorem of Wagner [34] yields $\log(U''(u)) \sim (1-\varepsilon)\kappa \cdot \frac{u}{(\log_{(k-2)}(u))^{\frac{1}{d}}}$, hence the assertion, if a corresponding Tauberian condition is satisfied. This is indeed true since for $g(s) = \exp_{k-1}\left((1-\varepsilon)\kappa\frac{1}{s^d}\right)$ we have $0 < g(s) \to \infty$ and for some fixed δ

$$0 < \delta \leq \frac{(g'(s))^2}{g(s)g''(s)} \leq 1$$

as $s \to 0$. ⊣

Let us now consider other models for enumerating ordinals. A classic partition function goes back to Mahler and it appears in our context via choosing a different norm function.

Definition 2. 1. $M0 := 0$.
2. $M\alpha := 2^{M\alpha_1} + \cdots + 2^{M\alpha_n}$ if $\alpha = \omega^{\alpha_1} + \cdots + \omega^{\alpha_n} \geq \alpha_1 \geq \cdots \geq \alpha_n$.

Let

$$c_\beta^M(n) := |\{\alpha < \beta : M\alpha = n\}|.$$

Then $c_{\omega^\omega}^M(n)$ is the classical Mahler partition function which satisfies the following identity

$$\sum_{n=0}^{\infty} c_{\omega^\omega}^M(n)z^n = \prod_{i=0}^{\infty} \frac{1}{1-z^{2^i}}.$$

Very good asymptotic bounds on $c^M_{\omega^\omega}(n)$ have been obtained for example by de Bruijn and Pennington in [9] and [28].

In general we obtain

$$\sum_{n=0}^{\infty} c^M_{\omega^\alpha}(n)z^n = \prod_{i=0}^{\infty} \frac{1}{(1-z^{2^i})^{c^M_\alpha(i)}}.$$

It seems to us that techniques of Flajolet and Dumas [12] are appropriate to classify the resulting asymptotics very accurately and we think it will be another worthwhile research project to carry this out. Nevertheless for most applications to phase transitions weak asymptotics suffice and these can be gotten by a classical Tauberian result of Parameswaran.

THEOREM 2. 1. $c^M_{\omega_1(d)} \sim \frac{1}{(d-1)!\prod_{l<d} 2^l} n^{d-1}$.

2. *There exists an explicitly calculable constant* C_2 *such that*

$$\log\left(c^M_{\omega_2(d)}(n)\right) \sim C_2 \cdot (\log(n))^{d+1}.$$

3. *If* $k \geq 2$ *then with the same constant as in the previous assertion*

$$\log_{(k-1)}\left(c^M_{\omega_k(d)}(n)\right) \sim C_2 \cdot \left(\log_{(k-1)}(n)\right)^{d+1}.$$

PROOF. The proof can be carried out by using Parameswaran's result [27]. Details are omitted and can be found in [37]. ⊣

Another aspect of the analytic combinatorics of the transfinite concerns the counting of ordinals with respect to their Gödel numbers. It turns out that the theory of multiplicative number systems as, for example, exposed in [8] applies very well to this situation. One specific natural coding goes back to Schütte 1977 (or even to Matula's coding of trees). For this purpose let $(p_i)_{i\geq 1}$ denote a listing of the prime numbers in increasing order.

DEFINITION 3. 1. $S0 := 1$.

2. $S\alpha := p_{S\alpha_1} \cdot \dots \cdot p_{S\alpha_n}$ if $\alpha = \omega^{\alpha_1} + \dots + \omega^{\alpha_n} \geq \alpha_1 \geq \dots \geq \alpha_n$.

This coding gives a bijection between positive integers and the ordinals below ε_0. Number theoretic aspects of this coding have been investigated in detail in [10, 17].

The corresponding local count function is defined as follows.

$$c^S_\beta(n) := |\{\alpha < \beta : S\alpha = n\}|$$

But in this multiplicative context we are mainly interested in the global count functions given by

$$c^S_\beta(\leq n) := |\{\alpha < \beta : S\alpha \leq n\}|.$$

The corresponding count function for prime elements is furthermore given by

$$p^S_\beta(n) := |\{\omega^\alpha < \beta : S(\omega^\alpha) = n\}|$$

and the fundamental identity for multiplicative number systems (Theorem 8.13 in [8]) yields

$$\sum_{n=0}^{\infty} c^S_{\omega^\alpha}(n)n^{-x} = \prod_{i=2}^{\infty} \frac{1}{(1-i^{-x})^{p^S_{\omega^\alpha}(i)}}$$

It can now easily be seen (by the elementary prime number theorem) that $p^S_{\omega^\alpha}(n)$ is very close to $c^S_\alpha(n)$ and so the asymptotics for the count functions can be obtained recursively from the fundamental identity using Tauberian theory analogously as before. In fact the Tauberian theorems used here are the same as the ones used in the proof of Corollary 1.

THEOREM 3. *Let*

$$\gamma := \frac{1}{d!\log(p_{2^0})\cdot\dots\cdot\log(p_{2^{d-1}})}$$

$$\eta := \left(1+\frac{1}{d}\right)\cdot\gamma^{\frac{1}{d+1}}\cdot(d\cdot d!\cdot\zeta(d+1))^{\frac{1}{d+1}}$$

$$\kappa := d\left(\frac{\eta}{d+1}\right)^{1+\frac{1}{d}}$$

1. $c^S_{\omega_1(d)}(\leq n) \sim \gamma\cdot(\log(n))^d$
2. $\log(c^S_{\omega_2(d)}(\leq n)) \sim \eta\cdot(\log(n)^{\frac{d}{d+1}})$
3. $\log(c^S_{\omega_3(d)}(\leq n)) \sim \kappa\cdot\frac{\log(n)}{\log(\log(n))^{\frac{1}{d}}}$
4. *If* $k\geq 3$ *then*

$$\log\left(c^S_{\omega_k(d)}(\leq n)\right) \sim \kappa\cdot\frac{\log(n)}{\log(\log_{(k-2)}(n))^{\frac{1}{d}}}$$

PROOF. Assertion 1 follows as usual from Karamata's theorem. Let

$$a(n) = \left|\left\{\alpha<\omega^d : S\alpha = n\right\}\right|.$$

Let

$$U(x) := \sum_{\log i\leq x} a(i).$$

We claim that

$$(*)\qquad \int e^{-sx}d(U(x))$$
$$= \sum a(n)n^{-s}$$
$$= \sum_{i_0\geq\dots\geq i_{d-1}} \left(p_{2^0}^{i_0}\cdot\dots\cdot p_{2^{d-1}}^{i_{d-1}}\right)^{-s}$$
$$\sim \gamma\cdot(d!)\cdot\frac{1}{s^d}.$$

This follows from

$$\left|\left\{(p_{2^k})^l \leq n\right\}\right| \sim \frac{\log(n)}{\log p_{2^k}}.$$

Indeed, put $U_k(x) := \sum_{\log i \leq x} a_k(i)$ with $a_k(i) := |\{\omega^k \cdot l : S(\omega^k \cdot l) = i\}|$. Then

$$U_k(x) \sim \frac{x}{\log p_{2^k}}$$

as $x \to \infty$. Therefore Karamata's theorem yields

$$\int e^{-sx} d(U_k(x)) = \sum_{k \geq 0} (\log(p_{2^k}))^{-s} \sim \frac{s}{\log p_{2^k}}$$

as $s \to 0$. By multiplying out the corresponding d Dirichlet series assertion (*) follows. Using Karamata's theorem again we obtain $U(n) = \frac{1}{d!} \cdot \gamma \cdot n^d$ as $n \to \infty$. Thus

$$c^S_{\omega^d}(\leq n) = \sum_{i \leq n} a(i) = U(\log(n)) \sim \frac{1}{d!}\gamma(\log(n))^d.$$

Ad 2. Let

$$a(n) := |\{\alpha < \omega_2(d) : S\alpha = n\}|$$

and

$$p(n) := \left|\{\beta < \omega_2(d) : S\beta = n \text{ and } \beta \text{ is additively indecomposable}\}\right|.$$

As already noticed we may consider the set $\{\alpha < \omega_2(d)\}$ as multiplicative number system equipped with multiplicative norm given by S. Note that $S(\alpha \# \beta) = S\alpha \cdot S\beta$.

By the fundamental identity we therefore have

$$\sum_{n=1}^{\infty} a_n n^{-s} = \prod_{i=2}^{\infty} \frac{1}{(1 - i^{-s})^{p(i)}}.$$

To this situation we apply again Kohlbecker's theorem. Actually an exponential Tauberian theorem by Hardy and Ramanujan [18] suffices already.

Note first that $c^S_{\omega_2(d)}(\leq n) = \sum_{i \leq n} a(i)$. We claim that $c^S_{\omega_1(d)}(n) \sim \sum_{i \leq n} p(i)$. This follows from the prime number theorem and the bounds for $c^S_{\omega_1(d)}$. Indeed, by classical results for the prime numbers we have the following system of weak inequalities [3].

$$\frac{1}{6}\frac{n}{\log(n)} < \left|\{p \leq n : p \text{ prime number}\}\right| < 6\frac{n}{\log(n)}$$

This yields

$$\begin{aligned}
&\sum_{i\leq n} p(i) \\
&= \sum_{p_{S\alpha}\leq n \ \&\ \alpha<\omega^d} 1 \\
&\leq \sum_{S\alpha\leq 6\cdot\frac{n}{\log(n)} \ \&\ \alpha<\omega^d} 1 \\
&= c^S_{\omega^d}\left(\leq \left\lfloor 6\frac{n}{\log(n)}\right\rfloor\right) \\
&\sim \gamma\cdot\left(\log\left(\left\lfloor 6\frac{n}{\log(n)}\right\rfloor\right)\right)^d \\
&\sim \gamma\cdot\log(n)^d.
\end{aligned}$$

Here the greatest integer smaller than or equal to a given real number x is denoted by $\lfloor x\rfloor$.

By the same argument we obtain

$$\begin{aligned}
&\sum_{i\leq n} p(i) \\
&= \sum_{p_{S\alpha}\leq n \ \&\ \alpha<\omega^d} 1 \\
&\geq \sum_{S\alpha\leq \frac{1}{6}\frac{n}{\log(n)} \ \&\ \alpha<\omega^d} 1 \\
&= c^S_{\omega^d}\left(\leq \left\lfloor \frac{1}{6}\frac{n}{\log(n)}\right\rfloor\right) \\
&\sim \gamma\left(\log\left(\left\lfloor \frac{1}{6}\frac{n}{\log(n)}\right\rfloor\right)\right)^d \\
&\sim \gamma\cdot(\log(n))^d.
\end{aligned}$$

Now define $\lambda_i := \log(i)$ and let $\Lambda := \{\lambda_i : i < \omega\}$. Let v_i enumerate the semi group generated by Λ and define q by

$$q(\log(n)) := \sum_{m_1\cdot\lambda_1+\cdots m_k\cdot\lambda_k=\log(n)} \prod_{k=1}^{n} \binom{p(k)+m_k-1}{m_k}$$

Then $a(n) = q(\log(n))$.

This gives the following Kohlbecker representation of the fundamental identity

$$\begin{aligned}
&\sum_{n=1}^{\infty} a(n)n^{-s} \\
&= \sum_{n=1}^{\infty} q(\log(n))e^{-s\log(n)} \\
&= \prod_{k=1}^{\infty}(1-e^{-s\lambda_k})^{-p(k)} \\
&= \prod_{i=2}^{\infty}\frac{1}{(1-i^{-s})^{p(i)}}.
\end{aligned}$$

Now let $n(u) := \sum_{\lambda_k \le u} p(k)$. Then $n(u) = \sum_{k \le e^u} p(k) \sim \gamma u^d$. Let

$$Q(u) := \sum_{\log(i)\le u} q(\log(i)).$$

Kohlbecker's result [cf. Corollary 1, p. 363 in [21]] yields

$$\log(Q(u)) \sim \eta u^{\frac{d}{d+1}}.$$

We finally obtain the desired result as follows.

$$c^S_{\omega_2(d)}(\le n) = \sum_{i\le n} a(i) = \sum_{i\le n} q(\lambda_i) = Q(\log(n)) \sim \eta(\log(n))^{\frac{d}{d+1}}.$$

Ad 3. The proof follows similar patterns. Kohlbecker's result [21] yields

$$\sum_{n=0}^{\infty} c^S_{\omega^{\omega^d}}(n)e^{-s\log(n)} \le \exp\left(\kappa(1+o(1))\frac{1}{s^d}\right). \tag{7}$$

Using the observation that

$$p^S_{\omega^{\omega^{\omega^d}}}(n) \sim c^S_{\omega^{\omega^d}}\left(\frac{n}{\log(n)}\right) = e^{\eta(1+o(1))\cdot(\log(n))^{\frac{d}{d+1}}}$$

we can rewrite this with (7) and the fundamental identity to

$$\sum_{n=0}^{\infty} p^S_{\omega^{\omega^{\omega^d}}}(n)e^{-s\log(n)} \le \exp\left(\kappa(1+o(1))\frac{1}{s^d}\right)$$

and using Proposition 8.22 in [8] we can now write

$$\sum_{n=0}^{\infty} c^S_{\omega^{\omega^{\omega^d}}}(n)e^{-s\log(n)} \le \exp\left(\exp\left(\kappa(1+o(1))\frac{1}{s^d}\right)\right).$$

(This is again more or less standard.) One might e.g. use estimates on geometric series as in [35] or the classical method of Laplace (cf. exercise 208 on page 80 in [30].) Now let $U(u) := \sum_{\log(i)\leq u} c^S_{\omega^{\omega^{\omega^d}}}(i)$. Then $\int_0^\infty e^{-sx} d(U(x)) = \sum_{n=0}^\infty c^S_{\omega^{\omega^{\omega^d}}}(n) e^{-s\log(n)}$. By the Tauberian theorem of Wagner [34] (cf. equation (10) in [34]) we obtain that

$$\log(U(u)) \leq (1+o(1))\kappa \cdot \frac{u}{\log(u)^{\frac{1}{d}}}.$$

This yields already $c^S_{\omega^{\omega^{\omega^d}}}(\leq u) = U(\log(u)) \leq (1+o(1))\kappa \cdot \frac{\log(u)}{(\log(\log(u)))^{\frac{1}{d}}}$. (The proof of this inequality is essentially based on Rankin's method and does not need any special Tauberian condition.)

The opposite inequality is harder to prove. Fix a small $\varepsilon > 0$. Let $f(s) = (1-\varepsilon)\kappa \frac{1}{s^d}$ and $h(v) = \sqrt[d+1]{\frac{(1-\varepsilon)\kappa d}{v}}$. Since

$$U(u) = \exp\left((1+o(1)) \cdot \eta \cdot u^{\frac{d}{d+1}}\right)$$

we can choose a subset of $T \subseteq \{\alpha < \omega^{\omega^{\omega^d}}\}$ such that

$$c^S_T(\leq \exp(v)) =: U'(v) \sim \frac{\exp(vh(v) + f(h(v)))}{h(v)\sqrt{2\pi f''(h(v))}}$$

as $v \to \infty$ (where f'' denotes the second derivative of f). Then Laplace's method (cf. exercise 208 on page 80 in [30]) yields

$$(8) \qquad s \cdot \int_0^\infty e^{-su} U'(u) du = \sum_{n=0}^\infty c_S(n) e^{-s\log(n)} \sim \exp\left(((1-\varepsilon))\kappa \frac{1}{s^d}\right).$$

Let $M(T)$ be the set of finite (normal form) sums (multisets) of elements from T. Then $M(T)$ is contained in $\{\alpha : \alpha < \omega^{\omega^{\omega^d}}\}$ so that $c^S_{\omega^{\omega^{\omega^d}}}(\leq n) \geq c^S_{M(T)}(\leq n)$. So it suffices to get a good lower bound on $c^S_{M(T)}(\leq n)$. Let

$$C^S_T(s) := \sum_{n=0}^\infty c^S_T(n) e^{-s\log(n)},$$

let

$$H(s) := \sum_{m=1}^\infty \frac{1}{m} C^S_T(sm)$$

and assume that $H(s) = \sum_{n=0}^\infty c'(n) e^{-s\log(n)}$. Then $c'(n) \geq c^S_T(n)$. Therefore the sum of the first n coefficients in $\exp(H(s))$ is not smaller than the sum of the first n coefficients, say $c''(i)$ $(i \leq n)$, in $\exp(C^S_T(s))$. Hence

$C^S_{M(T)}(\leq n) \geq c''(\leq n)$. We now have by (8)

$$\begin{aligned}
&\exp(C^S_T(s)) \\
&= \exp\left(\sum_{n=0}^{\infty} c^S_T(n)e^{-s\log(n)}\right) \\
&= \sum_{n=0}^{\infty} c''(n)e^{-s\log(n)} \\
&= \exp\left((1+o(1))\exp((1-\varepsilon))\kappa\frac{1}{s^d}\right).
\end{aligned}$$

Then we obtain by the Tauberian theorem of Wagner [34] that $\log(c''(\leq (u))) \sim (1-\varepsilon)\kappa \cdot \frac{u}{(\log(u))^{\frac{1}{d}}}$ if a corresponding Tauberian condition is satisfied. This is indeed so since for $g(s) = \exp((1+\varepsilon)\kappa\frac{1}{s^d})$ we have $0 < g(s) \to \infty$ and for some fixed δ

$$0 < \delta \leq \frac{(g'(s))^2}{g(s)g''(s)} \leq 1$$

as $s \to 0$.

Ad 4. Again we first obtain

$$\sum_{n=0}^{\infty} c^S_{\omega_k(d)}(n)e^{-s\log(n)} \leq \exp_{k-1}\left(\kappa(1+o(1))\frac{1}{s^d}\right).$$

This follows by induction on k using Proposition 2.33 in [8]. As before let $U(u) := \sum_{\log(i)\leq u} c^S_{\omega_k(d)}(i)$. Then $\int_0^\infty e^{-sx}d(U(x)) = \sum_{n=0}^\infty c^S_{\omega_k(d)}(n)e^{-s\log(n)}$. Again we obtain by the Tauberian theorem of Wagner [34] (cf. equation (10) in [34]) that $\log(U(u)) \leq (1+o(1))\kappa \cdot \frac{u}{\log_{(k-2)}(u)^{\frac{1}{d}}}$. This yields $\log(c^S_{\omega_k(d)}(\leq u)) \sim \log(U(\log(u))) \leq (1+o(1))\kappa \cdot \frac{\log(u)}{\left(\log_{(k-2)}(\log(u))\right)^{\frac{1}{d}}}$.

The opposite inequality is as usual harder to prove.

Fix a small $\varepsilon > 0$. Let $f(s) = \exp\left(\int_1^{\frac{1}{s^d}} \frac{d\exp_{k-1}((1-\varepsilon)\kappa t)}{t^{d+1}}dt\right)$ and $h(v) = \frac{\kappa(1-\varepsilon)}{\sqrt[d]{\log_{(k-1)}(v)}}$. Since by induction hypothesis

$$U(u) \geq \exp_{k-1}\left((1-o(1))\kappa\frac{u}{\log_{k-3}(u)}\right)$$

we can choose a subset of $T \supseteq \{\alpha < \omega_{k-1}\}$ such that

$$c^S_T(\leq \exp(v)) \sim \frac{\exp(vh(v) + f(h(v)))}{h(v)\sqrt{2\pi f''(h(v))}}.$$

Then Laplace's method yields

$$(9)\quad s\cdot\int_0^\infty e^{-su}c_T^S(\le u)du=\sum_{n=0}^\infty c_T^S(n)e^{-s\log(n)}\sim \exp_{k-1}\left((1-\varepsilon)\kappa\frac{1}{s^d}\right).$$

Let $M(T)$ be the set of finite (normal form) sums (multisets) of elements from T. Then $M(T)$ is contained in $\{\alpha : \alpha < \omega_k(d)\}$ so that $c^S_{\omega_k(d)}(\le n) \ge c^S_{M(T)}(\le n)$. So it suffices to get a good lower bound on $c^S_{M(T)}(n)$. Let $C_T^S(s) = \sum_{n=0}^\infty c_T^S(n)e^{-s\log(n)}$ and let $H(s) = \sum_{m=0}^\infty \frac{1}{m}C_T^S(sm)$ and assume that $H(s) = \sum_{n=0}^\infty c'(n)e^{-s\log(n)}$. Then $\sum_{i\le n} c'(i) \ge c_T^S(\le n)$. Therefore the sum of the first n coefficients in $\exp(H(s))$ is not smaller than sum of the first n coefficients, say $c''(i)$ $(i \le n)$, in $\exp(C_T^S(s))$. Hence $c^S_{M(T)}(\le n) \ge c''(\le n)$. We now have by (9)

$$\begin{aligned}
&\exp(C_T^S(s))\\
&=\exp\left(\sum_{n=0}^\infty c_T^S(n)e^{-s\log(n)}\right)\\
&=\sum_{n=0}^\infty c''(n)e^{-s\log(n)}\\
&=\exp\left((1+o(1))\exp_{k-1}\left(\kappa(1-\varepsilon)\frac{1}{s^d}\right)\right).
\end{aligned}$$

The Tauberian theorem of Wagner [34] yields $\log(c''(\le \exp(u))) \sim (1-\varepsilon)\kappa\cdot\frac{u}{(\log_{(k-2)}(u))^{\frac{1}{d}}}$ if a corresponding Tauberian condition is satisfied. This is indeed true since for $g(s) = \exp_{k-1}((1-\varepsilon)\kappa\frac{1}{s^d})$ we have $0 < g(s) \to \infty$ and for a fixed δ

$$0<\delta\le\frac{(g'(s))^2}{g(s)g''(s)}\le 1$$

as $s \to 0$.

So we obtain

$$\begin{aligned}
\log\left(c^S_{\omega_k(d)}(\le n)\right) &\sim \log(U(\log(u)))\\
&\sim (1-\varepsilon)\kappa\cdot\frac{\log(u)}{(\log_{(k-2)}(\log(u)))^{\frac{1}{d}}}.
\end{aligned}$$

⊣

Now we have a look at another Gödel coding which in fact goes back to Gödel's original exposition of his incompleteness results and it is widely used in logic, for example in textbooks on recursion theory.

Definition 4. 1. $G0 := 1$.
2. $G\alpha := p_1^{G\alpha_1}\cdot\dots\cdot p_n^{G\alpha_n}$ if $\alpha = \omega^{\alpha_1}+\dots+\omega^{\alpha_n} \ge \alpha_1 \ge \dots \ge \alpha_n$.

Let

$$c^G_\beta(\leq n) := |\{\alpha < \beta : G\alpha \leq n\}|$$

THEOREM 4. 1. *There exists an explicitly calculable constant C_1 such that*

$$c^G_{\omega_1(d)}(\leq n) \sim C_1\left(\frac{\log(n)}{\log(\log(n))}\right)^d.$$

2. *There exists an explicitly calculable constant C_2 such that*

$$\log(c^G_{\omega_2(d)}(\leq n)) \sim C_2\frac{(\log_{(2)}(n))^{d+1}}{(\log_{(3)}(n))^d}.$$

3. *For the constant C_2 from the previous assertion*

$$\log_{(k-1)}(c^G_{\omega_k(d)}(\leq n)) \sim \frac{(\log_{(k)}(n))^{d+1}}{(\log_{(k+1)}(n))^d}.$$

PROOF. The first assertion can be proved by an application of Karamata's theorem. A proof of the second and third assertion can be given by Parameswaran's Tauberian result [27]. Details are omitted and can be found, e.g., in [37]. ⊣

§3. A collection of further norms for ordinals. A further coding of ω_{l+1} (for $l \geq 1$) goes back to Ackermann 1940 [1] and makes use of coding ordinals in terms of exponentiation with respect to base 2 starting with an ordinal segment of length ω^2.

DEFINITION 5. 1. If $\alpha = \omega \cdot d + k < \omega^2$ then $A_0(\alpha) := 2^d(2 \cdot k + 1)$.
2. If $l \geq 1$ and $\alpha < \omega_{l+1}$ with $\alpha = 2^{\alpha_1} + \cdots + 2^{\alpha_n}$ where $\alpha_1 > \cdots > \alpha_n$ then let $A_l\alpha := 2^{A_{l-1}\alpha_1} + \cdots + 2^{A_{l-1}\alpha_n}$.

For $\beta < \omega_{l+1}$ put

$$c^{A_l}_\beta(n) := |\{\alpha < \beta : A_l\alpha \leq n\}|.$$

We guess that for $d > 2$ one should arrive at something similar to $\log(c^{A_1}_{\omega^d}) \sim C_d(\log(n))^2$ with C_d converging to a real number C. For $k > 1$ we believe that we should arrive at something similar to $\log_k(c^{A_k}_{\omega_k(d)}) \sim C'_{k,d}(\log_k(n))^2$ with $C'_{k,d}$ converging to a real number C'_d.

A corresponding additive Mahler type norm can be defined as follows.

DEFINITION 6. 1. If $\alpha = \omega \cdot d + k < \omega^2$ then $MA_0(\alpha) := 2d + k$.
2. If $l \geq 1$ and $\alpha < \omega_{l+1}$ with $\alpha = 2^{\alpha_1} + \cdots + 2^{\alpha_n}$ where $\alpha_1 > \cdots > \alpha_n$ then let $MA_l\alpha := 2^{MA_{l-1}\alpha_1} + \cdots + 2^{MA_{l-1}\alpha_n}$.

We guess that the asymptotic behavious of the resulting count function behaves similarly as the asymptotic of c^{A_l}.

Still another coding of ordinals less than ω_{l+1} for fixed l has roots going back to Hardy and Ramanujan [18].

DEFINITION 7. 1. If $\alpha = \omega^{k_1} + \cdots + \omega^{k_n} < \omega^\omega$ with $k_1 \geq \cdots \geq k_n$ then $HR_1\alpha := p_1^{k_1+1} \cdot \cdots \cdot p_n^{k_n+1}$.
2. For $\alpha = \omega^{\alpha_1} + \cdots + \omega^{\alpha_n} < \omega_{l+1}$ (where $l > 1$) with $\alpha_1 \geq \cdots \geq \alpha_n$ put $HR_l\alpha := p_1^{HR_{l-1}\alpha_1} \cdot \cdots \cdot p_n^{HR_{l-1}\alpha_n}$ if $\alpha = \omega^{\alpha_1} + \cdots + \omega^{\alpha_n} \geq \alpha_1 \geq \cdots \geq \alpha_n$.

For $\beta \leq \omega_{l+1}$ put

$$c_\beta^{HR_l}(n) := \left|\{\alpha < \beta : HR_l\alpha \leq n\}\right|.$$

THEOREM 5. 1. $c_{\omega^d}^{HR_1}(n) \sim \frac{1}{d!d!}\left(\frac{\log(n)}{\log(\log(n))}\right)^d$.
2. $\log(c_{\omega^\omega}^{HR_1}(n)) \sim \frac{2\pi}{\sqrt{3}}\sqrt{\frac{\log(n)}{\log(\log(n))}}$.

PROOF. The first assertion follows e.g. from Karamata's theorem as shown in [36]. The first assertion can also be proved by a more elementary argument of Schlage Puchta (private communication). The second assertion was proved by Hardy and Ramanujan [18]. ⊣

We expect that Parameswaran's theorem provides weak asympotics for the count functions related to higher ω-powers.

Finally we would like to mention a (perhaps somewhat artificial) norm function which leads to fractional exponents for partitions.

DEFINITION 8. 1. $N_k 0 := 0$
2. $N_k\alpha := (1 + N_k\alpha_1)^k + \cdots + (1 + N_k\alpha_n)^k$ if $\alpha = \omega^{\alpha_1} + \cdots + \omega^{\alpha_n} \geq \alpha_1 \geq \cdots \geq \alpha_n$.

We guess that the asymptotic for the resulting count functions can be determined using Petrogradsky's theorem starting at its base with

$$c_{\omega^d}^{N_k}(\leq n) \sim constant \cdot n^{\frac{d}{k}}.$$

The corresponding multiplicative analogue will be given by

DEFINITION 9. 1. $S_k 0 := 1$
2. $S_k\alpha := p_{S_k\alpha_1}^k \cdot \cdots \cdot p_{S_k\alpha_n}^k$ if $\alpha = \omega^{\alpha_1} + \cdots + \omega^{\alpha_n} \geq \alpha_1 \geq \cdots \geq \alpha_n$.

We expect that the asymptotic will start with something like

$$c_{\omega^d}^{S_k}(\leq n) \sim constant \cdot (\log n)^{\frac{d}{k}}.$$

The choice of N_k and S_k is of course somewhat artificial. But, and this seems to be of certain interest, if one uses N_k in place of N and S_k in place of S in Theorems 7 and 8 of the next section it will turn out that the threshold function will not change. So there will be a phenomenon of structural stability concerning applications of different norms.

§4. Applications to phase transition results for Gödel incompleteness. We start with some standard conventions. As usual the least integer greater than or equal to a given real number x is denoted by $\lceil x \rceil$ and similarly the greatest integer smaller than or equal to a given real number x is denoted by $\lfloor x \rfloor$. The binary length $|n|$ of a natural number n is defined by $|n| := \lceil \log(n+1) \rceil$. The d-times iterated length function $|\cdot|_d$ is defined recursively as follows $|x|_0 := x$ and $|x|_{d+1} := ||x|_d|$. (This function mainly serves as an L_{PA} definable version of $\log_{(d)}$.) As usual we assume that the ordinals less than ε_0 are available in PA via a standard coding. Recall that for $\alpha < \varepsilon_0$ the number $N\alpha$ denotes the number of occurrences of ω in the Cantor normal form of α. Let

$$\mathrm{CWO}(N, \alpha, f)$$

be the following principle

$$\forall K \exists M \; B(N, \alpha, f, K, M)$$

where $B(N, \alpha, f, K, M)$ is the assertion

$$(\forall \alpha_0, \dots, \alpha_M < \alpha)\big[(\forall i \le M)\big[N\alpha_i \le K + f(i)\big] \implies (\exists i, j < \omega)\big[0 \le i < j \le M \;\&\; \alpha_i \le \alpha_j\big]\big].$$

Theorem 6 (Friedman). *Let $d \ge 0$ be a fixed natural number. Then*

$$\mathrm{I}\Sigma_{d+1} \nvdash \mathrm{CWO}(N, \omega_{d+2}, f)$$

where f is the identity map.

Proof. See, for example, [33] for a proof. ⊣

Before we can sharpen this result we need to recall the definition of the standard system of canonical fundamental sequences and the Hardy hierarchy.

Definition 10. 1. If $\lambda = \omega^{\alpha_1} + \cdots + \omega^{\alpha_n} \ge \alpha_1 \ge \cdots \ge \alpha_n$ and $\alpha_n = \alpha_n' + 1$ then $\lambda[x] := \omega^{\alpha_1} + \cdots + \omega^{\alpha_n'} \cdot x$.
2. If $\lambda = \omega^{\alpha_1} + \cdots + \omega^{\alpha_n} \ge \alpha_1 \ge \cdots \ge \alpha_n$ and α_n is a limit then $\lambda[x] := \omega^{\alpha_1} + \cdots + \omega^{\alpha_n[x]}$.

Definition 11 (The Hardy hierarchy).

$$H_0(x) := x, \quad H_{\alpha+1}(x) := H_\alpha(x+1), \quad H_\lambda(x) := H_{\lambda[x]}(x) \text{ if } \lambda \text{ is a limit.}$$

The basic theory of the Hardy hierarchy is developed, for example, in [7] or [31]. The most fundamental result concerning this hierarchy is that $(H_\alpha)_{\alpha<\varepsilon_0}$ provides a classification of the provably recursive functions of PA. See, for example, [11] or [35] for recent expositions of this classical result (due to Kreisel and Ackermann). So if a function f is provably recursive in PA then there exists an $\alpha < \varepsilon_0$ such that $f(m) < H_\alpha(m)$ for all m. In particular the assertion $\forall x \exists y H_{\varepsilon_0}(x) = y$ is a prototype for an unprovable assertion in PA. Similarly the provably recursive functions of the theories $\mathrm{I}\Sigma_{d+1}$ are classified in terms of the hierarchy $(H_\alpha)_{\alpha<\omega_{d+2}}$. Note that typically the functions H_α

grow rather quickly. For example, the function H_{ω^ω} eventually dominates all primitive recursive functions.

For a given non decreasing unbounded function F we define its functional inverse F^{-1} by $F^{-1}(m) := \min\{n : F(n) > m\}$. Typically F^{-1} will be a rather slow growing when F is a fast growing function. In "practice" the inverse function of H_{ω^ω} will be almost equivalent to a constant function.

Now we can prove the first phase transition result.

THEOREM 7. *Let d be a fixed natural number.*

1. *If*

$$f(i) = |i| \cdot \sqrt[H^{-1}_{\omega_{d+2}}(i)]{|i|_d}$$

then

$$\mathrm{I}\Sigma_{d+1} \nvdash \mathrm{CWO}(N, \omega_{d+2}, f).$$

2. *If $\lambda < \omega_{d+2}$ and*

$$f(i) = |i| \cdot \sqrt[H^{-1}_{\lambda}(i)]{|i|_d}$$

then

$$\mathrm{I}\Sigma_{d+1} \vdash \mathrm{CWO}(N, \omega_{d+2}, f).$$

PROOF OF THEOREM 7. The proof is an adaptation of a corresponding proof in [4] and [35]. So we will be brief here. Fix $d \in \mathbb{N}$. Let $F(K)$ be the least number M such that for all sequences $\alpha_0, \alpha_1, \dots, \alpha_M$ with $N\alpha_i \leq K + i$ and $\alpha_0 < \omega_{d+1}$ we find $i < M$ with $\alpha_i \leq \alpha_{i+1}$. Then F is not provably recursive in $\mathrm{I}\Sigma_{d+1}$ and moreover one can verify with assertion c) of Theorem 4 in [7] that $H_{\omega_{d+2}}(i) \geq F(i)$ holds for all i. Let $M := F(K) - 1$. Then we find a long descending sequence $\alpha_0 > \alpha_1 > \cdots > \alpha_M$ with $N\alpha_i \leq K + i$ and $\alpha_0 < \omega_{d+2}$. Our aim is to transform this sequence into a strictly descending sequence $\beta_0 > \beta_1 > \cdots > \beta_M$ with $N\beta_i \leq p(K) + |i| \cdot \sqrt[H^{-1}_{\omega_{d+2}}(i)]{|i|_d}$ for $i \leq M$ where p is a primitive recursive function depending only on K. If this is achieved we are done since M depends on K in a way which is not provably recursive in $\mathrm{I}\Sigma_{d+1}$.

From $N\alpha_0 \leq K$ we conclude $\alpha_0 < \omega_{d+1}(K - d - 1)$. For the sequel assume that D depending primitive recursively on K is large enough so that the asymptotic bounds apply. For $i \geq D$ put

$$M_i := \left\{\alpha < \omega_{d+1}(K+3) : N\alpha \leq |i| \cdot \sqrt[K+1]{|i|_d}\right\}$$

and let $enum_i(l)$ be the l-th element of M_i. Then we define

$$\beta_i := \omega_{d+1}(K+3) \cdot \alpha_{|i|} + enum_i(2^{|i|} - i).$$

(This transformation might be interpreted as a renormalization operator on descending sequences.) The sequence β_i is strictly decreasing. Moreover we find for $i \geq D$

$$N\beta_i \leq (d+1+K+3)\cdot(K+|i|) + |i| \cdot \sqrt[K+1]{|i|_d} \leq K + |i| \cdot \sqrt[K]{|i|_d}.$$

Now we sharpen this slightly using a trick due to Arai [4]. For $i \leq F(K)$ we have $H^{-1}_{\omega_{d+2}}(i) \leq F^{-1}(i) \leq F^{-1}(F(K)) = K$. Therefore

$$N\beta_i \leq K + |i| \cdot \sqrt[H^{-1}_{\omega_{d+2}}(i)]{|i|_d}$$

holds for $i \geq D$.

Now we have to check the side conditions. For $i < D$ we simply put

$$\beta_i := \omega_{d+1}(K+3)\cdot\omega_{d+1}(K+3) + D - i.$$

We are nearly through but one last point has to be checked namely the well-definedness of β_i.

For this it suffices to check that M_i has at least $2^{|i|}$ many elements. This is the place where we need the bounds from analytic combinatorics.

CASE 1. $d = 0$.

We have by assertion 1 of Corollary 1 and obvious asymptotic calculations for a suitable and constant $C_{0,K}$ (which is primitive recursively computable with respect to K) and for all i larger than a suitable and primitive recursively computable constant $D_{0,K}$

$$\begin{aligned} |M_i| &\geq C_{0,K}\cdot\left(|i|\cdot\sqrt[K+1]{|i|_d}\right)^{K+2} \\ &\geq C_{0,K}\cdot\left(\sqrt[K+1]{i}\right)^{K+2} \\ &\geq C_{0,K}\cdot i^{\frac{K+2}{K+1}} \\ &\geq 2^{|i|}. \end{aligned}$$

CASE 2. $d = 1$.

We have by assertion 2 of Corollary 1 and obvious asymptotic calculations for a suitable and a constant $C_{1,K}$ (which is primitive recursively computable with respect to K) and for all i larger than a suitable and primitive recursively computable constant $D_{1,K}$

$$\begin{aligned} |M_i| &\geq 2^{C_{1,K}\left(|i|\cdot\sqrt[K+1]{|i|}\right)^{\frac{K+3}{K+4}}} \\ &\geq 2^{C_{1,K}^{\frac{K+3}{K+4}}|i|^{\left(1+\frac{1}{K+1}\right)\frac{K+3}{K+4}}} \\ &\geq 2^{C_{1,K}^{\frac{K+3}{K+4}}|i|^{\frac{K+2}{K+1}\cdot\frac{K+3}{K+4}}} \\ &\geq 2^{|i|}. \end{aligned}$$

CASE 3. $d \geq 2$.

We have by assertions 3 and 4 of Corollary 1 and obvious asymptotic calculations for a suitable and a constant $C_{d,K}$ (which is primitive recursively computable with respect to K) and for all i larger than a suitable and primitive recursively computable (with respect to K) constant $D_{d,K}$

$$|M_i| \geq 2^{C_{d,K}\left(\frac{|i|\cdot \sqrt[K+1]{|i|_d}}{\sqrt[K+3]{||i|\cdot \sqrt[K+1]{|i|_d}|_{d-1}}}\right)} \geq 2^{|i|}$$

since $\left(\sqrt[K+1]{|i|_d}\right)/\left(\sqrt[K+3]{||i|\cdot \sqrt[K+1]{|i|_d}|_{d-1}}\right) \to 0$ as $i \to \infty$.

Now we prove the second assertion. Fix $\lambda < \omega_{d+2}$. Given K let $M := 2^{H_\lambda(K\cdot 2)+1}$. This is well defined in $I\Sigma_{d+1}$. Moreover we may assume that M is large enough for asymptotic purposes. We claim that this M is sufficient for our purposes. Assume otherwise that $\alpha_0 > \alpha_1 > \cdots > \alpha_M$ is a sequence such that $N\alpha_i \leq K + |i| \cdot \sqrt[H_\lambda^{-1}(i)]{|i|_d}$ and $\alpha_0 < \omega_{d+2}$. Then $\alpha_0 < \omega_{d+1}(K-d-1)$. We have for $i \geq \frac{M}{2}$ that $H_\lambda^{-1}(i) \geq H_\lambda^{-1}(H_\lambda(2K)) = 2K$ and hence for $i \geq \frac{n}{2}$

$$\alpha_i \in \left\{\alpha < \omega_{d+1}(K) : N\alpha \leq |M| \sqrt[2K]{|M|_d}\right\}.$$

The number of elements of the latter set is at least $\frac{M}{2}$ as witnessed by the α_i. We obtain the contradiction by calculating bounds on the number of elements on this set. To this end we distinguish three cases.

CASE 1. $d = 0$.

By choice of M we may assume that the bounds of assertion 1 of Corollary 1 apply and that for a certain constant $C_{0,K}$ we have

$$\begin{aligned}\frac{M}{2} &< C_{0,K} \cdot \left(|M| \sqrt[2K]{M}\right)^K \\ &= C_{0,K} \cdot (|M|)^K \sqrt{M} \\ &< \frac{M}{2}.\end{aligned}$$

CASE 2. $d = 1$.

By choice of M we may assume that the bounds from assertion 2 of Corollary 1 apply and that for a certain constant $C_{1,K}$ we have

$$\begin{aligned}\frac{M}{2} &< 2^{C_{1,K}\cdot\left(|M| \sqrt[2K]{|M|}\right)^{\frac{K}{K+1}}} \\ &= 2^{C_{1,K}\cdot(|M|)^{\frac{2K+1}{2K}\frac{K}{K+1}}} \\ &< \frac{M}{2}.\end{aligned}$$

CASE 3. $d \geq 2$.

By choice of M we may assume that the bounds from assertions 3 and 4 of Corollary 1 apply and that for a certain constant $C_{d,K}$ we have

$$\frac{M}{2} < 2^{\frac{C_{d,K} \cdot \left(|M| \sqrt[2K]{|M|_d}\right)}{\sqrt[K]{\left|\left(|M| \sqrt[2K]{|M|_{d-1}}\right)\right|_{d-1}}}} < \frac{M}{2}.$$

since $C_{d,K} \cdot \frac{\sqrt[2K]{|M|}}{\sqrt[K]{|M|}} \to 0$ as $M \to \infty$. ⊣

THEOREM 8. *Fix $d \geq 0$. Let $f_\alpha(i) := 2^{\left(|i| \cdot \sqrt[H_\alpha^{-1}(i)]{|i|_d}\right)}$.*

1. *If $\alpha = \omega_{d+2}$ then $\mathrm{I}\Sigma_{d+1} \nvdash \mathrm{CWO}(S, \omega_{d+2}, f_\alpha)$.*
2. *If $\alpha < \omega_{d+2}$ then $\mathrm{I}\Sigma_{d+1} \vdash \mathrm{CWO}(S, \omega_{d+2}, f_\alpha)$.*

PROOF. The theorem follows by mimicking the proof of Theorem 7 using Theorem 3 in place of Corollary 1. ⊣

The second assertion and the cases $d = 1, 2$ of the first assertion follow alternatively also from results by Gutman and Ivic on Matula numbers. From their article [17] we namely know that

$$\frac{\log(S\alpha)}{\log(\log(S\alpha))} \leq N\alpha \leq \frac{3}{\log(5)} \log(S\alpha) \tag{10}$$

if we identify $S\alpha$ with the Matula number of the tree $T\alpha$ which is canonically associated to α (see the beginning of section 2). Therefore the partial results follows directly from the corresponding result about the $\mathrm{I}\Sigma_{d+1}$-(un-)provability of $\mathrm{CWO}(N, \omega_{d+2}, |f_\alpha|)$.

Note that (10) does not directly yield the independence results related to $\mathrm{I}\Sigma_d$ for $d \geq 3$.

REMARKS. 1. For all norm functions mentioned in Section 2 there will be related phase transition results for the fragments of PA. Results related to M and G can be found in [37].

2. Let us remark that the analytic combinatorics of the transfinite also provides intriguing problems for research in combinatorial probability theory. There are, for example, known results relating the contour process of simply generated trees to the Brownian excursion process in terms of stochastic convergence [16]. In our context it would be natural to ask what the nature of the contour process for ordinals below ε_0 is.This process will not have the Markov property and is therefore different from the usual tree processes.
3. We believe that a general sandwich result in the style of Bell and Burris [5] can be proved for additive and multiplicative number systems by one

corresponding result on Laplace transforms. The hope is to get a uniform way to prove logical limit laws simultaneously for (structures based on) additive and multiplicative number systems. The author is grateful for J. P. Bell for discussions of related issues.

4. We believe that for all norms considered in this article there will be related logical limit laws for ordinal segments at least for first order logic. (See, for example, Burris's book [8] for an excellent exposition of this field.) This will be part of a future research project with Alan R. Woods.

Acknowledgements. The author is very grateful for Jaap Korevaar for his helpful comments on the Tauberian aspects of the subject. Further thanks go to Jan Christoph Schlage Puchta for proofreading an early draft of this paper and for severeal useful suggestions. The author acknowledges gratefully financial support of the John Templeton Foundation (grand ID 13396) and of the FWO (grant ID 3G009308).

REFERENCES

[1] W. Ackermann, *Zur Widerspruchsfreiheit der Zahlentheorie*, ***Mathematische Annalen***, vol. 117 (1940), pp. 162–194.

[2] G. E. Andrews, ***The Theory of Partitions***, Encyclopedia of Mathematics and Its Applications, vol. 2, Addison-Wesley, Reading, Mass.-London-Amsterdam, 1976.

[3] T. M. Apostol, ***Introduction to Analytic Number Theory***, Undergraduate Texts in Mathematics, Springer-Verlag, New York, 1976.

[4] T. Arai, *On the slowly well orderedness of* ε_0, ***Mathematical Logic Quarterly***, vol. 48 (2002), no. 1, pp. 125–130.

[5] P. Bell and S. Burris, *Partition identities. I. Sandwich theorems and logical 0-1 laws*, ***Electronic Journal of Combinatorics***, vol. 11 (2004), no. 1, pp. Research Paper 49, 25 pp. (electronic).

[6] N. H. Bingham, C. M. Goldie, and J. L. Teugels, ***Regular Variation***, Encyclopedia of Mathematics and Its Applications, vol. 27, Cambridge University Press, Cambridge, 1989.

[7] W. Buchholz, A. Cichon, and A. Weiermann, *A uniform approach to fundamental sequences and hierarchies*, ***Mathematical Logic Quarterly***, vol. 40 (1994), no. 2, pp. 273–286.

[8] S. N. Burris, ***Number Theoretic Density and Logical Limit Laws***, Mathematical Surveys and Monographs, vol. 86, American Mathematical Society, Providence, RI, 2001.

[9] N. G. de Bruijn, *On Mahler's partition problem*, ***Nederl. Akad. Wetensch., Proc.***, vol. 51 (1948), pp. 659–669 = Indagationes Math. 10, 210–220 (1948).

[10] R. de la Bretèche and G. Tenenbaum, *Sur certaines équations fonctionnelles arithmétiques*, ***Université de Grenoble. Annales de l'Institut Fourier***, vol. 50 (2000), no. 5, pp. 1445–1505.

[11] A. den Boer, *An independence result for PA using multiplicative number theory*, `http://cage.rug.ac.be/~weierman/`.

[12] P. Dumas and P. Flajolet, *Asymptotique des récurrences mahlériennes: le cas cyclotomique*, ***Journal de Théorie des Nombres de Bordeaux***, vol. 8 (1996), no. 1, pp. 1–30.

[13] P. Erdős, *On an elementary proof of some asymptotic formulas in the theory of partitions*, ***Annals of Mathematics. Second Series***, vol. 43 (1942), pp. 437–450.

[14] P. Flajolet and R. Sedgewick, ***Analytic Combinatorics***, Cambridge University Press, Cambridge, 2009.

[15] J. L. Geluk, L. de Haan, and U. Stadtmüller, *A Tauberian theorem of exponential type*, ***Canadian Journal of Mathematics. Journal Canadien de Mathématiques***, vol. 38 (1986), no. 3, pp. 697–718.

[16] B. Gittenberger, *On the contour of random trees*, ***SIAM Journal on Discrete Mathematics***, vol. 12 (1999), no. 4, pp. 434–458 (electronic).

[17] I. Gutman and A. Ivić, *On Matula numbers*, ***Discrete Mathematics***, vol. 150 (1996), no. 1-3, pp. 131–142, Selected papers in honour of Paul Erdős on the occasion of his 80th birthday (Keszthely, 1993).

[18] G. H. Hardy and S. Ramanujan, *Asymptotic formulæ for the distribution of integers of various types [Proc. London Math. Soc. (2)* **16** *(1917), 112–132]*, ***Collected Papers of Srinivasa Ramanujan***, AMS Chelsea, Providence, RI, 2000, pp. 245–261.

[19] A. E. Ingham, *A Tauberian theorem for partitions*, ***Annals of Mathematics. Second Series***, vol. 42 (1941), pp. 1075–1090.

[20] D. E. Knuth, ***The Art of Computer Programming. Volume 1: Fundamental Algorithms***, Addison-Wesley Series in Computer Science and Information Processing, Addison-Wesley Publishing, Reading, Mass.-London-Don Mills, Ont., 1973.

[21] E. E. Kohlbecker, *Weak asymptotic properties of partitions*, ***Transactions of the American Mathematical Society***, vol. 88 (1958), pp. 346–365.

[22] J. Korevaar, ***Tauberian Theory: A Century of Developments***, Grundlehren der Mathematischen Wissenschaften [Fundamental Principles of Mathematical Sciences], vol. 329, Springer-Verlag, Berlin, 2004.

[23] L. Mutafchiev, *Asymptotic Enumeration of Plane Partitions of Large Integers and Hayman's Theorem for Admissible Generating Functions*, unpublished preprint.

[24] A. M. Odlyzko, *Asymptotic enumeration methods*, ***Handbook of Combinatorics, Vol. 1, 2*** (R. L. Graham, M. Groetschel, and L. Lovasz, editors), Elsevier, Amsterdam, 1995, pp. 1063–1229.

[25] A. M. Odlyzko and L. B. Richmond, *Asymptotic expansions for the coefficients of analytic generating functions*, ***Aequationes Mathematicae***, vol. 28 (1985), no. 1-2, pp. 50–63.

[26] H.-H. Ostmann, ***Additive Zahlentheorie. Erster Teil: Allgemeine Untersuchungen. Zweiter Teil: Spezielle Zahlenmengen***, Ergebnisse der Mathematik und ihrer Grenzgebiete (N.F.), Hefte 7, vol. 11, Springer-Verlag, Berlin, 1956.

[27] S. Parameswaran, *Partition functions whose logarithms are slowly oscillating*, ***Transactions of the American Mathematical Society***, vol. 100 (1961), pp. 217–240.

[28] W. B. Pennington, *On Mahler's partition problem*, ***Annals of Mathematics. Second Series***, vol. 57 (1953), pp. 531–546.

[29] V. M. Petrogradsky, *Growth of finitely generated polynilpotent Lie algebras and groups, generalized partitions, and functions analytic in the unit circle*, ***International Journal of Algebra and Computation***, vol. 9 (1999), no. 2, pp. 179–212.

[30] G. Pólya and G. Szegö, ***Aufgaben und Lehrsätze der Analysis (Problems and Theorems in Analysis)***, Springer Grundlehren, Berlin, 1928.

[31] H. E. Rose, ***Subrecursion: Functions and Hierarchies***, Oxford Logic Guides, vol. 9, The Clarendon Press Oxford University Press, New York, 1984.

[32] W. Schwarz, *Schwache asymptotische Eigenschaften von Partitionen*, ***Journal für die Reine und Angewandte Mathematik***, vol. 232 (1968), pp. 1–16.

[33] R. Smith, *The consistency strengths of some finite forms of the Higman and Kruskal theorems*, ***Harvey Friedman's Research on the Foundations of Mathematics***, Studies in Logic and the Foundations of Mathematics, vol. 117, North-Holland, Amsterdam, 1985, pp. 119–136.

[34] E. Wagner, *Ein reeller Tauberscher Satz für die Laplace-Transformation*, ***Mathematische Nachrichten***, vol. 36 (1968), pp. 323–331.

[35] A. Weiermann, *An application of graphical enumeration to PA*, ***The Journal of Symbolic Logic***, vol. 68 (2003), no. 1, pp. 5–16.

[36] ———, *An application of results by Hardy, Ramanujan and Karamata to Ackermannian functions*, ***Discrete Mathematics & Theoretical Computer Science. DMTCS.***, vol. 6 (2003), no. 1, pp. 133–141 (electronic).

[37] ———, *Phase transitions for Gödel incompleteness*, ***Annals of Pure and Applied Logic***, vol. 157 (2009), pp. 281–296.

[38] M. Yamashita, *Asymptotic estimation of the number of trees*, ***The Transactions of the Institute of Electronics and Communication Engineers of Japan***, vol. 62-A (1979), pp. 128–135 (in Japanese).

GHENT UNIVERSITY
VAKGROEP ZUIVERE WISKUNDE EN COMPUTERALGEBRA
KRIJGSLAAN 281 GEBOUW S22
9000 GHENT
BELGIUM
E-mail: Andreas.Weiermann@ugent.be

Printed in the United States
By Bookmasters